Biology of the Adipocyte: Research Approaches

Van Nostrand Reinhold Advanced Cell Biology Series
Miguel M. Azar, Series Editor

CONTROL OF CELL GROWTH AND PROLIFERATION, edited by Carlo M. Veneziale

NEW METHODOLOGIES IN STUDIES OF PROTEIN CONFIGURATION, edited by Tai Te Wu

CHROMOSOME STRUCTURE AND FUNCTION, edited by Michael S. Risley

MITOCHONDRIAL PHYSIOLOGY AND PATHOLOGY, edited by Gary Fiskum

VERTEBRATE LECTINS, edited by Kenneth Olden and James Brian Parent

BIOLOGY OF THE ADIPOCYTE: Research Approaches, edited by Gary J. Hausman and Roy J. Martin

Biology of the Adipocyte: Research Approaches

GARY J. HAUSMAN, Editor
USDA-ARS, Richard Russell Research Center
Athens, Georgia

ROY J. MARTIN, Editor
University of Georgia
Athens, Georgia

Van Nostrand Reinhold
Advanced Cell Biology Series

VNR VAN NOSTRAND REINHOLD COMPANY
New York

Library of Congress Catalog Card Number 86-26705
ISBN 0-442-23227-6

Printed in the United States of America

Van Nostrand Reinhold Company Inc.
115 Fifth Avenue
New York, New York 10003

Van Nostrand Reinhold Company Limited
Molly Millars Lane
Wokingham, Berkshire RG11 2PY, England

Van Nostrand Reinhold
480 La Trobe Street
Melbourne, Victoria 3000, Australia

Macmillan of Canada
Division of Canada Publishing Corporation
164 Commander Boulevard
Agincourt, Ontario MIS 3C7, Canada

16 15 14 13 12 11 10 9 8 7 6 5 4 3 2 1

Library of Congress Cataloging-in-Publication Data
Biology of the adipocyte.
(Van Nostrand Reinhold advanced cell biology series)
Includes index.
1. Fat cells Research Technique. I. Hausman, Gary J. II. Martin, Roy J. III. Series. [DNLM: 1. Adipose Tissue physiology. QS 532.5.A3 B615]
QP88.15.B56 1987 611′.0182 86-26705
ISBN 0-442-23227-6

Contents

Foreword

A few years ago we collected the thoughts of a number of distinguished investigators to write *Advanced Cell Biology,* which was published by Van Nostrand Reinhold. Seventy-five experts shared their insight in cell biology in what became eighty-four clear topics presented with an unprecedented timeliness. Although during the early stages it seemed too large a task to complete, the combined efforts of the talented participants resulted in that most excellent book.

The late Lazar M. Schwartz, M.D., was mainly responsible for the coordination and the success of the book, which reached a sizeable readership. Confronted with a wide variety of topics, a fast moving field, and a demanding audience, we decided next to propose to Van Nostrand Reinhold to extend the scope of the basic *Advanced Cell Biology* book by publishing a series of timely, concise, and didactic monographs. This Advanced Cell Biology Series will bring to investigators in the scientific community, as well as to students, an up-to-date presentation of select topics of great significance. Now we are pleased to bring our increasing number of readers a new volume in this series.

The *Biology of the Adipocyte* crystallizes the hard, combined effort of a selected nucleus of investigators. Gary J. Hausman and Roy J. Martin have realized the fruits of many hours of intense labor to bring together the chapters that put such an important topic into a clear perspective.

At the time this subject was selected, we did not anticipate in totality the depth and significance of the new research. Our aim is to share with our readers this most valuable monograph.

MIGUEL M. AZAR, M.D., Ph.D.

Preface

Fat cells, once considered metabolically inert entities, have been the focus of intense study by scores of basic and clinical investigators during the last 20 years. The biology of the fat cell, or adipocyte, is of active concern to researchers in the fields of obesity, diabetes, and animal production. The adipocyte has also been used by many researchers as a model cell to study a wide range of cell biology topics, which range from cell growth and differentiation to the molecular aspects of hormone action. The collective results of these research efforts have shown that the development and function of adipocytes and adipose tissue is complex, intriguing and not fully understood. An eventual description of the full physiological role of adipocytes in normal and pathological conditions is dependent on the integration of a wide variety of techniques used by researchers from various disciplines.

Biology of the Adipocyte: Research Approaches has been designed to demonstrate the very divergent techniques that represent state of the art in the study of adipocyte biology. Each of the ten chapters describes a unique research approach complemented with a literature review and/or description of original contributions from authors' laboratories. Two chapters describe cell culture approaches, three chapters detail short term in vitro techniques and five chapters describe in vivo techniques. The authors and their contributions represent a wide range of disciplines that include biochemistry, anatomy, nutrition, physiology, endocrinology, and histochemistry. The end result is a monograph that should be useful to current and future graduate students, postdoctoral fellows, and seasoned investigators.

GARY J. HAUSMAN
ROY J. MARTIN

Contributors

Dr. Margaret Ashwell
Dunn Nutrition Laboratory, Downhams Lane, Milton Road, Cambridge, CB4 1XJ, England

Dr. Basant Bhandari
Metabolism Section, Department of Medicine, Veterans Administration Medical Center, Case Western Reserve University, 10701 East Boulevard, Cleveland, Ohio 44106

Dr. Douglas M. Burns
Mineral Metabolism Section, Medical Research Service, American Lake VA Medical Center, Tacoma, Washington 98493

Dr. Barbara Cannon
The Wenner-Gren Institute, University of Stockholm, Norrtullogatan 26, S-113 45 Stockholm, Sweden

Dr. Aubrey L. Cartwright
USDA-ARS, Poultry Research Lab, Route 2, Box 600, Georgetown, Delaware 19947

Dr. Mario DiGirolamo
Department of Physiology, Emory University, 69 Butler Street, Atlanta, Georgia 30303

Dr. Terry D. Etherton
The Pennsylvania State University, 301 Animal Industries Building, University Park, Pennsylvania 16802

Dr. Susan Fried
Department of Physiology, Emory University, 69 Butler Street, Atlanta, Georgia 30303

Dr. Gary J. Hausman
USDA-ARS, Richard Russell Research Center, Athens, Georgia 30601

Dr. Alayne Makula
Department of Foods and Nutrition, University of Georgia, Dawson Hall, Athens, Georgia 30602

Dr. Roy Martin
Department of Foods and Nutrition, University of Georgia, Dawson Hall, Athens, Georgia 30602

Dr. Richard E. Miller
Metabolism Section, Department of Medicine, Veterans Administration Medical Center, Case Western Reserve University, 10701 East Boulevard, Cleveland, Ohio 44106

Dr. Jan Nedergaard
The Wenner-Gren Institute, University of Stockholm, Norrtullogatan 26, S-113 45 Stockholm, Sweden.

Dr. Jan E. Novakofski
205 Meat Science Lab, 1503 S. Maryland Drive, Urbana, Illinois 61801-4737

Dr. Bernard G. Slavin
University of Southern California, School of Medicine, 1333 San Pablo Street, Los Angeles, California 90033

1

The Developmental Morphology of the Adipocyte

Gary J. Hausman
Richard Russell Research Center

Morphological studies of developing, presumptive adipose depots in the fetus or very young animal have led to new perspectives on the anatomy and physiology of adipose tissue (1, 2). Wassermann was the first to demonstrate that adipose tissue depots are first delineated by the development of stromal-vascular elements (1). Developing blood cells were located in well-developed vascular beds in the clear absence of mature adipocytes (1). Therefore, "primitive fat organs" were considered sites of blood cell formation for the fetus (1). The significance of blood cell formation in immature adipose tissue has yet to be determined. Nevertheless, the concept of the "fat cell cluster" was initially demonstrated by Wassermann (1). The immature fat cell cluster is characterized by cells with either no lipid or little lipid conspicuously arranged around small blood vessels. Therefore, adipocyte development takes place only in clusters of cells and blood vessels. The degree of adipose tissue development can be assessed by measuring the size and number of cell clusters (3, 4). For instance, in the subcutaneous depot of fetal pigs, fat cell cluster number and size increase throughout fetal development (3, 5). Fat lobule (cluster) size and number were quantitated in a recent study of the buccal fat pad from a large number of human embryos and fetuses (4). Fat lobule number increases between the 15th and 23rd weeks, whereas adipose tissue growth between the 23rd and 29th weeks is primarily due to an increase in lobule size (4). In order to elucidate and understand the true "growth" of adipose tissue, it is necessary to examine the complete morphological aspects of depot development (1, 2). The following review demonstrates the importance and utility of developmental morphological and histochemical studies.

IDENTIFICATION OF ADIPOSE TISSUE PREMORDIA AND THE PREADIPOCYTE

Study of the stromal-vascular components of adipose depots permits the identification and verification of presumptive adipose depots (3, 5). The dorsal subcutaneous adipose tissue in the pig fetus displays structural complexity (Figure 1-1). Four distinct depots can be identified within a given cross section of the tissue (Fig. 1-1). Inner, middle, and outer layers, and areas around hair follicles (hair follicle lobules) develop during the last 64 days of fetal development (3, 5–8). The thickness of fetal subcutaneous tissue at different areas over the back (neck, shoulder, etc.), showed the same pattern as that for backfat thickness of mature pigs. This characteristic was present before adipocyte lipid deposition, indicating that the differences in thickness can be due only to either differences in numbers of preadipocytes or to the potential of preadipocyte proliferation (3, 5). This characteristic represents a positive verification of the subcutaneous depot in fetal pigs despite few or no ''adipocytes.''

The temporal and spatial relationships between the development of tissue components and the appearance of fat cell cluster primordia have been analyzed (6, 9–11). The downward and lateral growth of hair follicles into the rat dermis and hypodermis may physically concentrate and flatten stromal cells (9). Fat cell cluster primordia in the rat hypodermis were limited to small areas immediately around descended hair follicles (9). Apparently, the concentrated stromal cells developed into the primordial tissue in the hypodermis (9). The growth and development of hair follicles, sweat glands, large blood vessels, large nerves, and mammary gland ducts in the pig and rat are also associated temporally and spatially with adipose tissue development (6, 10, 11). Lipid and enzyme cytochemistry demonstrates physical continuity between the flattened cells (directly adjacent to tissue structures) and fat cell cluster primordia (Fig. 1-2). The wide distribution of the aforementioned structures throughout the body could explain the extensive distribution of adipose tissue.

The presence of a histochemically detectable basal lamina (PAS-positive) has been used to identify very small adipocytes (6, 11–13). Stromal cells such as fibroblasts and macrophages lack basal laminae and are easily distinguished from small adipocytes (11–13). The histochemical nature of the connective tissue matrix has been used to localize sites of adipose tissue

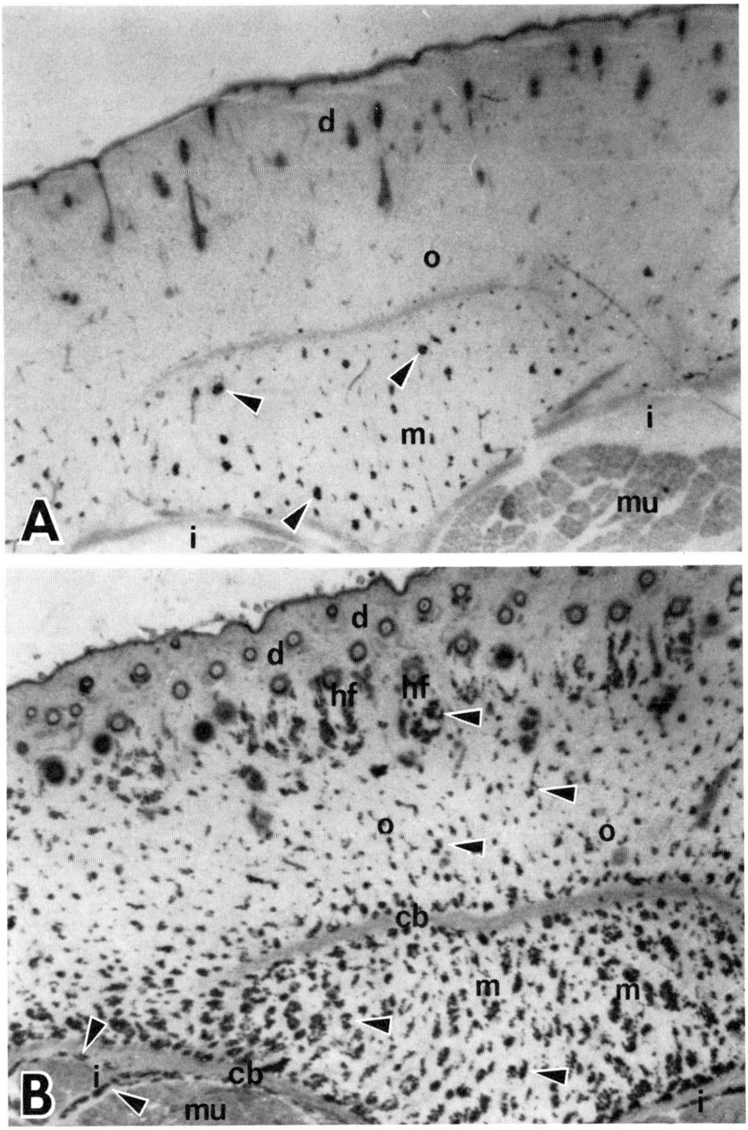

Figure 1-1. The anatomy of dorsal-most subcutaneous adipose tissue (shoulder region) from 70-day and 110-day-old fetuses (A and B, respectively). Lipid-containing fat cell clusters (arrows) are present in four distinct areas in the sample (B) from the 110-day-old fetus: hair follicle (hf) areas, outer (o) layer, middle layer (m) and the innermost layer (i). Note the connective tissue bands (cb) that separate the outer from the middle layer and the middle from the innermost layer. Adipocytes are not present in hf areas, the outer and innermost layers in the 70-day-old fetus (A). Also indicated are dermis (d) and muscle (mu). Cryostat sections (50 microns) stained with oil red O and Harris hematoxylin, × 15.

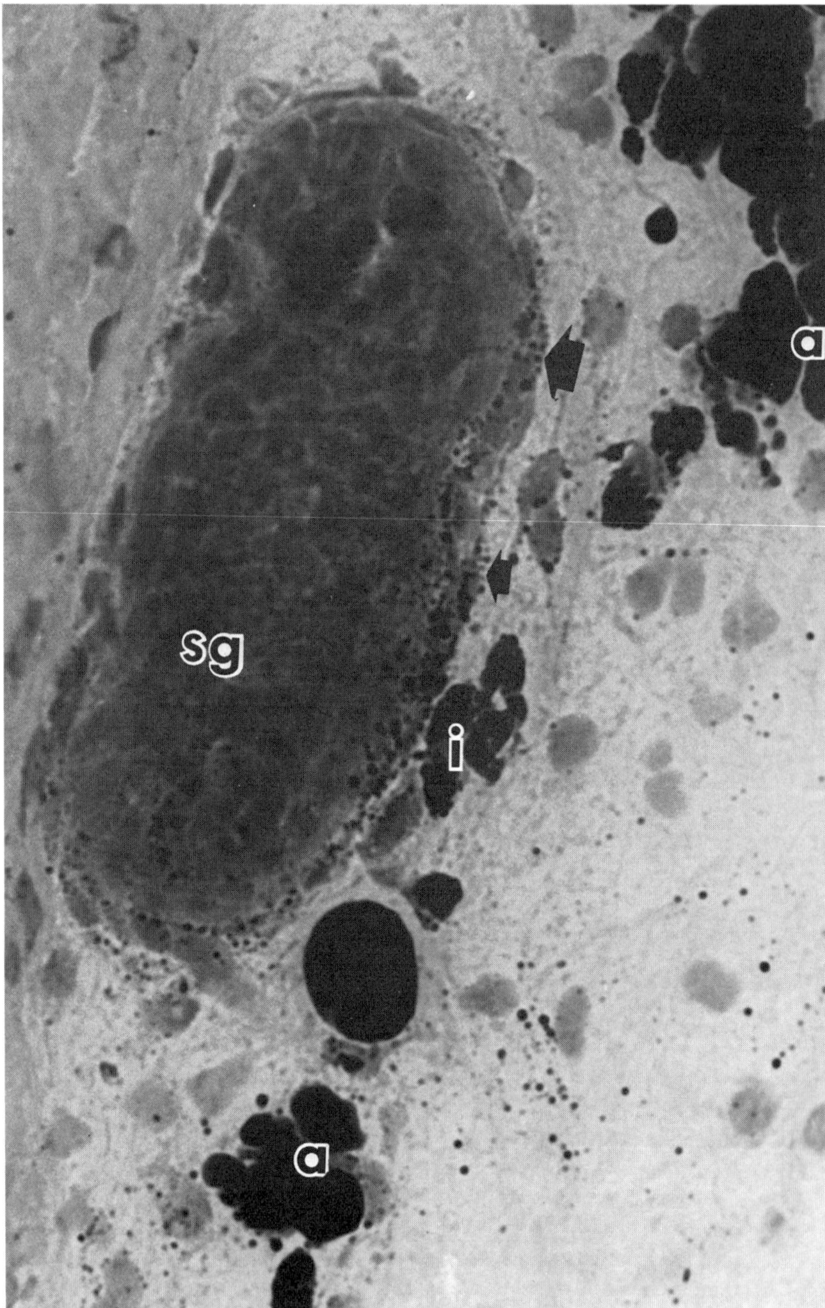
a
sg
i
a

differentiation (10, 14). Toluidine blue metachromasia was present around developing adipocytes, whereas adjacent connective tissue was not metachromatic (10). Simple cellular histochemical tests have been used to discriminate between stromal cells and adipose primordia (11, 13, 15, 16). For instance, primordial cells but not stromal cells react for Nicotinamide Adenine Dinucleotide (NADH) and Nicotinamide Adenine Dinucleotide 3^1-Phosphate (NADPH) tetrazoleum reductases (TR), which indicates the capacity to oxidize these nucleotides (11, 13, 15, 16). In the tests (NADH-TR, NADPH-TR), sections are incubated in a medium that contains a simple buffer with the nucleotide and nitro blue tetrazoleum. A biochemical similarity between precursor stromal cells and adipose tissue primordia can be demonstrated by eliminating the electron acceptor from NADP-dependent dehydrogenase histochemical assays. Color production in these assays depends on the presence of both the NADPH-TR activity and the particular dehydrogenase itself.

HORMONAL AND DIETARY EFFECTS ON DIFFERENTIATING ADIPOSE TISSUE AND PREADIPOCYTES

Reduction in the level of pituitary hormones in the pig fetus (pituitary removed through decapitation) was associated with impaired development of the outer layer of subcutaneous adipose tissue (7, 15). Fat cell cluster numbers were reduced in decapitated fetuses (outer layer), and adipocytes were larger and contained more lipogenic enzymes than adipocytes in control fetuses (7, 15). In decapitated fetuses, skin development was severely retarded and sweat glands and hair follicle adipocyte lobules were nearly absent (7, 15). In an ontogeny study, Ossabaw (fetal, obese) pig fetuses were compared with a strain of contemporary fetuses (Hausman, 1984, unpublished observations). The skin and associated structures in Ossabaw fetuses were differentiated earlier than observed in contemporary fetuses. Hair follicle adipocyte lobules were obvious at 70 days of gestation (term is 114 days) in Ossabaw fetuses, whereas they were not yet present in 70-day con-

Figure 1-2. Histochemistry of connective tissue cells around a sweat gland (sg) in subcutaneous tissue from a fetal pig. Lipid-laden connective tissue cells (arrows) are around a portion of a sweat gland (sg). Indicated are adipocytes (a). Note the physical proximity of these connective tissue cells (lipid +) and immature adipocytes (i). Cryostat section from a 112-day-old fetus stained with oil red O and Hematoxylin, ×420. (*Adapted from Hausman, 1984, p. 149*)

temporary fetuses. Collectively, these studies (7, 15; Hausman, 1984, unpublished observations) indicate that fat cell development around hair follicles is an integral aspect of skin development in the pig fetus.

Fat cell size and the subcutaneous tissue thickness were increased in fetuses obtained from diabetic sows (alloxan-induced diabetes; 17). Fat cell cluster concentrations in the subcutaneous layers were similar in fetuses from control sows and in fetuses from diabetic dams. Therefore, the thicker subcutaneous tissue in fetuses of diabetic sows was associated with higher numbers of fat cell clusters. Thus, maternal diabetes in pigs produces an increased adipocyte size and number (cell clusters) in fetal adipose tissue (17). In a comparison of fetuses (110 days of gestation) from genetically lean and obese sows and from Ossabaw sows, fat cell lipoprotein lipase (LPL) histochemistry completely distinguished obese and Ossabaw fetuses (all cells positive) from lean fetuses (all cells negative for LPL) (18). Fat cell size and adipose tissue morphology (subcutaneous) were essentially unaffected by phenotype (18). Altering the level of fat in the maternal diet (0, 2, 10, and 30%) during the last 30 days of gestation had no effect on fetal pig adipose tissue morphology (subcutaneous tissue) (Hausman, 1984, unpublished observations).

In ontogeny studies of obese fetuses (Hausman, 1984, unpublished observations) examined at 70, 90, and 110 days, and decapitated fetuses (15; decapitated at day 45 and examined at 65, 85, and 110 days), the time period in which fetal pig adipose tissue is susceptible to altered hormonal profiles has been tentatively identified. The major morphological and enzyme-histochemical alterations associated with either obesity or decapitation take place after 70 days of gestation. Since adipocytes are detectable in subcutaneous tissue by 60 days of gestation (3), adipose tissue is apparently refractory to normal stimuli during the first 10–15 days of existence.

SITE TO SITE VARIATIONS IN FETAL ADIPOSE DEPOTS

Fat cell cluster development in the middle and outer layers of fetal pig subcutaneous tissue takes place along a ventral to dorsal gradient (3). The development of cell clusters in hair follicle lobules (3) and in the inner layer (8) was temporally unrelated to that in the middle and outer layers. Enzyme and glycogen histochemistry and blood vessel distribution and histochemistry also clearly distinguished hair follicle and inner layer cells (3, 8, 19). In recent experiments, the fetal perirenal depot has been analyzed and several distinctive developmental characteristics have been recorded (Hausman, 1984, unpublished observations). Larger arterioles are more numerous in the perirenal depot than in subcutaneous tissues. Alkaline

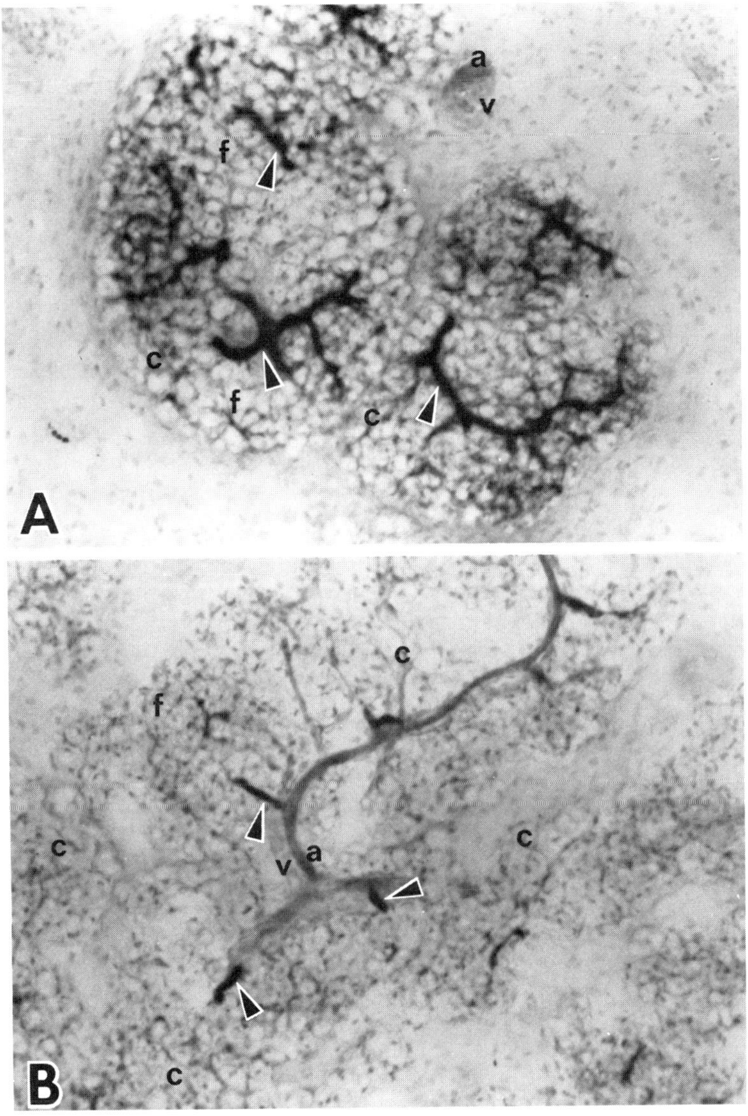

Figure 1-3. A comparison of alkaline phosphatase reactive vasculature (arrows) in samples of perirenal (*A*) and subcutaneous (*B*) adipose tissues from the same 110-day-old fetus. Arterioles (a) and associated venules (v) are very lightly stained, whereas the most reactive vessels (arrows) are located between capillaries (c) and arterioles (a). Note that the perirenal tissues (*A*) contain the highest concentration of reactive vessels (arrows) despite similar fat cell (f) concentrations in perirenal (*A*) and subcutaneous (*B*) tissues. Cryostat sections (50 μ), ×120.

phosphatase reactivity in the adipocyte-associated vessels sharply distinguishes the perirenal and subcutaneous depots (Fig. 1-3). Perirenal fat cell clusters also have higher levels of lipogenic enzymes than cell clusters in subcutaneous tissues. The percentage of esterase-positive cells varies according to depot in the following order (more to less): perirenal > hair follicle lobules > middle > outer layers of subcutaneous tissues (Hausman, 1984, unpublished observations). Adipocyte size in all four fetal depots (middle, outer layers, perirenal, etc.) is very similar (20–25 microns) despite differences in vascularity and enzyme cytochemistry. Therefore, inherent differences between particular sites (blood vessel and nerve distribution) probably account for the metabolic and cellular differences between depots in older animals (20, 21).

STROMAL-VASCULAR CHARACTERISTICS AND ADIPOCYTE CELLULARITY

In several studies, blood vessels in developing adipose depots have been histochemically and structurally analyzed (12, 22). In subcutaneous tissues of young pigs and rats, capillary lumen diameters range between 4 and 10 microns (12, 22). In mature subcutaneous depots of both species, capillaries have lumen diameters between 2 and 4 microns (12, 22). In contrast with the larger capillaries, the small capillaries (2–4 microns) are straighter, have fewer nuclei (per unit length), and have thinner walls (12, 22). Lipoprotein lipase histochemistry and ultrastructural analysis indicated that the development of the narrow lumen capillaries may impede LPL-mediated adipocyte hypertrophy in the very young rat (12).The inner layer of subcutaneous tissue in fetal pigs has stromal-vascular characteristics that distinguish it from the middle layer of tissue (8). The collagen matrix is denser and better organized in the inner layer (8). The adipocyte-associated capillaries in the inner layer react strongly for adenosine triphosphate-ase (ATPase) activity, whereas capillaries in the middle layer are either lightly stained or are not stained for ATPase activity (8). The inner layer of tissue was most affected by fetal decapitation as compared with the middle layer (8). A poorly developed collagen matrix was associated with larger fat cell clusters in decapitated fetuses, and most capillaries (adipocyte-associated) reacted little for ATPase activity (8). Apparently the loss of ATPase reactivity in the blood vessels (decapitated fetuses) was associated with the maturation of the fat cell cluster capillary bed. Thus, these studies demonstrate that the amount and organization of the collagen matrix was inversely associated with the size of fat cell clusters and fat cells and with the maturity of the adipocyte-associated capillaries (8).

A seemingly simple relationship between collagen fiber organization and cell morphology has been observed (6, 9, 10, 14, 22). Typical fibroblasts (polar, spindle-shaped, basophilic cells) were present within organized collagen fibers. Cells with little basophilia, irregular shapes, no particular polarity, and large, pale staining nuclei were located in unorganized fibers. Small and immature adipocytes were present only in the unorganized collagen matrix. Delaying the maturation of the connective tissue matrix may result in conditions suitable for adipocyte development.

Connective tissue organization and vascularization take place in the middle and outer layers (fetal pig subcutaneous) along a ventral to dorsal gradient (3, 5). The appearance of fat cell clusters follows the same ventral to dorsal gradient but at a slightly later time (3, 5). The use of short sampling intervals (15–20 days) in ontogeny studies has revealed that the stromal-vascular development of fetal adipose depots is progressive with age (3, 5). However, fat cell cluster development is not as age dependent, since very immature cell clusters can be found at any fetal age. Possibly, the differentiation of the stromal-vascular aspects of adipose depots is more tightly coupled to overall growth than is fat cell cluster development.

BIOCHEMICAL AND HISTOLOGICAL DIFFERENTIATION OF WHITE ADIPOSE TISSUE

In the subcutaneous depot of the fetal pig the appearance of cytochemically detectable enzymes was always along the ventral to dorsal gradient of fat cell cluster development (Hausman, 1984, unpublished observations). Reactions for NADH-TR (16), NADPH-TR (15), and several dehydrogenases were present in the earliest lipid-positive cells. Cytochemical detection of lipoprotein lipase activity was evident at a later time than the dehydrogenases and NADH-TR and NADPH-TR activities (19). In normal fetuses very few adipocytes were esterase positive (16), but in decapitated fetuses esterase activity was induced along the ventral to dorsal gradient (15). Therefore, along the ventral to dorsal gradient there were some significant temporal lags between the appearance of enzyme activities in the subcutaneous layers. The appearance of LPL and esterase activities was associated with the development of larger fat cell clusters; no other histological changes were associated with the appearance of these enzymes.

In decapitated fetuses the temporal lag between the appearance of esterase activity and other enzymes is virtually abolished (15). Lipoprotein lipase activity was significantly elevated in adipose tissue from decapitated fetuses as revealed by biochemical (7) and histochemical (19) assays. In contrast with adipocytes from control fetuses, adipocytes in decapitated fetuses were larger and in much tighter clusters (7, 15, 19). Experiments in vitro indicate that confluence of adipocyte precursor cells is associated with marked increases in LPL and lipolytic enzymes (23, 24, 25). Experiments with decapitated fetuses may indicate a similar relationship in vivo (i.e., more tightly packed cells associated with greater LPL and esterase activities).

Adipocytes located around hair follicles (hf) in the fetal pig are histochemically distinct (greater esterase and LPL) from other subcutaneous cells (6, 19). These adipocytes (hf) are located in the largest lobules in the entire subcutaneous depot (6). In ontogeny studies it is clear that between 90 and 110 days of gestation areas around hair follicles have the highest concentrations of adipocytes at similar levels of differentiation (6). The close association of cells at similar stages of differentiation may therefore be associated with a tighter coupling of lipid deposition and the appearance of enzyme activities.

MATERIALS AND METHODS

Utilizing a histological approach in adipose tissue studies offers several obvious and distinct advantages over other techniques. For instance, cell diameters can be measured in frozen sections and histochemical tests can be conducted for a large range of enzymes, metabolites, antibodies, and other tissue components. Within minutes after sample collection, frozen sections can be cut, stained, and analyzed. Possibly no other approach or technique has such potential for a short time interval between sample collection and analysis.

Preparing fixed and wax- or resin-embedded blocks of tissue also allows some versatility. New sections can be cut from tissue blocks obtained in studies done years ago and analyzed immunocytochemically with newly developed antibodies. With many techniques it is simply impossible to go back and reanalyze tissues from a past study.

The histological approach does not allow generation of high sample numbers as is possible with electronic measurement of the site and number of osmium-fixed fat cells. An objective approach is easier to achieve with techniques that involve large numbers of cells. However, in these procedures (Coulter counting, etc.) tissue disruption is an integral part of the experimental protocol. Obviously, the tissues are not disrupted as such in a histological-histochemical approach, and this offers several advantages. The biochemistry of the tissue can be localized to either adipocytes or stromal cells with enzyme histochemical tests of frozen sections and immunocytochemical tests of wax-embedded sections. Many studies have demonstrated distinct variations in cell size within small areas of a depot.

The continued use of histological-histochemical approaches is necessary to identify factors responsible for these variations.

Enzyme histochemical studies have been useful in elucidating the metabolic heterogeneity of muscle cells within a given muscle. These studies have demonstrated that adjacent muscle cells can be metabolically and physiologically distinct (26). The ontogeny of various muscle cell types has been thoroughly investigated with histochemical analyses of frozen sections (26). The general approach used by muscle histochemists has been easily adopted to the study of immature fat depots.

Tissue Handling Procedures

The subcutaneous depot of fetal swine is removed from fetuses to optimize sample number (18). The dorsal-most subcutaneous tissue from the neck and shoulder area is removed with some underlying muscle still attached for purposes of orientation. This tissue is removed in a manner that results in a rectangular piece of tissue, the length of which is equal to the length extending from a point just caudal to the head to the fifth and sixth ribs. The width of this sample ranges from 2 to 4 cm, depending on fetal age. The excised tissue is then divided into 3- to 5-mm sections cut perpendicular to the long axis of the body. One-half to one-fourth of these sections are fixed in Bouin's fixative, and the remaining sections are frozen in isopentane cooled to liquid nitrogen temperature, and are stored in an upright ultracold freezer (−80° C) until they are analyzed. Fetal and young rats are cut into three equal-sized cross sections (cranial, middle, and caudal) (12). One cut is made near the umbilicus and the other is made caudal to the shoulders. The front and back legs and tail are removed from the sections. For frozen section analysis, the body sections are frozen as described above in cold isopentane. To obtain paraffin sections, the rat body sections and the fetal pig sections are placed in Bouin's fixative for 12–18 hours. The sections are then removed and sliced into 1- to 2-mm whole-body cross sections (rats) or 1- to 2-mm tissue sections (fetal pigs). These sections (1–2 mm) are fixed for an additional 12–24 hours in Bouin's fixative and then are routinely processed into paraffin blocks (27). The time in fixative prior to "cutting down" is just long enough to harden the tissue sufficiently to allow cutting with minimal disruption along the cutting edge.

The use of Bouin's fixative (27) for histological samples is recommended since it preserves the connective tissue element and resultant sections are completely intact. The relevance of fixed sections in the study of adipose cellularity has been indicated by Ashwell and co-workers (28). For frozen section analysis, tissue blocks are frozen almost instantaneously in isopentane cooled to liquid nitrogen temperature (29, 30). These tissue blocks are not fixed by any means prior to freezing (29, 30).

Section Preparation and Staining

Mounting of tissue sections or whole-body cross sections is done to ensure that histological sections will be perpendicular to the skin surface. Specific procedures for handling and staining fresh-frozen cryostat (−15 to −20° C) sections are reported elsewhere (30). When cryostat sections are cut, the face of the tissue block is often cut down to allow routine collection of serial sections (30). To do so, a scalpel, small scissors, very fine pointed forceps, and a larger forceps are kept in the cryostat so that they are cold and ready to use at all times. Air-dried cryostat sections of 25 to 50 microns in thickness (30) are routinely used. Air-drying for at least 5 minutes allows permanent adherence of a section to a slide even during prolonged staining or incubation. The glass slides are not coated or dipped with anything prior to use (30). The use of thick sections (25–50 microns as opposed to 10–20 microns) allows the collection of serial sections that are completely intact. Sections from fixed and embedded blocks are routinely cut at 7 or 10 microns thick.

Some of the routine staining procedures used for sections from paraffin-embedded blocks are the PAS procedure (27) with and without Harris hematoxylin (HH) counterstaining, the Picro Ponceau procedure (27) with and without HH, and the Lillies allochrome procedure (31) without hematoxylin counterstaining.

The Picro Ponceau procedure is useful for staining collagen in the tissue matrix, and the PAS and Lillies allochrome procedures delineate basal laminae around differentiated cell types. The PAS procedure is also useful for glycogen staining when diastase treated control sections are used for comparison (27). A Phloxine-methylene blue procedure described by Humason (27) is very useful for identifying small adipocytes in sections from wax-embedded blocks that have been fixed with neutral buffered formalin (27). This staining procedure cannot be used on Bouin's fixed tissues, since the picric acid interferes with the stain.

In this laboratory 200-ml plastic staining dishes or 300-ml glass dishes and slide racks that hold 25 slides are routinely used for staining and histochemistry. This procedure allows a large number of slides (as high as 50 with two 25-slide racks in a 300-ml dish) to be stained or reacted simultaneously in the same media or stain. However, with dehydrogenase and tetrazoleum reductase histochemistry small amounts of media are placed directly over the sections on slides laid horizontally in an incubator.

Several routine staining procedures are used on cryostat sections (30). A toluidine blue procedure (27, p. 350) and the Picro Ponceau procedure (27) are used directly on air-dried cryostat sections (30). A fixative per se is never used with these procedures, and fixation artifacts should therefore be minimal (30). The chemistry (toluidine blue) and structure (Picro Ponceau) of the connective tissues can be evaluated with these procedures. Adipocyte diameters and other histological traits can also be studied in sections stained with these procedures (toluidine blue and Picro Ponceau). Lipids are routinely stained with oil red (30) preceded by fixation with cold (4° C) Baker's formalin (27).

ENZYME HISTOCHEMISTRY

As a general rule it is recommended that no fixative be used on air-dried cryostat sections prior to enzyme incubation (29). The usefulness and need for a fixation step should be verified by controlled experimentation before a fixation step is routinely used. Nucleotide tetrazoleum reductases are simple reactions and are very useful for staining immature adipocytes (Fig. 1-4;6,13,15). The capability to reduce NADH (NADH-TR) or NADPH (NADPH-TR) can be assessed in unfixed cryostat sections with the following media as suggested by Chayen and co-workers (29): To 12.5 ml of 0.2 M tris buffer (pH 7.4) add 12.5 mg nitro blue tetrazoleum and either 10

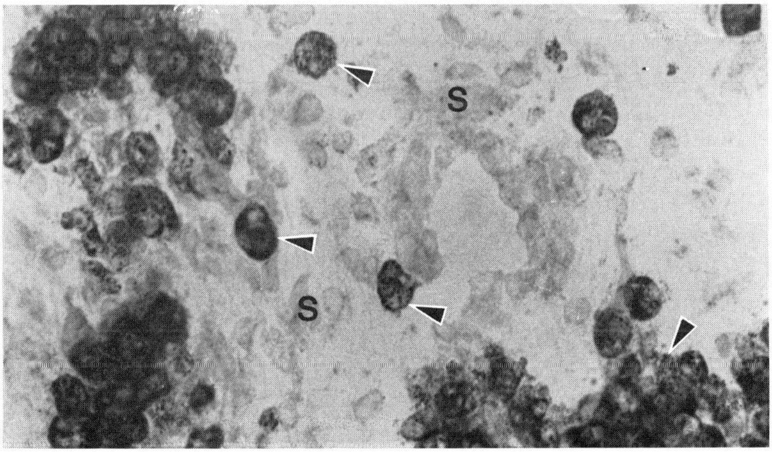

Figure 1-4. Histochemistry for NADH-TR in immature adipocytes (arrows) in a whole-body cross section of a newborn unsuckled rat. These cells are located in the inguinal area. Note the absence of positive reactions in stromal cells (S) and the dark reactions in the small (10 micron diameters) adipocytes (arrows). Cryostat (25 μ) section, ×320.

mg of NADH (NADH-TR) or 10 mg of NADPH (NADPH-TR). The reduction of NADH and NADPH is dependent on the pH of the medium and on the concentration of nucleotide. Incubation is conducted at 37° C for 10–15 minutes and is followed by a 15-minute, cold (4° C) Baker's formalin fixation (27). Sections are mounted in glycerol gel after several water rinses. Esterase is another useful marker enzyme of adipocyte differentiation (6, 13, 15, 30) and is thought to indicate the presence of the enzyme glycerol phosphate dehydrogenase (32). Esterase media is used as detailed by Barka and Anderson (33, p. 120) with a 15-minute cold (4° C) Baker's formalin (27) fixation prior to incubation. A counterstain is not routinely used to allow better visualization of reaction product.

DEHYDROGENASE HISTOCHEMISTRY

The NADP-dependent dehydrogenases, malate, isocitrate, glucose-6-phosphate, and 6-phosphogluconate dehydrogenases can be used to study cytosolic differentiation of adipocytes (13, 15, 34). Histochemical procedures for these enzymes are precisely detailed by Rieder and co-workers (35). Phenazine methosulfate (PMS) is used in these procedures (35) as an electron acceptor, and sodium azide is used to inhibit cytochrome oxidase. If cells have demonstrable NADPH-TR activity, the PMS can be eliminated in the NADP-dependent dehydrogenase procedures (35) so that the particular dehydrogenase and NADPH-TR activities can be demonstrated simultaneously (13, 34). We usually make one medium with PMS and another without PMS (13, 34) and compare reactions from the two media. Another marker of cytosolic differentiation is glycerol phosphate dehydrogenase, which is used as described by Chayen and co-workers (29).

A number of dehydrogenase activities can be used for mitochondrial markers. We determine mitochondrial malate and isocitrate dehydrogenase activities by a modified procedure of Rieder and co-workers (35). The modification consists of replacing NADP with NAD at the same concentration, and the polyvinyl alcohol (PVA) solution is eliminated and replaced with the same amount of buffer (13, 34). The PVA is used to retain cytosolic enzymes within the cell during enzyme incubation. We eliminated PVA in these reactions since we were interested only in the mitochondrial reactions of these enzymes. Glutamate and succinate dehydrogenases are determined with and without PMS concentration used by Rieder and co-workers (35) in procedures suggested by Chayen (29). Alpha glycerol phosphate dehydrogenase is determined essentially in the same way as succinate dehydrogenase but with the addition of PMS and azide at concentrations used by Rieder and co-workers (35).

A procedure for demonstrating the presence of lactate dehydrogenase has been devised by modifying the procedure of Rieder and co-workers (35) as follows: The magnesium chloride $MgCl_2$ solution is eliminated, 1 ml of 200 mg sodium lactate in 5 ml distilled water is used as a substrate and 0.2 ml of 63 mg Na-NAD in 2 ml of distilled water is used as a cofactor. All dehydrogenase reactions are conducted at 37° C for 15 to 30 minutes, and Baker's formalin is used to post-fix the sections prior to water rinsing and mounting in glycerol gel.

PHOSPHATASE HISTOCHEMISTRY

Nucleosidediphosphatase (NDPase), with thiamine pyrophosphate (TPP) or inosine diphosphate (IDP) as substrates, is useful as a marker enzyme for recticular organelles (Golgi and endoplasmic reticulum, 32). Histochemistry for NDPase clearly demonstrated the ontogeny of the Golgi and endoplasmic reticulum in differentiating 3T3-LI preadipocytes (32). The media described by Novikoff and co-workers (32) with IDP substrate have been successfully used in several studies (13, 34). In rat and pig tissues, NDPase histochemistry demonstrates specific adipocyte reactions (13, 34) and clearly distinguishes arterioles (positive reactions) from venules (negative reactions, 36). In fact, the ontogeny of the arteriolar system in adipose tissue can be demonstrated with inosine diphosphatase (IDPase) histochemistry (36). We have used the NDPase procedure for demonstrating the presence of ATPase by substituting ATP for IDP at the same molar concentration. The NDPase and ATPase enzymes are distributed similarly in blood vessels (36), but stromal preadipocyte cells in fetal pig perirenal depots are reactive for ATPase but not for NDPase activities (Hausman, 1984, unpublished observations). In primary cultures of rat adipose tissue stromal cells, the Golgi apparatus of differentiating adipocytes is clearly demonstrated with NDPase reactions (IDP) but not with ATPase reactions (Fig. 1-5; Hausman, 1984, unpublished observations). For NDPase and ATPase enzymes, cryostat sections are fixed in cold (4° C) Baker's formalin (27) for 15 minutes, and then are rinsed several times in a tris-maleate buffer (pH 7.2) prior to incubation in a medium described by Novikoff and co-workers (32). Sucrose is added to the medium as suggested (32).

An acid phosphatase (AcPase) procedure (32) can be used to locate lysosomes in differentiating rat adipocytes in vivo (13) and in vitro (Hausman, 1984, unpublished observations). The medium used is detailed by Novikoff et al. (32), and fixation and rinsing is done as described for ATPase and NDPase. Optimal times of incubation for AcPase, ATPase, and NDPase have to be determined empirically for a specific tissue of a species. In fetal pig adipose tissue, AcPase activity has not been demonstrated despite the use of several techniques (16).

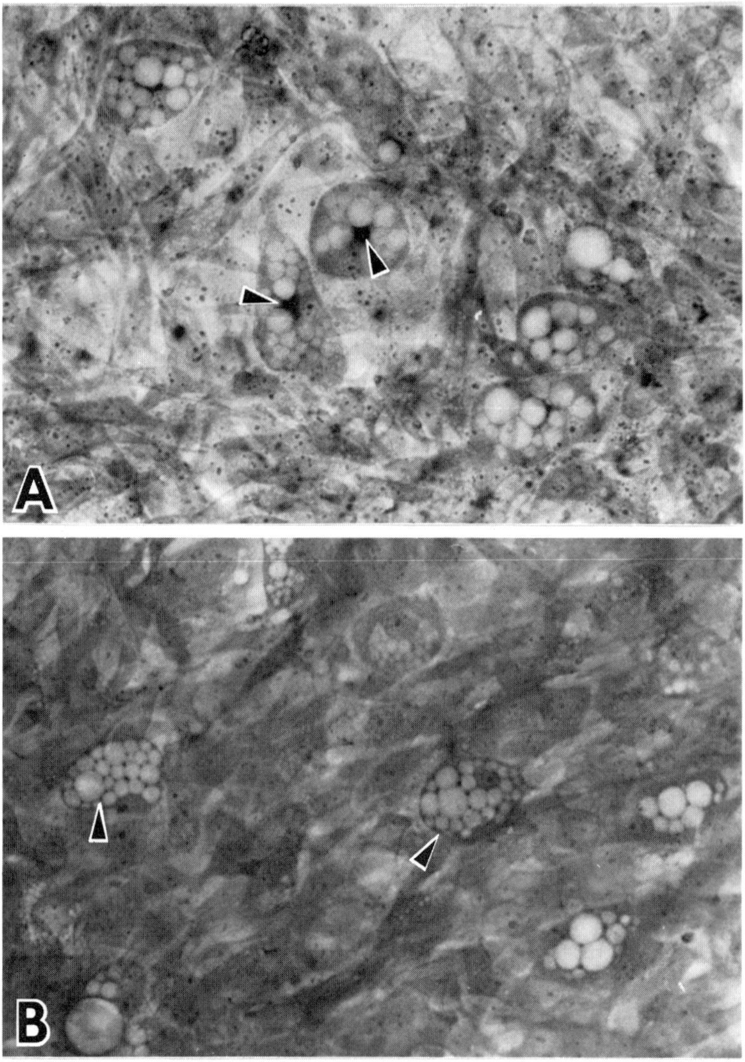

Figure 1-5 Nucleosidediphosphatase (NDPase, *A*) adenosine triphosphatase (ATPase, *B*) histochemistry in cultured adipocytes (arrows). A distinct Golgi region (arrows) is delineated with NDPase reaction product (*A*), but no delineation (Golgi) is evident with ATPase (*B*) histochemistry. Some distinct foci of reaction product for ATPase are evident between lipid droplets, but most cells contain no reaction product (ATPase) and no cells have a reactive Golgi zone (ATPase). These are primary cultures of rat adipose tissue stromal vascular cells exposed to a medium that promotes differentiation. Cells were grown on glass coverslips for 10 days prior to staining on the same coverslips. Cell membranes were permeabilized by exposure to cold (−20°C) methanol for 4 minutes prior to Baker's formalin fixation and routine IDPase and ATPase staining, ×320.

Alkaline phosphatase histochemistry can be used to demonstrate age-related changes in vascular beds of adipose tissues (36). Vascular traits specific for various adipose depots can also be clearly documented with alkaline phosphatase reactions (37). The presence of the enzyme may indicate physiologically active vascular transport because the enzyme is thought to be involved in active transport mechanisms (29). A procedure detailed by Humason (27) is used on unfixed cryostat sections. Sigma's 104 phosphatase substrate is routinely used (27), but several other substrates (i.e., glycerolphosphate) work as well.

LIPOPROTEIN LIPASE AND LIPASE HISTOCHEMISTRY

A simple cytochemical assay (38) has been used to study the ontogeny of LPL activity in adipose depots of fetal pigs (18, 19) and young rats (12). The assay medium (38) has no substrate per se but relies on the presence of endogenous substrate (VLDL, chylomicrons) in the tissues. Validation of the assay has been reported for pig adipose tissue (18). Fresh-frozen cryostat sections are incubated in a 0.1 M tris buffer (pH 8.6) that contains 0.02 M CaCl2. Incubation for 3 to 5 hours at 37° C is followed by incubation in (15 minutes; room temperature) a lead nitrate solution (1%) followed by 1 to 2 minutes in an ammonium sulfide (0.5%) solution (38). We have tried several procedures (27, 29, 38) for lipase demonstration, but positive cellular reactions are always associated with heavy artifacts and nuclei staining (in pig adipose tissues).

OTHER CYTOCHEMICAL APPLICATIONS

Autonomic nerve varicosities can be studied in cryostat sections with a technique described by de la Torre and Surgeon (39). This technique is an improved version (39) of techniques that utilize catecholamine fluorescence to visualize the nerve varicosities. This laboratory routinely uses 50-micron-thick sections with this technique (39) to maintain tissue integrity. The thicker sections do not significantly limit resolution. Immunohistochemistry for antitubulin can also be used to visualize nerve varicosities since tubulin is present in the microtubules of nerve varicosities. A large number of antibodies raised against major cellular components, hormones, and enzymes are commercially available and are suitable for immunohistochemistry. Complete staining kits for a wide range of antibodies are also available from several companies (Biomeda Corp., Foster City, Calif.; Bio Genex Laboratories, Dublin, Calif.; Cambridge Medical Diagnostics, Billerica,

Mass; Dako Corporation, Santa Barbara, Calif.). Excellent technical publications dealing with immunohistochemistry techniques utilizing paraffin and cryostat sections are available from most of these companies.

INTEGRATION AND APPLICATION OF THESE METHODOLOGIES

In fetal pig studies (3, 6, 8, 11, 15, 19, 34, 36) the use of at least three sampling periods (equally spaced during fetal development) has demonstrated the existence of distinct developmental gradients for stromal vascular and adipocyte components of adipose tissue. Anatomical markers are used to locate a similar point in subcutaneous tissue from fetuses of different ages. This can be done only if a constant sectioning plane is maintained throughout tissue and section preparation. The use of a wide range of enzyme and metabolite cytochemical assays allows a detailed analysis of age-associated aspects of adipocyte development. This type of analysis (various enzymes, etc.) combined with morphological analysis can be used to postulate the existence of distinct stages of adipose tissue differentiation (13, 34).

The developmental gradients and stages of adipocyte differentiation can be used to gauge development in a particular experiment. An alteration in a developmental gradient can be reported as indirect evidence of changes in fat cell size or number.

REFERENCES

1. F. Wasserman, 1965, The development of adipose tissue, in *Handbook of Physiology, Section 5: Adipose Tissue,* A. E. Renold, and G. F. Cahill, Jr., eds., American Physiological Society, Washington, D.C., pp. 87–100.
2. G. J. Hausman, D. R. Campion, and R. J. Martin, 1980, Search for the adipocyte precursor cell and factors that promote differentiation: A review, *J. Lipid Res.* **21**:657–670.
3. G. J. Hausman, 1978, Cellular and histological aspects of developing adipose tissue, *Proc. Reciprocal Meat Conf.* **31**:35–52.
4. C. M. Poissonnet, A. R. Burdi, and F. L. Bookstein, 1983, Growth and development of human adipose tissue during early gestation, *Early Hum. Dev.* **8**:1–11.
5. G. J. Hausman, 1977, Cellular and histological aspects of developing porcine adipose tissue, Ph.D. thesis, University of Wisconsin.
6. G. J. Hausman and R. J. Martin, 1982, The development of adipocytes located around hair follicles in the fetal pig, *J. Anim. Sci.* **54**:1286–1296.
7. G. J. Hausman, D. R. Campion, J. P. McNamara, R. L. Richardson, and R. J. Martin, 1981, Adipose tissue development in the fetal pig after decapitation, *J. Anim. Sci.* 1634–1644.
8. G. J. Hausman and G. B. Thomas, 1984, The development of the inner layer of backfat in fetal and young pigs, *J. Anim. Sci.* **58**:1550–1560.

9. G. J. Hausman, D. R. Campion, R. L. Richardson, and R. J. Martin, 1981, Adipocyte development in the rat hypodermis, *Am. J. Anat.* **161**:85–100.
10. G. J. Hausman, 1982, Adipocyte development in subcutaneous tissues of the young rat, *Acta Anat.* **112**:185–196.
11. G. J. Hausman, 1984, Histochemistry of connective tissue cells in subcutaneous adipose tissue of normal and decapitated pig fetuses, *Acta Anat.* **118**:147–152.
12. G. J. Hausman and R. L. Richardson, 1983, Cellular and vascular development in immature rat adipose tissue, *J. Lipid Res.* **24**:522–532.
13. G. J. Hausman and G. B. Thomas, 1984, Enzyme histochemical differentiation of white adipose tissue in the rat, *Am. J. Anat.* **169**:315–326.
14. G. J. Hausman and D. R. Campion, 1982, Histology of the stroma in developing rat subcutaneous adipose tissue, *J. Anim. Sci.* **55**:1336–1342.
15. G. J. Hausman and G. B. Thomas, 1984, Histochemical and cellular aspects of adipose tissue in decapitated fetuses: An ontogeny study, *J. Anim. Sci.* **58**:1540–1549.
16. G. J. Hausman and R. L. Richardson, 1982, Ultrastructural and histochemical aspects of adipocyte development in the fetal pig, *Acta Anat.* **114**:228–247.
17. G. J. Hausman, T. R. Kasser, and R. J. Martin, 1982, The effect of maternal diabetes and fasting on fetal adipose tissue histochemistry in the pig, *J. Anim. Sci.* **55**:1343–1350.
18. G. J. Hausman, D. R. Campion, and G. B. Thomas, 1983, Adipose tissue cellularity and histochemistry in fetal swine as affected by genetic selection for high or low backfat, *J. Lipid Res.* **24**:223–228.
19. G. J. Hausman, 1982, Histochemically detectable lipoprotein lipase activity in adipose tissue of pigs and normal and decapitated pig fetuses, *Acta Anat.* **114**:281–290.
20. D. B. Anderson, R. G. Kauffman, and L. L. Kastenschmidt, 1972, Lipogenic enzyme activities and cellularity of porcine adipose tissue from various anatomical locations, *J. Lipid Res.* **13**:593–599.
21. D. B. Anderson and R. G. Kauffman, 1973, Cellular and enzymatic changes in porcine adipose tissue during growth, *J. Lipid Res.* **14**:160–168.
22. G. J. Hausman and R. J. Martin, 1981, Subcutaneous adipose tissue development in (lean) Yorkshire and (obese) Ossabaw pigs, *J. Anim. Sci.* **42**:1442–1449.
23. R. Negrel, P. Grimaldi, and G. Ailhaud, 1978, Establishment of preadipocyte clonal line from epididymal fat pad of ob/ob mouse that responds to insulin and to lipolytic hormones, *Proc. Natl. Acad. Sci. (USA)* **75**:-6054–6058.
24. P. Grimaldi, R. Negrel, J. P. Vincent, and G. Ailhaud, 1979, Differentiation of ob 17 preadipocytes to adipocytes, *J. Biol. Chem.* **254**:6849–6852.
25. P. Bjorntorp, M. Karlsson, P. Petterson, and G. Sypniewaka, 1980, Differentiation and function of rat adipocyte precursor cells in primary culture, *J. Lipid Res.* **21**:714–723.
26. R. G. Cassens, 1970, The morphology of muscle as food, in *Physiology and Biochemistry of Muscle as a Food,* E. J. Briskey, R. G. Cassens, and B. B. Marsh, eds. The University of Wisconsin Press, Madison, Wisconsin, pp. 679–704.
27. G. L. Humason, 1972, *Animal Tissue Techniques,* W. H. Freeman and Company, San Francisco, California.
28. M. A. Ashwell, P. Priest, and C. Sowter, 1975, Importance of fixed sections in the study of adipose tissue cellularity, *Nature* **256**:724–725.
29. J. Chayen, L. Bitensky, and R. G. Butcher, 1973, *Practical Histochemistry,* John Wiley and Sons, London.
30. G. J. Hausman, 1981, Techniques for studying adipocytes, *Stain Tech.* **56**:149–154.
31. R. D. Lillie, 1951, The allochrome procedure, *Am. J. Clin. Pathol.* **21**:484–488.
32. A. B. Novikoff, P. M. Novikoff, O. M. Rosen, and C. S. Rubin, 1980, Organelle relationships in cultured 3T3-L1 preadipocytes, *J. Cell Biol.* **87**:180–190.

33. T. Barka and P. J. Anderson, 1963, *Histochemistry,* Harper and Row, New York, p. 263.
34. G. J. Hausman, 1985, Cellular and enzyme-histochemical aspects of adipose tissue development in obese (Ossabaw) and lean (Crossbred) pig fetuses: An ontogeny study, *J. Anim. Sci.* **60:**1539–1552.
35. H. Rieder, H. F. Teutsch, and D. Sasse, 1978, NADP-dependent dehydrogenases in rat liver parenchyma, *Histochemistry* **56:**283–297.
36. G. J. Hausman and G. B. Thomas, 1985, Differentiation of blood vessels in the adipose tissue of lean and obese fetal pigs, studies by differential enzyme histochemistry, *Acta Anat.* **123:**137–144.
37. D. W. Fawcett, 1952, A comparison of the histological organization and cytochemical reactions of brown and white adipose tissue, *J. Morphol.* **90:**363–386.
38. M. S. Moskowitz, and A. A. Moskowitz, 1965, Lipase: Localization in adipose tissue, *Science* **149:**72–73.
39. J. C. de la Torre and J. W. Surgeon, 1976, A methodological approach to rapid and sensitive monoamine histofluorescence using a modified glyoxylic acid technique: The SPG method, *Histochemistry* **49:**81–93.

2

Adipocytes, Preadipocytes, and Mitochondria from Brown Adipose Tissue

Barbara Cannon and Jan Nedergaard
University of Stockholm

BROWN ADIPOSE TISSUE

Brown Adipose Tissue

It has taken a long time for brown adipose tissue to be recognized as a tissue in its own right. As has been pointed out by Hausman and co-workers (1), earlier studies on the ontogenic development of adipose tissue were not always strictly defined as to whether the "normal" white adipose tissue or brown adipose tissue was studied; this omission has led to some confusion in the area. Since brown adipose tissue typically is well developed in the perinatal state, it was often the subject of these investigations.

The present view is that two types of adipose tissues exist; certain precursors have the potential to develop only into brown fat cells, while others develop only into white fat cells (2, 3). This clear distinction between the tissues is understandable because of the very different, if not opposing, roles of the two tissues. In principle white adipose tissue is an energy-storing tissue, whereas brown fat is an energy-consuming tissue. Brown fat is therefore well developed in animals and man when there is a need for combustion of food. Regarding the benefits for animal homeostasis, there are two products of this food combustion: One is heat as such, the other is the "excretion" of water, carbon dioxide, and heat.

In the traditional view, the heat produced is the homeostatic product, and this is why brown fat is prominent in newborn mammals (4), and in cold-acclimated mammals (5).

However, the elimination of ingested food can also be the homeostatic "product" of brown fat activity, as is seen in "cafeteria diet" experiments. Compared to animals fed ordinary laboratory food, animals fed high-energy (but low-nutrition) foods convert significantly more of the ingested food to heat (6, 7), and this conversion takes place in brown fat. The physiological significance of this decrease in metabolic efficiency may be that it allows the animals to extract the essential components from a "poor" diet, without having to store the extra energy ingested and therefore become obese.

However, the simple functional delineation between brown and white fat is anatomically complicated. Although in a given animal species a given fat depot (which must be believed to have its origin in a single precursor cell) is generally either white or brown, there are inconsistencies between species. Thus, the perirenal site may be considered to be "typical" white in many species, but it is "typical" brown in lambs, (8). Furthermore, within a single depot—for example—the parametrial depot, from an animal in a given state, certain patches are typical white, whereas some are typical brown (9). Finally, unrecruited brown adipose tissue (i.e., the tissue in a nonactivated state) tends to take on an appearance similar to white fat.

A possible (although cumbersome and technically difficult) definition of a brown fat depot would be that it is a depot of fat containing precursors that, when grown in culture or when physiologically activated, [*can* develop the characteristics of brown fat. These characteristics include the high degree of vascularization, the tendency to a multilocular appearance of the fat droplets, and most selectively, the ability to express the gene for the brown fat-specific protein: thermogenin.

The Mitochondrial "Uncoupling" Protein: Thermogenin

To a large extent, the functional characteristics of brown adipose tissue can be understood through investigation of a protein unique to the mitochondria of this tissue. It is *the regulated uncoupling property of thermogenin* that endows the tissue with its ability to combust substrate and as a result to produce heat at a rate unsurpassed in mammalian tissue, somewhere around 300 watts/kg brown fat (10).

Thermogenin (for more specialized reviews see Fig. 2-1 and refs. 11–13) is a polypeptide of monomeric molecular weight 32,000 that is found in the inner membranes of mitochondria in brown adipose tissue. It binds purine di- and trinucleotides (adenosine triphosphate, ATP; adenosine diphosphate, ADP; guanosine 5[1]-triphosphate, GTP; guanosine 5[1]-diphosphate, GDP) with approximately equal affinity (on the order of 1 μM). The affinity for ATP is probably highest, but thermogenin is, for historical reasons, known as "the GDP-binding protein." The presence of thermogenin also endows the mitochondria with a high permeability for chloride ion (Cl^-), and this permeability is regulated by purine nucleotides; that is, the addition of GDP, for example, diminishes the Cl^- permeability of brown-fat mitochondria to a level comparable to that found in mitochondria from other tissues. The high Cl^- permeability is generally believed to be physiologically irrelevant and merely a biochemically observable effect associated with the presence of thermogenin.

The presence of thermogenin is also responsible for the high "proton" permeability that renders the mitochondria thermogenic. This permeability

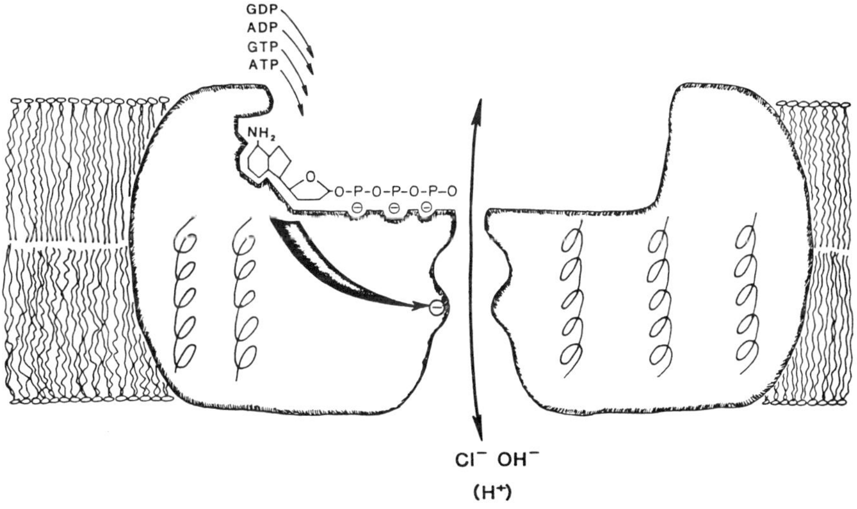

Figure 2-1. The characteristics of the brown fat-specific protein thermogenin. The figure is a visualization of some of the characteristics of thermogenin, as revealed from studies with isolated brown-fat mitochondria and with the isolated protein. In the mitochondrial membrane, thermogenin probably exists as a dimer. It probably has quite a large part of its tertiary structure as alpha helices, shown here to be perpendicular to the membrane. Thermogenin functions experimentally as a channel for Cl^-, and functions physiologically as a channel for hydroxyl ions (OH^-) or the experimentally indistiguishable hydrogen ion (H^+), leading to uncoupling and heat production. However, when nucleotides bind to the site on thermogenin, the proton channel is closed and thermogenesis is inhibited.

is regulated by for example, GDP, since the addition of GDP "couples" the mitochondria; that is, it renders them functionally similar to mitochondria from other tissues. The added GDP acts as an allosteric modifier; it is not metabolized. Whether the "proton" permeability is in fact this or in reality is a hydroxyl (OH^-) permeability as presently discussed, OH^- and Cl^- permeability may be two aspects of the same anion conductance property of thermogenin (11). Such an identity of Cl^- and "proton" permeability is feasible, but it may not be consistent with all present data.

Thermogenin was originally identified by photoaffinity labeling of isolated brown fat mitochondria with ^{32}P-ATP (in the presence of atractylate to block possible labeling of the adenine nucleotide carrier) (14). The monomeric molecular weight was found to be 32,000, which was in agreement with that observed earlier for the only polypeptide that increased in relative concentration during cold acclimation (15). Studies performed after the isolation of the molecule tend to suggest that it exists as a dimer in the mitochondria, with only one nucleotide-binding site per dimer (16). The amino acid sequence has recently been analyzed both by direct amino acid analysis (17) and from cloned MRNA (18). There appear to be many similarities between thermogenin and the adenine nucleotide carrier, with respect to both size and nucleotide-binding properties, and also some direct homologies between the amino acid sequences of the two proteins. Several regions of thermogenin could have an alpha helix structure and consequently be membrane-spanning species (17, 19), but as yet the configuration within the membrane is unknown.

The above information indicates that the passage of protons through thermogenin is the molecular basis for thermogenesis. This proton permeability is physiologically regulated, and this is accomplished within the cell as summarized below.

β-adrenergic Response and Acute Regulation of Tissue Activity

Norepinephrine released from sympathetic nerve vesicles surrounding the brown fat cells elicits the intracellular processes that finally lead to the production of heat in the tissue. Most of this heat production (about 80%) results from the stimulation of β-adrenergic pathways (20, 21). The β-adrenergic pathway is summarized in Figure 2-2.

β_1-Adrenergic receptors are found on brown fat cells (22), and it has generally been assumed that the stimulation of these receptors leads to heat production. However, a novel subtype of β-receptors (sometimes called "β_3"-receptors) has recently been suggested as being involved instead (23).

Stimulation of β-receptors leads to activation of adenylate cyclase, but the coupling between receptor occupancy and adenylate cyclase activity is influenced by the physiological state of the animal. Physiological states associated with stimulated brown adipose tissue—such as chronic cold exposure—diminish the degree of coupling (24).

The cyclic adenosine 5^1-monophosphate (AMP) produced by the adenylate cyclase activates protein kinase(s) (25). Among the proteins phosphorylated, a hormone-sensitive lipase is undoubtedly found, although this enzyme has never been directly characterized in brown fat. The fatty acids produced by the lipase are the substrate for heat production. As only a limited amount of triglycerides is found in the tissue, prolonged periods of thermogenesis necessitate the uptake of more fatty acids from the circulation. In agreement with this, β-adrenergic stimulation also stimulates lipoprotein lipase activity in brown adipose tissue (26) (in contrast to the case in white adipose tissue). Interestingly, this stimulation seems to be associated with an increased synthesis of the lipase itself (27), and not with an activation of existing enzyme, as is the case for the hormone-sensitive lipase.

It has, however, long been recognized that brown-fat mitochondria are not substrate limited; thus, the mere procuration of substrate (fatty acids) for thermogenesis is not sufficient to elicit thermogenesis (28). It must therefore be concluded that an "activator" of thermogenin must also exist (see Fig. 2-1 and 2-2), but the molecular nature of this activator of thermogenesis has not been fully established. The mere addition of fatty acids to brown fat cells induces a thermogenic reaction that seems to be identical to that seen after norepinephrine stimulation (29, 30, 31). Therefore, it is generally accepted that the physiological activator must be fatty acids or some of their derivatives. The physiological activator is thus thought to activate thermogenin in the mitochondria in the presence of cytosolic nucleotides (ATP), and when proton permeability is increased, substrate oxidation is no longer inhibited and thermogenesis is evoked.

α_1-Adrenergic Response and Tissue Recruitment

Besides the β-adrenergic receptors discussed above, brown fat cells also possess α_1-adrenergic receptors (32). Stimulation of these receptors is responsible only for a minor part of heat production. The α_1-receptors, however, are associated with quite marked ionic effects.

Adrenergic stimulation leads to a plasma membrane depolarization, down from about 60 mV to about 30 mV (33), and this is probably, at least in part, due to an increased (Na^+) permeability (34, 35). α_1-adrenergic stimulation leads to a mobilization of (Ca^{2+}) from intracellular stores (probably

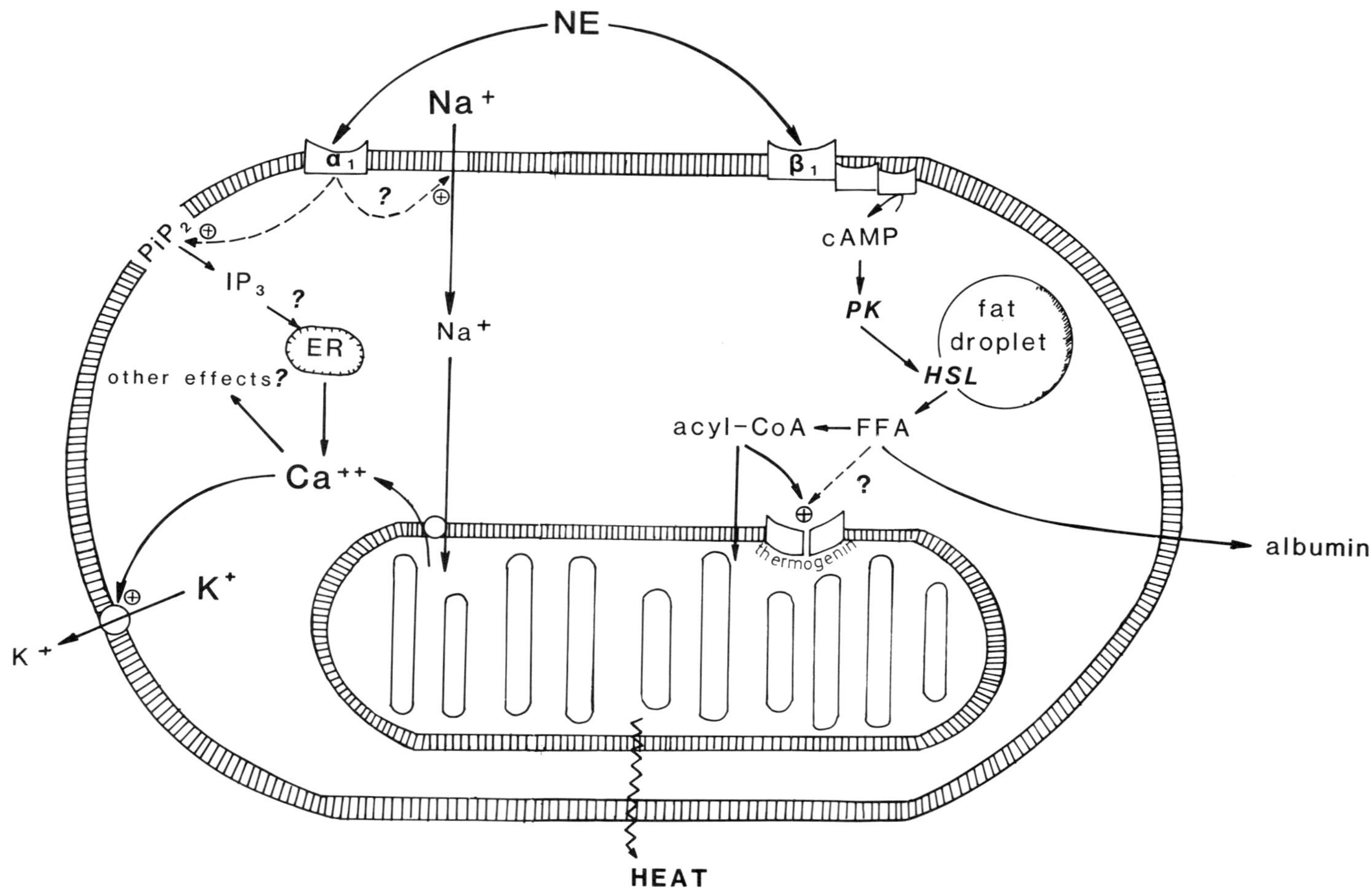
NE
Na+
α1
β1
?
PiP2
IP3
?
ER
other effects?
Na+
cAMP
PK
fat droplet
HSL
acyl-CoA
FFA
?
Ca++
K+
K+
thermogenin
albumin
HEAT

mitochondria) (36), and the increased intracellular Ca^{2+} concentration apparently stimulates a Ca^{2+}-dependent potassium ion (K^+) channel (37, 38) (as one of its effects). Perhaps cyclic guanosine 5'-monophosphate (GMP) levels are also elevated (39). However, these are not the "final" effects of α_1-activation.

The above events—possible increases in cytosolic Na^+ and Ca^{2+}, and an increase in cyclic GMP—are events which are often discussed as being associated with cell proliferation and growth. Interestingly, the number of α_1-adrenergic receptors is markedly increased under conditions where brown fat is activated in vivo, such as cold acclimation (40) and cafeteria-diet feeding (41). Thus, the above pathways of ion fluxes are probably augmented under these conditions.

Very little is known about the regulation of tissue proliferation, and about cell growth and differentiation in general. An impressive increase in cell proliferation can be elicited in brown fat by such a simple stimulus as exposure of the animal to cold (an increase in mitotic index to over 70 by 2 days' cold exposure, (42). This characteristic makes the tissue highly interesting as a model system for further studies on control of cell growth in peripheral tissues of adult animals.

Physiological Recruitment of Brown Adipose Tissue

Neonates. Brown adipose tissue is known to occur in an active state in neonates of all mammalian species. The degree of activity of the tissue at the time of birth relates to the state of development at birth (4). Conse-

Figure 2-2. The adrenergic pathways for stimulation of brown adipocytes. In the β-*adrenergic pathway* (to the right) it is shown how norepinephrine (NE), working through a traditional cyclic (adenosine 3' 5'-monophosphate) AMP cascade with activation of a protein kinase (PK), leads to the activation of hormone-sensitive lipase (HSL) and through this to the release of free fatty acids (FFA) from the intracellular fat droplets. The free fatty acids (or their acyl-coenzyme A, acyl-CoA, derivatives) may activate thermogenin in such a way that uncoupling occurs. The fatty acid derivatives are transported into the mitochondria, where they are combusted and heat is released. In the α_1-adrenergic pathway (to the left), norepinephrine (NE) binds to receptors in the membrane. As a consequence of this, there is an increased turnover of phosphatidyl inositols (e.g., PIP_2), and a production of inositol *tris*-phosphate (IP_3). There is also a release of Ca^{2+} from intracellular stores (mitochondria or part of the endoplasmic reticulum (ER)). The released Ca^{2+} has as one of its effects activation of K^+ channels in the membrane, but may also have other effects.

quently, relatively mature ("precocial") animals such as guinea pigs have abundant, active brown fat immediately at birth (43), whereas the highly immature and poikilothermic newborn of the golden hamster first demonstrate active brown fat at around 2 weeks, when they also begin to regulate body temperature (44). Rat pups are "altricial," that is, they recruit their brown fat in the first postnatal days (45).

Adult Rodents

Cold Acclimated. Small adult rodents (rats, mice, hamsters) may demonstrate active brown fat. Traditionally, these animals are exposed to chronic cold (+5° C) for periods of about 3 weeks, after which notable deposits of brown fat can be found in the animal where the tissue was found in the neonate.

Short Photoperiod. In a natural environment at least two cues are involved in the initiation of brown fat development in preparation for winter. These cues are a lowered environmental temperature, as indicated above, but also a short photoperiod. Consequently, animals will develop partially activated brown fat when days are short (46, 47).

Cafeteria Diets. When rats and mice are offered palatable, but low-protein diets (cafeteria diets), there is a considerable increase in the amount of brown fat in the animals and the brown fat has increased thermogenic activity. These facts are witnessed by an increased protein content and, most important, an increased thermogenin content (48, 49).

Hibernators. A variety of small rodents and insectivores spend winter in deep hibernation, with a body temperature close to that of the nest, in order to conserve energy. These animals probably utilize their brown fat in order to initiate the arousal process. As in the case in cold-acclimating adult rodents that remain homeothermic during winter, the hibernators develop brown fat during autumn in response to environmental cues and can thus also be used as an experimental source of active brown fat.

Adult Humans. In at least two pathological states, human adults can redevelop brown fat that both morphologically and biochemically resembles that which occurs in the neonatal state. One condition is a benign tumor of fat tissue known as *hibernoma.* The stimulus activating the tissue is not understood. The tissue is usually only defined as brown fat postoperatively; after a pathological study, and is consequently seldom accessible for biochemical study. However, electron micrograph studies leave little doubt that it is in fact active brown fat that has regenerated (50).

The other condition is an adrenal tumor, pheochromocytoma. In this condition enormous quantities of circulating catecholamines are found in the blood, and it is believed that this is a sufficient stimulus to reinitiate proliferation and differentiation of the tissue. Apparently even large adult mammals are thus capable of regaining brown fat, providing the stimulus is sufficiently strong. Since pheochromocytoma is diagnosed preoperatively, it is possible to obtain biopsy samples of the brown fat that can be used for both morphological and biochemical studies (51, 52).

Brown fat has also been described in healthy adult humans (53), but the quantities found are in all cases small and its potential role in energy balance is still unclear.

Pharmacological Activation

Catecholamines. The incidence of brown fat in patients with pheochromocytoma indicates that high circulating levels of catecholamines are a sufficient stimulus for brown fat development. As expected, infusion of norepinephrine over prolonged periods seems to entirely mimic cold acclimation, with respect to brown fat development (54). The site of action of the catecholamines—whether it is directly on the tissue or via central mechanisms—has not yet been clarified.

Melatonin. Peripheral implants of melatonin entirely mimic the action of a short photoperiod on brown fat development. The action of melatonin would appear to be central, as implants of much smaller doses of melatonin (which are inactive peripherally) into the hypothalamus allow the same brown fat development to take place (55).

METHODOLOGIES

The preceding account of present knowledge of the metabolism of brown adipose tissue is largely based on a synthesis of the results of experiments performed with isolated brown fat cells and isolated brown fat mitochondria. Brown fat cell culture is a most suitable experimental system for studying the most challenging current question, the control of the proliferation and differentiation of the tissue.

The following sections describe in detail the preparation of the biological material normally used for in vitro studies of brown-fat metabolism: the mature brown adipocytes, the differentiating brown preadipocytes, and the thermogenic brown fat mitochondria. Experiments typically performed with these preparations are also described (although in less detail), and the properties of each isolated system are characterized.

Brown Adipocytes

Brown adipocytes are, of course, the smallest units still capable of hormonally regulated thermogenesis. They have been in experimental use since they were first isolated in the laboratories of Fain (56) and Lindberg (28, 29) in the 1960s.

Sources. Functional brown adipocytes are routinely prepared from control hamsters and rats, but have also been prepared from guinea pigs (57).

It could be envisaged that a larger yield of more thermogenic brown fat cells could be obtained by the use of cold-acclimated animals, but it is generally difficult to obtain cells from such animals; the yield may be lower than that obtained with control animals, and the cells may have a lower quality and may also exhibit refractoriness to norepinephrine stimulation (24, 58). Although cells have been isolated from cafeteria-fed animals (59), their ability to respond to hormonal stimulation has not been demonstrated. Only a few reports concern cells from fetal rats (60), fetal (61) or newborn lambs (9), and rabbits (62, 63).

We shall, therefore, describe only the isolation of brown adipocytes from control hamsters.

Preparation of Isolated Mature Brown Adipocytes. Hamsters are killed by decapitation and brown adipose tissue is dissected from the interscapular, cervical, subscapular, axillary, and periaortic sites. The tissue is dissected free of adhering white adipose tissue and muscle and is washed in a fortified Krebs-Ringer phosphate buffer (KRP) pH 7.4, containing 10 mM glucose and fructose and 4% crude bovine serum albumin. The approximate yield of brown fat from one adult control hamster is 1–1.2 g tissue. The tissue is agitated in the buffer on a vortex mixer for a few seconds to remove blood cells and is then filtered onto wet silk cloth (Joymar Scientific Inc., Hicksville, N.Y.). The tissue (from one or two hamsters) is then placed in a plastic vial (as used for scintillation counting) and minced moderately with scissors to facilitate the digestion process. Six milliliters of KRP buffer containing 2 mg/ml collagenase (e.g., type I, Sigma, but batches may vary and products of other companies may also yield good results) is added, and the suspension is vortex-mixed at maximum speed for 5 sec. The tissue is incubated at 37° C for 8 min with vortex mixing every 2 min.

The material in the vial is then filtered onto wet silk cloth, and the filtrate, which contains mainly broken fat cells, fat droplets and red blood cells but very few brown fat cells, is discarded. The tissue is taken from the filter and placed into a plastic vial with 6 ml of KRP buffer containing 1 mg/ml collagenase, vortex-mixed as before, and incubated at 37° C for

30 min with vortex mixing every 5 min. The vigorous vortex mixing of the vial is essential for good digestion of the tissue.

The tissue/cell suspension is then filtered through wet silk cloth and rinsed thoroughly with 10 ml of KRP buffer to wash through all the cells. The remaining tissue is placed into a new plastic vial with 6 ml of KRP buffer containing 0.5 mg/ml collagenase. (When the tissue from only one hamster is used, often very little material is left after the first 30 min digestion.) The tissue is incubated as before at 37° C until digestion is complete (15–30 min). Often there is a residue of material that is undigestible and is very red in color.

The filtrates containing the brown adipocytes are poured into 15-ml plastic centrifuge tubes, which contain a length of plastic tubing, and the tubes are centrifuged at 500 rpm (20 *g*) for 10 min. After centrifugation, the tube contains a pellet of red blood cells, sometimes overlaid with a layer of small, dense brown adipocytes ("sinking cells"), an infranatant of buffer containing the collagenase and debris material, and a thick layer of compacted brown adipocytes at the surface (Fig. 2-3).

The filtrate from the first collected digestion usually contains the most cells but is heavily contaminated with red blood cells. The second collected digestion usually gives a smaller, but cleaner, cell population.

The pellet of red blood cells and the infranatant are aspirated from below the cells with the plastic tube, 10 ml of fresh buffer is added, and the cells are resuspended. The tube is then placed in the refrigerator to allow the cells to float. The buffer is aspirated again, and the cells from the first and second digestions are pooled, resuspended in a further 10 ml of buffer, and placed in the refrigerator for flotation.

The buffer is removed to leave the final cell suspension at a density of about 2.10 cells/ml. The cells are then ready for use in experiments. Cells are counted in a Bürker chamber.

Cells may be stored overnight at 4° C and used the next day. However, before use, a further filtration of the cells is recommended to remove aggregates that appear overnight.

Yields are generally from 8 to 10.10 cells from the tissue of one hamster.

Experimentation. The most significant property of isolated brown adipocytes is their ability to vastly increase the rate of heat production when they are stimulated by norepinephrine. This heat production can be monitored directly in a microcalorimeter (64), but as this technique is not in common use, measurement of respiratory rate is routinely used. Heat production can be calculated from respiratory rates by the use of the factors 500 Rilajoules (kJ) per mol oxygen O_2 or 8 kilowatts (kW) per mol oxygen per minute (64).

Stimulation with catecholamines also induces increases in the rates of release of fatty acids and changes in ion fluxes and plasma membrane potentials. The adrenergic receptors responsible for these effects can also be directly characterized on the cells.

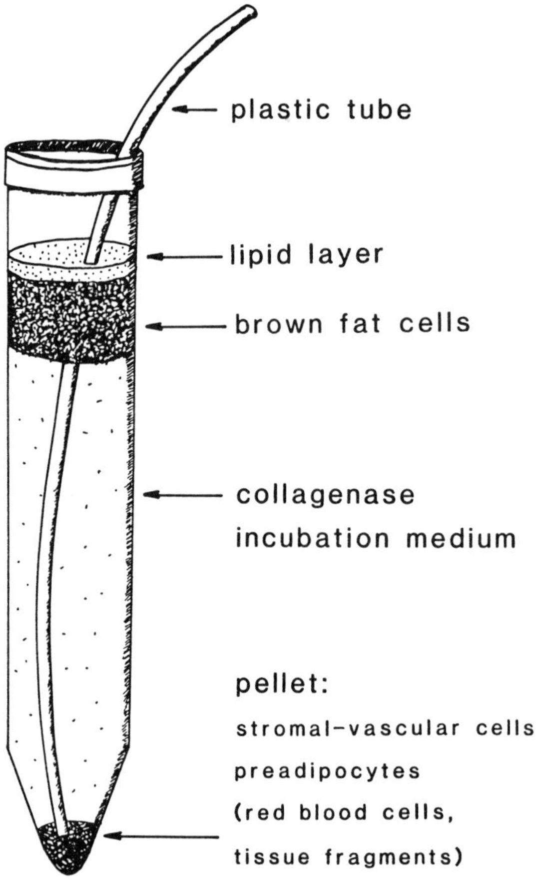

Figure 2-3. Separation of isolated cells. In the plastic tube where the cells are collected after collagenase digestion and filtration, the following layers can be distinguished: on top often a lipid layer, formed from coalescent triglycerides from broken cells; below that the true mature brown fat cells. Below the collagenase incubation medium, a pellet is found; it consists mainly of red blood cells and tissue fragments, but in this pellet the lipid-free preadipocytes are found. This pellet is thus the starting material for in vitro culture of brown preadipocytes.

Respiratory Studies. Although it is possible to use manometric techniques, more physiologically valid responses are probably obtained with polarographic techniques. The isolated brown fat cells are incubated in a bicarbonate buffer, which also contains 10 mM glucose (and possibly fructose) and 2–4% albumin, preferably fatty acid-free.

Traditionally, many studies were performed in Krebs-Ringer phosphate buffer. However, in this buffer, the norepinephrine-induced respiration was transient (29), but the presence of CO_2 was later observed to change the transient stimulatory effect of norepinephrine to a persistent one (31). This change has been demonstrated to be due to the internal anaplerotic effects of pyruvate carboxylase, which, by reaction of CO_2 with pyruvate (formed from the added glucose) forms oxaloacetate. This reaction increases the capacity of the citric acid cycle to receive acetyl moieties from β-oxidation of fatty acids (65).

The high concentration of phosphate in the (CO_2-bubbled) Krebs-Ringer phosphate buffer has, however, in itself an inhibitory effect on the cells, and considerably higher rates of thermogenesis are observed in a Krebs-Ringer bicarbonate buffer (66). As the lipolytic capacity of the cells is increased under these circumstances, they may exhibit feedback inhibition of lipolysis (67), but this is prevented by the addition of fatty acid-free bovine serum albumin (66). Crude albumin cannot be recommended, as it may contain insulinlike substances with an inhibitory effect on the cells (56, 66).

In hamster cells, it is possible to induce respiration hormonally only by the addition of catecholamines. The sensitivity pattern of the main part of the response demonstrates a typical β-adrenergic response (68). However, careful analyses of the response also demonstrate the existence of α_1-adrenergic responses, which at least in cells from control hamsters may be responsible for 20% of the total thermogenic response (20, 21). In rat cells (but not in hamster cells), glucagon can also induce thermogenesis in vitro (69), but the physiological significance of this is not known, and extremely high doses are needed in comparison to what is found in the blood of cold-acclimated rats (70).

Besides hormones, substrates added to brown fat cells may also induce their own respiration, but great care should be taken in the interpretation of these responses. Thus, it is clear that in these cellular preparations, flavoprotein-linked substrates such as glycerol-3-phosphate and succinate are oxidized predominantly by mitochondrial fragments present in the cell suspension, and this respiration can be considered to be an artifact (71). Pyruvate, being a nicotinamide adenine dinucleotide NADH-linked substrate, is not oxidized by the fragments, but enters the cells. The addition of an artificial uncoupler such as carbonylcyanide *p*-trifluoromethoxyphenylhydrazone elicits pyruvate oxidation, demonstrating that respiration in brown fat cells is normally not substrate limited but is under control of the mitochondrial proton motive force (71). The addition of free fatty acids also induces respiration, a very important phenomenon known for more than 15 years (29, 30). The interpretation of this phenomenon is that the addition of free fatty acids mimics the processes normally taking place after catecholamine addition (29, 30), which leads to stimulation of lipolysis and subsequently to thermogenin activation. This observation is the basis for the generally accepted "minimal" theory that an increase in the level of intracellular fatty acids or their derivatives is a sufficient signal for the induction of thermogenesis. The fatty acids (or their derivatives) thus have a dual role in being both the physiological intracellular activator of thermogenesis and the substrate for the thermogenesis (72, 73).

Fatty Acid Metabolism. The stimulation of β-receptors also leads to a release of fatty acids from isolated cells (and probably also in situ to the animal's circulation) (66, 72). This release is not just an overshoot of fatty acids released after the capacity of the mitochondria for combustion has been reached, but rather is seen also at suboptimal norepinephrine concentrations (74). The release may be followed with Ho's ^{63}Ni method (75). In this method an aliquot (e.g., 1 ml) of the medium with cells is added to 5 ml Dole's extraction solution (isopropanol: heptane 0.5 M H_2SO_4; 40:10:1; v:v:v). After addition of 3 ml heptane and 2 ml H_2O, 3 ml of the resulting upper phase is evaporated to dryness in a vacuum oven (100° C). The dried sample is diluted in 300 μl chloroform: methanol (2:1; v:v), and 250 μl is added with 40 μl Ni reagent (^{63}NiCi in 1 M triethanolamine buffer) into an Eppendorf tube and centrifuged for 1 min. The upper phase is discarded and 150 μl of the lower (organic) phase (to which $^{63}Ni^{++}$ has been transferred in complex with the fatty acids) is added to scintillation vials.

Ion Flux Studies. For these studies, the cells must be isolated from the incubation buffer in an intact form, still containing cellular fluid. The methods developed for this utilize centrifugation through oil to obtain this gentle separation, as described in Figure 2-4 (36, 37, 38).

With this method it has been possible to show that α_1-adrenergic stimulation leads to Ca^{2+} mobilization within the cell (36) and, through this, to the activation of Ca^{2+}-dependent K^+ channels (38), and also that β-adrenergic stimulation leads to increased Na^+ influx (76).

Cellular Membrane Potentials. Today it is not possible to measure the plasma membrane potential electrically in isolated cells, so the information

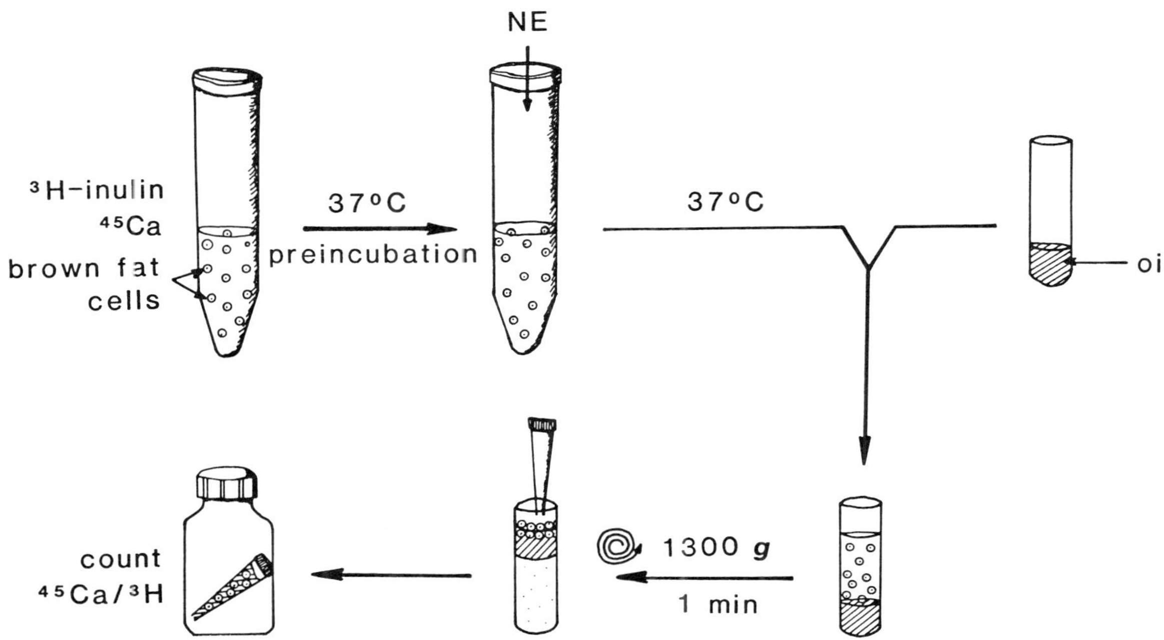

Figure 2-4. The oil centrifugation technique. As the mature brown fat cells, due to their lipid content, are lighter than the incubation solution, normal centrifugation techniques for separation of cells from medium cannot be used. Instead, the cellular incubation is poured on top of a phthalate oil mixture, and when this mixture passes (during centrifugation) through the incubation solution, the cells are also pressed through the solution and finally are found on top of the tube. They can then be collected and their intracellular contents can be examined.

presently available results from experimentation either with intact tissue in the anesthetized rat (77, 35) or with dissected tissue (34, 33).

The use of microelectrodes has revealed that the resting plasma membrane potential is about −60 mV, and that α-adrenergic stimulation may decrease this some 10 mV (33). Observed effects of β-adrenergic stimulation have been more difficult to analyze.

Receptor Studies. Through binding studies with radioactively labeled adrenergic ligands, the receptors responsible for the effects of adrenergic stimulation have been identified on isolated cells. Studies of this type are traditionally performed by filtration of a cell suspension incubated with radioactive ligand through filter paper.

This technique has made it possible to substantiate the existence on isolated brown fat cells of β_1-adrenergic receptors (identified with (^{3}H)alprenolol) (22). It has been suggested that atypical β-receptors, 'β_3'-receptors, are really the receptors responsible for stimulation of lipolysis (and thermogenesis), but this type of receptor has not been identified with radioligand binding techniques (23).

α_1-Receptors are also found on the brown fat cells from hamsters (32) and rats (30), but α_2-receptors have not been identified with radioligand techniques, although there are pharmacological indications of their existence (61, 78).

Brown Preadipocytes

Sources. Brown preadipocytes have been isolated from fetal (79, 80) or young rats (3, 81, 82), and also from guinea pigs (83), but the method described below apparently does not yield brown preadipocytes from hamsters.

Preparation of Isolated Brown Preadipocytes. The rats used are 20- to 25-day-old males of the Sprague-Dawley strain. They are weaned to ad libitum, onto ordinary chow, kept at room temperature (21 ± 2° C), and killed by a blow on the head, followed by cervical dislocation.

From each rat, the interscapular plus cervical brown fat depots, and (if desired for comparison) epididymal white fat, are carefully dissected out under sterile conditions. The tissues are then cut into small pieces and treated to isolate the stromal-vascular fraction, essentially as described by Björntorp and co-workers for white adipose tissue preadipocytes (84). Each tissue from each rat is first incubated in the presence of 0.2% (w/v) collagenase in 4 ml isolation buffer (see below) in siliconized scintillation vials at 37°

C for 30 min in a shading water bath, with vortexing of the vials every 3–5 min. After incubation, the tissue remnants are removed by filtration through a 250-μm nylon screen into plastic test tubes. The mature adipocytes and the fat droplets from broken cells are then allowed to float to the surface of the cell suspension for about 30 min; the infranatants are then collected through a siliconized needle and filtered through a 25-μm nylon screen to remove cell aggregates. The cells passing the filter are finally pelleted in test tubes by centrifugation at 700 g_{max} for 10 min. The concentrated cells are then resuspended in 500 μl culture medium (see below).

Numerous red blood cells are contained in the final brown fat cell suspension. These may—if desired—be eliminated by a hypoosmotic shock: 1 ml Eagle's medium (see below) + 2 ml H_2O for 20 sec at 4° C. After addition of 5 ml Eagle's medium, the remaining cells are pelleted and washed once with Eagle's medium. The cells are finally resuspended in culture medium (see below).

Cell Culture. Brown preadipocytes may be cultured in tissue-culture flasks (e.g., Nuňclon Delta, 25 cm^2) containing 5 ml culture medium (see below). Each flask may be inoculated with about 0.5×10^6 brown- or white fat cells, obtained from one rat, and the cultures can be kept at 36° C in an atmosphere of 5% CO_2 in air with 80% humidity in an incubator. The culture medium is changed the day after inoculation of the cells into the flasks, and then every second day. At the low inoculation densities used (when individual rats are used) the plating efficiency is only 10–20%, but this efficiency can be improved by increasing the number of cells in the inoculum. The doubling time under these conditions is about 20 h.

The cells can be grown until they—due to lipid deposition—detach from the bottom of the culture flasks. This normally happens after 12–14 days.

Cell Harvesting. The cultured cells are first incubated with 5 ml Ca, Mg-free Earle's balanced salt solution (Flow Laboratories) for 10 min at 36° C. The cells are then removed from the bottom of the flasks by incubation with 3 ml 0.2% collagenase in isolation buffer (see below) in a shaking water bath at 37° C for 5–30 min (depending on the state of development of the cells; i.e., on their fat content and their tendency to float). The flask is washed with 3 ml isolation buffer, and the combined suspension obtained is further incubated in a test tube for 15–30 min in order to dissociate cell aggregates. The culture flasks may be reincubated as above with an additional 3 ml 0.2% collagenase in isolation buffer in order to remove the remaining cells. The suspensions are pooled and after addition of 4 ml isolation buffer, the cell suspension centrifuged at 400 g_{max} for 10 min. However, only those cells that are pelleted can be collected and investigated by this method; floating cells cannot be concentrated sufficiently (contrast the preparation of mature brown adipocytes described above). Cells may alternatively be removed from the culture flasks by treatment with a rubber policeman, and although this method is more gentle than collagenase digestion, it is more difficult to separate cell aggregates into single-cell suspensions.

After resuspension in a minimum volume of isolation buffer, the cells may be counted in a Bürker chamber.

The isolation buffer consists of 123 mM Na^+, 5 mM K^+, 1.3 mM Ca^{2+}, 131 mM Cl^-, 5 mM glucose, 1.5% (w/v) crude bovine serum albumin and 100 mM N-2-hydroxyethylpiperizine-N'-2-ethanesulfonic acid (HEPES) (pH adjusted with NaOH to 7.4). The buffer is sterile filtered (Millipore 0.45 μm-HA; 0.22 μm Millex-GS) before use, after the addition of collagenase.

The Eagle's medium to be used is Eagle's Minimum Essential Medium with added nonessential amino acids, antibiotics (200 IU penicillin and 100 μg streptomycin per ml). This may be obtained as a final sterile solution (Flow Laboratories).

The culture medium consists of this Eagle's medium to which is added 20% newborn calf serum (Flow Laboratories), 40 mIU insulin (Vitrum, Sweden) per ml, 0.05% (w/v) Intralipid solution (Vitrum, Sweden), and 20 mM HEPES buffer (Flow Laboratories).

Characteristics, Prognosis. The cells that develop in culture as a result of the above procedure retain multilocularity (3, 85), although they contain only few, characteristic mitochondria, compared with mature brown adipocytes (81). They are hormone sensitive (80), demonstrating β-adrenergically induced fatty acid release (82, 83), but little or no norepinephrine-stimulated increase in oxygen consumption, perhaps because of the low complement of mitochondria (3). The cells are therefore highly interesting as a potential model system to study control of proliferation and of differentiation, particularly with respect to mitochondriogenesis.

Brown Adipose Tissue Mitochondria

Mitochondria from brown adipose tissue are of interest because they are the site of nonshivering thermogenesis—their study illuminates both theoretical and possibly clinical considerations; that is, questions related to both bioenergetics and obesity control. In addition, during the rapid proliferation of the tissue there is an extensive mitochondriogenesis, which can probably be studied better here than in any other tissue. A further unique and exciting feature of the mitochondria is that they possess the brown fat-specific "uncoupling" protein thermogenin, which is the molecular manifestation of their ability to produce heat.

As the mitochondria are readily isolated, they have been studied in animals and in humans under a variety of very different physiological states, and this source of variation must always be considered. Results from these studies can provide insight into the degree of adaptation of the animal to a specific physiological state with respect to energy balance, and such studies are therefore of considerable comparative interest. However, in many cases they also necessitate caution in directly comparing the work of different laboratories, as these have seldom been performed on animals under identical conditions of age, diet, temperature, photoperiod, and sex.

Sources. Brown fat mitochondria may in principle be isolated from all the conditions where brown fat is activated, described in Section 1.5. However, for functional investigations of regulation of thermogenesis and other factors, it is most advantageous to use brown adipose tissue from cold-acclimated rats or hamsters.

Preparation of Isolated Brown Fat Mitochondria. Mitochondria from brown fat are most readily prepared from tissue in the activated state, as this tissue gives the highest yield of mitochondria. However, as has been pointed out above, it is essential to always bear in mind that this is a physiological state vastly different from the control state; comparisons therefore must be performed with caution. From small rodents, a routine preparation can be performed in the manner described below (see also Ref. 86).

The interscapular, cervical, periaortic, and perhaps axillary brown adipose tissue depots are dissected out and immediately placed in ice-cold 250 mM sucrose. The pieces (generally about 1 g of tissue) are then added to about 30 ml sucrose solution, and the tissue is homogenized on an ice bath with six strokes in a Potter-Elvehjem homogenizer. The resulting homogenate is filtered through one layer of cheesecloth. All subsequent steps are carried out at 4° C. The brown fat homogenate is first centrifuged at 8500 *g*. The supernatant, including the fat cake, is discarded, and the pellet is resuspended in about 30 ml sucrose in the homogenizer. The homogenate is then centrifuged at 800 *g* (2,600 rpm) for 10 min. The pellet, containing nuclei and cellular debris, is discarded. The supernatant is centrifuged at 8,500 *g* for 10 min, and the new pellet is resuspended in 10 ml of a 250 mM sucrose solution, which may contain 2% bovine serum albumin, and recentrifuged. The pellet is finally suspended in 10 ml sucrose and recentrifuged. The resulting albumin-washed mitochondria are taken up in a small amount of sucrose solution and used for protein concentration determination.

Mitochondria suspended in sucrose in this way may be used for a variety of different experimental procedures, but for respiratory studies it is essential that the mitochondria have an expanded matrix, which is not the case when they are prepared as described above (see below for explanation). The most convenient way to obtain mitochondria with an expanded matrix volume is to suspend the pellet after the albumin wash in a solution of 100 mM KCl and 20 mM K-Tes at pH 7.2. After the final centrifugation in this medium (N=tris [hydroymethyl] methyl-2-aminoethane sulfuric acid) the mitochondria should be suspended in a small volume of the same solution for storage.

Whether sucrose or a KCl solution is used for the final suspension, it is often advantageous to add fatty acid-free albumin to a final concentration of 2% to the final mitochondrial preparation. The albumin facilitates stability and therefore reproducibility during the course of an experimental series.

Experimentation

Respiratory Studies. When conventionally prepared as described above, brown fat mitochondria are normally in an energy-dissipating state and are unable to demonstrate respiratory control, ATP synthesis, and other energy-requiring processes. In this state they are suitable for investigation of substrate specificities, although it is generally advisable nonetheless to add a mitochondrial uncoupler (such as FCCP) to the experimental medium (see below) in order to be certain that no rate limitation other than the capacity for uncontrolled respiration of the substrate under investigation is being observed. For typical respiratory traces, see Figure 2-5.

Experimental medium: A suitable ionic medium for respiratory studies may have the following composition: 100 mM KCl, 20 mM K-TES, 1 mM ethylenediaminetetraacetate (EDTA) 2 mM $MgCl_2$, 4 mM KH_2PO_4, pH 7.2.

Respiratory Substrates. Fatty acids and their derivatives as substrates. For brown fat mitochondria from all species studied, the highest respiration is always found with activated fatty acid derivatives, such as palmitoyl-L-carnitine. Added in this way, the fatty acids are in their least detergent form and can be added in concentrations that will elicit very high rates of respiration. Average values for measurements at 37° C are 600 nmol/mg protein/min. These values agree well with those found in isolated cells and correlate also with estimates that can be derived from the tissue in vivo. Thus, the respiratory capacity that can be observed in vitro probably reflects that found in the animal.

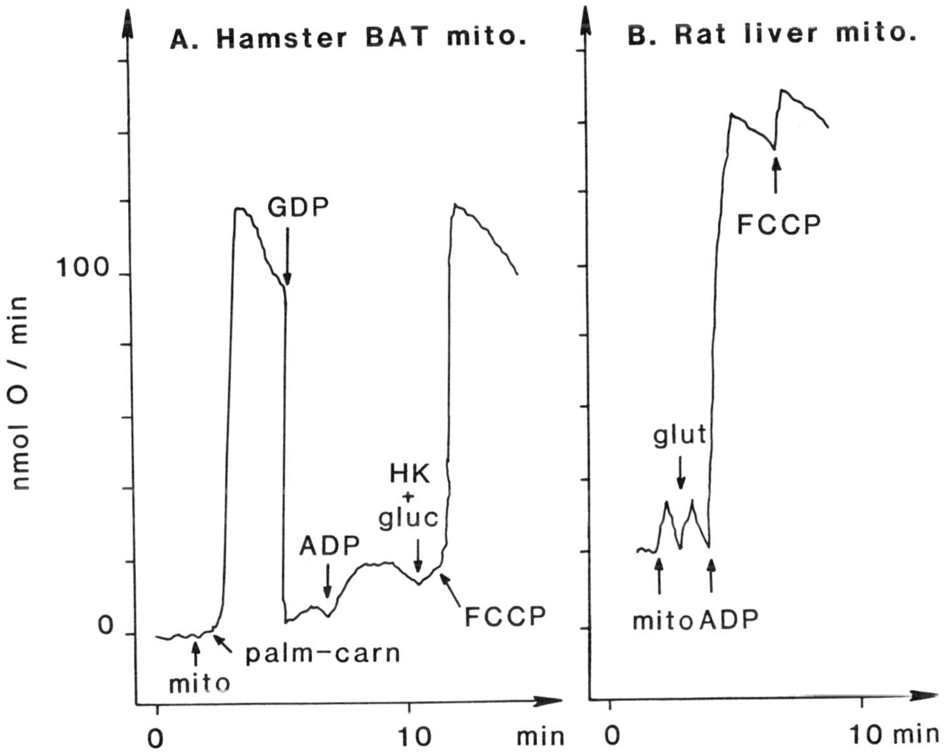

Figure 2-5. Induction of respiratory control in brown fat mitochondria. Isolated brown fat mitochondria behave characteristically when followed with an oxygen electrode. Thus, as seen in *A*, the mere addition of substrate (here palm-carn: palmitoyl-carnitine) to *brown fat mitochondria* immediately leads to a vast increase in oxygen consumption. However, when a purine nucleotide (here GDP) is added, respiration immediately ceases. When ADP is added, only a relatively small increase in respiration can be seen, and the addition of an ATP trap (in the form of glucose + hexokinase) leads to no further increase in respiration. However, after the addition of an artificial uncoupler (here FCCP), respiration is again stimulated to the high level seen from the start.
In contrast, *liver mitochondria* (*B*) do not respond with a stimulated respiration to the mere addition of substrate (here, glut: 2-oxo-glutarate in the presence of malate). There is no effect of GDP (not shown), but the addition of ADP leads to a vast increase in respiration, fully up to the level found after the further addition of an artificial uncoupler (here FCCP). Thus, whereas brown fat mitochondria are optimized for heat production and are poor ATP producers, liver mitochondria do not possess the pathway for heat production, but are optimized for ATP production. (*Adapted from Lindberg et al., 1976 (110)*)

Glycerol-3-phosphate as a substrate. The flavoprotein dehydrogenase for this substrate is found on the outside of the mitochondrial inner membrane and is extremely well developed in all brown fat mitochondria studied. There is also an active cytoplasmic dehydrogenase, and therefore a well-developed shuttle for reducing equivalents exists in the tissue (87). A further function of the dehydrogenase may be the oxidation of glycerol-3-phosphate in order to prevent triglyceride resynthesis in catabolic situations (88).

Citric acid cycle intermediates as substrates. Although the dehydrogenases for these substrates are clearly well developed, as is evidenced by the high rates of respiration on fatty acid substrates, their oxidation when they are added exogenously varies depending on the species studied. The reason for this variation is that the substrate permeases have very different capacities. In hamsters, for example, the permeases are extremely poorly developed (89) whereas lambs apparently have much higher permease capacities (9), and the rat is somewhere in between.

Pyruvate as a substrate. Pyruvate is a good substrate for brown fat mitochondria from all species tested. The permease is active and the dehydrogenase is also very active. The fate of pyruvate in the mitochondria may be either to be degraded to acetyl coenzyme A (acetyl-CoA) or to be converted to oxaloacetate through the anaplerotic activity of pyruvate carboxylase under conditions of high rates of β-oxidation (65).

Induction of Energy Conservation. Although brown-fat mitochondria function in vivo to produce heat by a dissipation of respiration-derived energy through a loose coupling pathway thermogenin, in the resting state they are normally well coupled. The physiological regulation of the energy dissipation pathway is not entirely clarified, but evidently the possession of this pathway yields uncoupled mitochondria after isolation; that is, when the regulatory factors are removed.

In order to introduce energy conservation, it is necessary to add to the medium both serum albumin (fatty acid-free, at about 0.2% concentration) and a purine nucleotide; for example, ATP, ADP, GDP, or GTP (at about 0.5 mM concentration). The function of the albumin appears to be to remove fatty acids (or their derivatives) that are either bound to the mitochondria endogenously or become bound during isolation. The fatty acids may either uncouple as "classical" uncouplers, as they are weak lipophilic acids, and/or they (or their derivatives) may in some way interact with thermogenin in a more specific manner. The nucleotide binds to thermogenin in the inner mitochondrial membrane (on the outside) and in so doing closes the ion (or "proton") channel; when open, this channel allows dissipation of the proton motive force. It is generally suggested that ATP is the natural inhibitor of the ion channel, but in studies with isolated mitochondria it is generally more convenient to use GDP, as GDP does not interfere with energy-conserving processes. Its action therefore can be more easily studied.

In the energy-conserving state induced by albumin and GDP, brown fat mitochondria function as other mitochondria. The proton motive force is as high as that in other mitochondria (90), and the mitochondria are able to perform energy-requiring processes such as calcium transport (91). The capacity to synthesize ATP, however, is not equally well developed in all species of animals studied. Although lamb brown fat mitochondria have an ATP synthase capacity that is almost as high as the capacity of the respiratory chain (9), brown fat mitochondria from hamsters possess very little ATP synthase (92). Consequently, they demonstrate very low rates of ADP-stimulated respiration. This observation gave early support to the hypothesis that the mechanism for heat production in brown fat involves an uncoupling process rather than being based on an ATPase activity (93).

Ion Conductance Studies. When purine nucleotides are removed from BAT mitochondria during isolation, the regulated ion conductance channel is left open and matrix ions can leak out. This unusual permeability is the reason for the collapse of the matrix during isolation and the very condensed matrix volume that is then observed (94). In the condensed state, the mitochondria are virtually unable to respire on substrates that are oxidized within the matrix (94). This inability explains the requirement for KCl or similar permeant ions in the storage medium in order to expand the matrix prior to respiratory studies.

The physiologically significant permeability found in the absence of purine nucleotides is for protons (or hydroxide ions), because this allows the dissipation (or neutralization) of the proton motive force, derived from respiration. High permeabilities are also observed for chloride, bromide, and nitrateions. All of these permeabilities return to levels normally observed in, for example, liver mitochondria, when purine nucleotides are added (95).

Ion conductance studies are most easily performed as measurements of changes in light scattering by the mitochondria, which reflect changes in the mitochondrial volume. Light scattering can be followed as a decrease in absorbance at 520 nm, but may also be followed in nephelometers or in fluorometers as an increase in scattered light. These changes result from the uptake of ions from the incubation medium. It is essential that only one ion under study is rate limiting; this is usually achieved by using either K^+ with an ionophore (valinomycin) or ammonium as the positive counter ion for the studies. These measurements are only semiquantitative and do not readily give rates. When more quantitative ion transport measurements are required, studies can be performed with radioactive isotopes of the ion of interest; $^{36}Cl^-$, for example (96).

Mitochondrial Membrane Potential and Proton Motive Force. The electrical component of the proton motive force, that is, the mitochondrial membrane potential, can be estimated with different techniques. It may be estimated as the distribution of a small amount of rubidium ($^{86}Rb^+$) over the mitochondrial membrane, but this method necessitates the presence of the K^+(Rb^+) ionophore valinomycin (90); it is therefore possible to induce the membrane potential by the measurement itself. This is not the case when a tetraphenylphosphonium (TPP^+) electrode (97) is used, but the matrix volume and the activity coefficient of TPP^+ in the matrix must then be determined. It is also possible to use the lipophilic cation safranine (91). The absorbance of safranine is inversely proportional to the membrane potential, but it is necessary to calibrate the response with a K^+ gradient and valinomycin (91).

In the coupled state, that is, in the presence of albumin and GDP, brown fat mitochondria exhibit "normal" membrane potentials of about 160 mV (90, 91). The pH component of the membrane potential can be estimated by distribution studies with acetate (90), and amounts to a further 30–50 mV, so in the resting condition a proton motive force of about 220 mV is developed across the mitochondrial membrane (90).

Uncoupling through activation of thermogenin results only in a very limited membrane potential decrease—about 15 mV. This means that physiologically activated brown fat mitochondria should not be considered to be in the same "uncoupled" state as that which is seen after the addition of an excess of an artificial uncoupler such as FCCP or DNP (2, 4=dinetrophenol); rather they should be considered equivalent to, for example, liver or muscle mitochondria actively synthesizing ATP; and they thus remain capable of among other things Ca^{2+} control (91).

(^{3}H)GDP Binding Studies. Because of the great interest in the recruitment of brown adipose tissue in different physiological situations (see previous section on physiological recruitment of brown adipose tissue), measurements of (^{3}H)GDP-binding capacities of brown fat mitochondria isolated from animals under different conditions have been used considerably. Such measurements were initially performed by filtration of isolated mitochondria, incubated with a nucleotide such as 10 μM (^{3}H)GDP, and with (^{14}C)sucrose as an extramitochondrial space marker (98, 44, 99). Recently, however, many studies have instead been performed using a centrifugation technique in Eppendorf tubes (43, 100, 101). It is not known to what extent the two methods yield similar results, but it seems that the centrifugation technique yields lower values than the filtration technique (102).

It may be stressed that the physiologically important parameter probably is not the specific amount of GDP-binding per milligram of mitochondrial protein (the parameter generally measured), but rather the total amount of thermogenin in the entire animal. It is this amount that is rate-limiting for the nonshivering thermogenesis capacity (44, 45, 103).

Brown fat Mitochondria as a Potential Source for Thermogenin. By a fairly simple technique, developed from the isolation of the ATP/ADP carrier, it has been possible to isolate thermogenin from brown fat mitochondria in a pure form (104, 19).

One advantage of this technique is the possibility of producing antibodies against thermogenin. These antibodies have been used as an independent parameter for determining thermogenin content in investigations concerning the possible existence of a "masked" form of thermogenin and an "unmasking" process (105, 106, 107, 108). A recent report demonstrates reconstitution of GDP-sensitive H^+-permeability by incorporation of thermogenin into liposomes (109).

ACKNOWLEDGMENTS

Work reported from the authors' laboratory has been supported by the Swedish Natural Science Research Council and the Swedish Central Committee for Laboratory Animals. The authors would like to thank Eamonn Connolly and the rest of the brown fat group for valuable discussions.

REFERENCES

1. G. J. Hausman, D. R. Campion, and R. J. Martin, 1980, Search for the adipocyte precursor cell and factors that promote its differentiation, *J. Lipid Res.* **21**:657–670.
2. B. Cannon, and J. Nedergaard, 1985, Brown adipose tissue. The molecular mechanisms controlling activity and thermogenesis, in *New Perspectives in Adipose Tissue,* R. Van, and A. Cryer, eds., Butterworth, London, pp. 233–270.
3. M. Nechad, P. Kuusela, C. Carneheim, P. Bjorntorp, J. Nedergaard, and B. Cannon, 1983, Development of brown fat cells in monolayer culture. I. Morphological and biochemical distinction from white fat cells in culture, *Exp. Cell Res.* **149**:105–118.
4. J. Nedergaard, E. Connolly, and B. Cannon, 1985, Brown adipose tissue in the mammalian neonate, in *Brown Adipose Tissue,* P. Trayhurn, and D. G. Nicholls, eds., Edward Arnold Ltd., London.
5. J. Himms-Hagen, 1986, Brown adipose tissue in cold-acclimated mammals, in *Brown Adipose Tissue,* P. Trayhurn, and D. G. Nicholls, eds., Edward Arnold, London.
6. A. Sclafani, and D. Springer, 1976, Dietary obesity in adult rats: Similarities to hypothalamic and human obesity syndromes, *Physiol. Behav.* **17**:461–471.
7. N. J. Rothwell, and M. J. Stock, 1979, A role for brown adipose tissue in diet-induced thermogenesis, *Nature* **281**:31–35.
8. B. Cannon, L. Romert, U. Sundin, and T. Barnard, 1977, Morphology and biochemical properties of perirenal adipose tissue from lamb (*Ovis aries*). A comparison with brown adipose tissue, *Comp. Biochem. Physiol.* **56B**:87–99.

9. P. Young, J. R. S. Arch, and M. Ashwell, 1984, Brown adipose tissue in the parametrial fat pad of the mouse, *FEBS Lett.* **167**:10–14.
10. J. Nedergaard and O. Lindberg, 1982, The brown fat cell, *Int. Rev. Cytol.* **74**:187–286.
11. D. G. Nicholls, 1979, Brown adipose tissue mitochondria, *Biochim. Biophys. Acta* **549**:1–29.
12. M. Klingerberg, 1984, Characteristics of the uncoupling protein from brown-fat mitochondria, *Biochem. Soc. Trans.* **12**:390–393.
13. J. Nedergaard and B. Cannon, Thermogenic mitochondria, in *New Comprehensive Biochemistry (Bioenergetics),* L. Ernster, eds., Elsevier, Amsterdam, pp. 291–314.
14. G. M. Heaton, R. J. Wagenvoord, A.J.R. Kemp and D. G. Nicholls, 1978, Brown-adipose-tissue mitochondria: Photoaffinity labelling of the regulatory site of energy dissipation, *Eur. J. Biochem.* **82**:515–521.
15. D. Ricquier and J. C. Kader, 1976, Mitochondrial protein alteration in active brown fat, A sodium dodecyl sulfate-polyacrylamide gel electrophoretic study, *Biochem. Biophys. Res. Commun.* **73**:577–583.
16. C. S. Lin, H. Hackenberg, and E. M. Klingerberg, 1980, The uncoupling protein from brown adipose tissue mitochondria is a dimer. A hydrodynamic study, *FEBS Lett.* **113**:304–306.
17. H. Aquilla, T. A. Link, and M. Klingerberg, 1985, The uncoupling protein from brown fat mitochondria is related to the mitochondrial ADP/ATP carrier. Analysis of sequence homologies and of the folding protein in the membrane, *EMBO Journal.* **4**:2369–2376.
18. F. Bouillaud, J. Weisenbach and D. Ricquier, 1986, Complete DNA-derived amino acid sequence of rat brown fat uncoupling protein. *J. Biol. Chem.* **261**:1487–1490.
19. C. S. Lin and M. Klingenberg, 1982, Characteristics of the isolated purine nucleotide binding protein from brown fat mitochondria, *Biochemistry* **21**:2950–2956.
20. N. Mohell, J. Nedergaard, and B. Cannon, 1983, Quantitative differentiation of alpha- and beta-adrenergic respiratory responses in isolated hamster brown fat cells: Evidence for the presence of an $alpha_1$-adrenergic component, *Eur. J. Pharmacol.* **93**:183–193.
21. R. J. Schimmel, L. McCarthy, and K. K. McMahon, 1983, $Alpha_1$-adrenergic stimulation of hamster brown adipose respiration, *Am. J. Physiol.* **244**:C362–C368.
22. P. Svoboda, J. Svartengren, M. Snochowski, J. Houstek, and B. Cannon, 1979, High number of high-affinity binding sites for (−)-(^{3}H)dihydroalprenolol on isolated hamster brown-fat cells, *Eur. J. Biochem.* **102**:203–210.
23. J. R. S. Arch, A. T. Ainsworth, M. A. Cawthorne, V. Piercy, M. V. Sennitt, V. E. Thody, C. Wilson, and S. Wilson, 1984, Atypical beta-adrenoceptor on brown adipocytes as target for anti-obesity drugs, *Nature* **309**:163–165.
24. J. Svartengren, P. Svoboda, and B. Cannon, 1982, Desensitisation of beta-adrenergic responsiveness in-vivo. Decreased coupling between receptors and adenylate cyclase in isolated brown fat cells, *Eur. J. Biochem.* **128**:481–488.
25. J. P. Skala and B. L. Knight, 1977, Protein kinases in brown adipose tissue of developing rats, *J. Biol. Chem.* **252**:1064–1070.
26. C. Carneheim, J. Nedergaard, and B. Cannon, 1984, Beta-adrenergic stimulation of lipoprotein lipase activity in rat brown adipose tissue during acclimation to cold, *Am. J. Physiol.* **246**:E327–E333.
27. C. Carneheim, J. Nedergaard, and B. Cannon, 1985, Beta-adrenergic stimulation of lipoprotein lipase activity in rat brown adipose tissue is due to CAMP-dependent mRNA synthesis. *Acta Physiol. Scand.* 48, 124, 542.
28. S. B. Prusiner, R. H. Eisenhart, E. Rylander, and O. Lindberg, 1968, The regulation of oxidative metabolism of isolated brown fat cells, *Biochem. Biophys. Res. Commun.* **30**:508–515.

29. S. B. Prusiner, B. Cannon, and O. Lindberg, 1968, Oxidative metabolism in cells isolated from brown adipose tissue. I. Catecholamine and fatty acid stimulation of respiration, *Eur. J. Biochem.* **6:**15–22.
30. N. Reed and J. N. Fain, 1968, Potassium-dependent stimulation of respiration in brown fat cells by fatty acids and lipolytic agents, *J. Biol. Chem.* **243:**6077–6083.
31. B. Pettersson, 1977, CO_2-mediated control of fatty acid metabolism in isolated hamster brown-fat cells during norepinephrine stimulation, *Eur. J. Biochem.* **72:**235–240.
32. N. Mohell, J. Svartengren, and B. Cannon, 1983, Identification of (^{3}H)prazosin binding sites in brown adipose tissue as alpha$_1$-adrenergic receptors, *Eur. J. Pharmacol.* **92:** 15–25.
33. L. Girardier and G. Schneider-Picard, 1983, Alpha- and beta-adrenergic mediation of membrane potential changes and metabolism in rat brown adipose tissue, *J. Physiol.* **335:**629–641.
34. L. Girardier, J. Seydoux, and T. Clausen, 1968, Membrane potential of brown adipose tissue. A suggested mechanism for the regulation of thermogenesis, *J. Gen. Physiol.* **52:**925–940.
35. J. M. Horowitz, B. A. Horwitz, and R. E. Smith, 1971, Effect in vivo of norepinephrine on the membrane resistance of brown fat cells, *Experientia* **27:**1419–1421.
36. E. Connolly, E. Nånberg, and J. Nedergaard, 1984, Na^+ dependent, alpha-adrenergic mobilization of intracellular (mitochondrial) Ca^{2+} in brown adipocytes, *Eur. J. Biochem.* **141:**187–193.
37. E. Nanberg, J. Nedergaard, and B. Cannon, 1984, Alpha-adrenergic effects on $^{86}Rb^+$ (K^+) potentials and fluxes in brown fat cells, *Biochim. Biophys. Acta* **804:**291–300.
38. E. Nanberg, E. Connolly, and J. Nedergaard, 1985, Alpha$_1$-adrenergic stimulation in brown adipocytes: Presence and possible role of a Ca^{2+}-dependent K^+ channel, *Biochim. Biophys. Acta* **804:**291–300.
39. J. P. Skala and B. L. Knight, 1979, Cyclic GMP and cyclic GMP-dependent protein kinase in brown adipose tissue of developing rats, *Biochim. Biophys. Acta* **582:**122–131.
40. A. Raasmaja, N. Mohell, and J. Nedergaard, 1985, Increased alpha$_1$-adrenergic receptor density in brown adipose tissue of cold-acclimated rats and hamsters, *Eur. J. Pharmacol.* **106:**489–498.
41. A. Raasmaja, N. Mohell, and J. Nedergaard, 1984, Increased alpha$_1$-adrenergic receptor density in brown adipose tissue of cafeteria-fed rats, *Biosci. Rep.* **4:**8851–8859.
42. I. L. Cameron, and R. E. Smith, 1964, Cytological responses of brown fat tissue in cold-exposed rats, *J. Cell Biol.* **23:**89–100.
43. J. Rafael and H. W. Heldt, 1976, Binding of guanine nucleotides to the outer surface of the inner membrane of guinea pig mitochondria in correlation with the thermogenic activity of the tissue, *FEBS Lett.* **63:**304–308.
44. U. Sundin, D. Herron, and B. Cannon, 1981, Brown fat thermoregulation in developing hamsters (*Mesocricetus auratus*). A GDP-binding study, *Biol. Neonate* **39:**141–149.
45. U. Sundin and B. Cannon, 1980, GDP-binding to the brown fat mitochondria of developing and cold-adapted rats, *Comp. Biochem. Physiol.* **65B:**463–471.
46. G. Heldmaier and K. Hoffman, 1974, Melatonin stimulates growth of brown adipose tissue, *Nature* **247:**224–225.
47. G. Heldmaier, S. Steinlechner, J. Rafael, and P. Vsiansky, 1981, Photoperiodic control and effects of melatonin on nonshivering thermogenesis and brown adipose tissue, *Science* **212:**917–919.
48. S. L. Brooks, N. J. Rothwell, M. J. Stock, A. E. Goodbody, and P. Trayhurn, 1980, Increased proton conductance pathway in brown adipose tissue mitochondria of rats exhibiting diet-induced thermogenesis, *Nature* **286:**274–276.

49. J. Nedergaard, A. Raasmaja, and B. Cannon, 1984, Parallel increases of amount of (^{3}H)GDP binding and thermogenin antigen in brown-adipose-tissue mitochondria of cafeteria-fed rats, *Biochem. Biophys. Res. Commun.* **122**:1328–1336.
50. I. Dardick, 1978, Hibernoma: A possible model of brown fat histogenesis, *Hum. Pathol.* **9**:321–329.
51. D. Ricquier, M. Nechad, and G. Mory, 1982, Ultrastructural and biochemical characterization of human brown adipose tissue in pheochromocytoma, *J. Clin. Endocrinol. Metab.* **54**:803–807.
52. F. Bouillaud, M. Combes-George, and D. Ricquier, 1983, Mitochondria of adult human brown adipose tissue contain a 32000-Mr uncoupling protein, *Biosci. Rep.* **3**:775–780.
53. S. Cunningham, E. Rial, and D. G. Nicholls, 1984, Bioenergetic characterization of human brown fat mitochondria, *Eur. Bioenerg. Conf.* **3**:477–478.
54. G. Mory, F. Bouillaud, M. Combes-George, and D. Ricquier, 1984, Noradrenaline controls the concentration of the uncoupling protein in brown adipose tissue, *FEBS Lett.* **166**:393–396.
55. G. Heldmaier, 1985, Seasonal acclimatization and thermogenesis, Circulation, Respiration, and Metabolism, R. Gilles, ed., Springer-Verlag, Berlin, pp. 490–501.
56. J. N. Fain, N. Reed, and R. Saperstein, 1967, The isolation and metabolism of brown fat cells, *J. Biol. Chem.* **242**:1887–1894.
57. R. M. Locke, E. Rial, and D. G. Nicholls, 1982, The acute regulation of mitochondrial proton conductance in cells and mitochondria from the brown fat of cold-adapted and warm-adapted guinea pigs, *Eur. J. Biochem.* **129**:381–387.
58. J. Nedergaard, 1982, Catecholamine sensitivity in brown fat cells from cold-adapted hamsters and rats, *Am. J. Physiol.* **242**:C250–C257.
59. L. Bukowiecki, A. J. Collet, N. Follea, G. Guay, and L. Jahjah, 1982, Brown adipose tissue hyperplasia: A fundamental mechanism of adaptation to cold and hyperphagia, *Am. J. Physiol.* **242**:E353–E359.
60. K. R. F. Elliot, C. Roncero, M. Fernandez, and B. Benito, 1984, Preparation of hormone-sensitive adipocytes from fetal rat brown adipose-tissue, *Biosci. Rep.* **4**:397–401.
61. J. N. Fain, N. Mohell, M. A. Wallace, and I. Mills, 1984, Metabolic effects of beta, $alpha_1$ and $alpha_2$ receptor activation on brown adipocytes isolated from the perirenal adipose tissue of fetal lambs, *Metabolism* **33**:289–294.
62. A. H. Klein, J. J. Jenkins, A. Reviczky, and D. A. Fisher, 1981, Thyroid hormone sensitive brown adipose tissue respiration in the newborn rabbit, *Am. J. Physiol.* **24**:E449–E453.
63. A. H. Klein and D. A. Fisher, 1983, Effects of changes in cell size and ouabain sensitive respiration on development of brown adipose tissue thermogenesis in the rabbit, *Pediatr. Res.* **17**:220–223.
64. J. Nedergaard, B. Cannon, and O. Lindberg, 1977, Microcalorimetry of isolated mammalian cells, *Nature* **267**:518–520.
65. B. Cannon and J. Nedergaard, 1979, The physiological role of pyruvate carboxylation in hamster brown adipose tissue, *Eur. J. Biochem.* **94**:419–426.
66. J. Nedergaard and O. Lindberg, 1979, Norepinephrine-stimulated fatty-acid release and oxygen consumption in isolated hamster brown fat cells, *Eur. J. Biochem.* **95**:139–145.
67. N. Reed and J. N. Fain, 1968, Stimulation of respiration in brown fat cells by epinephrine dibutyryl-3',5'-adenosine monophosphate, and m-chloro-(carbonyl cyanide) phenylhydrazone, *J. Biol. Chem.* **243**:2843–2848.
68. B. Pettersson and I. Vallin, 1976, Norepinephrine-induced shift in levels of adenosine 3':5'-monophosphate and ATP parallel to increased respiratory rate and lipolysis in isolated hamster brown fat cells, *Eur. J. Biochem.* **62**:383–390.
69. A. Kuroshima and T. Yahata, 1979, Thermogenic responses of brown adipocytes to

noradrenaline and glucagon in heat-acclimated and cold-acclimated rats, *Jap. J. Physiol.* **29**:683–690.

70. A. Kuroshima, K. Doi, and T. Ohno, 1978, Role of glucagon in metabolic acclimation to cold and heat, *Life Sci.* **23**:1405–1410.
71. V. S. M. Bernson, P. Lundberg, and B. Pettersson, 1979, Flavoprotein-linked substrate oxidation in preparations of hamster brown adipocytes. A discrimination between internally and externally oxidized substrates, *Biochim. Biophys. Acta* **587**:353–361.
72. L. L. Bieber, B. Pettersson, and O. Lindberg, 1975, Studies on norepinephrine-induced efflux of free fatty acid from hamster brown adipose tissue cells, *Eur. J. Biochem.* **58**:375–381.
73. S. B. Prusiner, B. Cannon, T. M. Ching, and O. Lindberg, 1968, Oxidative metabolism in cells isolated from brown adipose tissue. 2. Catecholamine-regulated respiratory control, *Eur. J. Biochem.* **7**:51–57.
74. B. Cannon, J. Nedergaard, L. Romert, U. Sundin, and J. Svartengren, 1978, The biochemical mechanism of thermogenesis in brown adipose tissue, in *Strategies in Cold: Natural Torpidity and Thermogenesis,* L. L. Wang, and J. Hudson, eds. Academic Press, New York, pp. 567–594.
75. J. Ho, 1970, Radiochemical assay of long chain fatty acids using ^{63}Ni as tracer, *Anat. Biochem.* **36**:105–113.
76. E. Connolly, E. Nånberg, and J. Nedergaard, 1986, Norepinephrine-induced Na^+ influx in brown adipocytes is cyclic-AMP mediated. *J. Biol. Chem.* In press.
77. B. A. Horwitz, J. M. Horowitz, and R. E. Smith, 1969, Norepinephrine-induced depolarisation of brown fat cells, *Proc. Nat. Acad. Sci. (USA)* **64**:113–120.
78. U. Sundin and J. N. Fain, 1983, Alpha$_2$-adrenergic inhibition of lipolysis in rat brown adipocytes, *Biochem. Pharmacol.* **32**:3117–3120.
79. H. Sugihara, N. Yonemitsu, K. Ohta, S. Miyabara, and A. Nagayama, 1983, Immunocytochemistry of the microtubules of fat-laden cells. Brown fat cells and adrenocortical cells in primary monolayer culture, *Histochemistry* **79**:227–236.
80. H. Sugihara, S. Miyabara, N Yonemits, and K. Ohta, 1983, Hormonal sensitivity of brown fat cells of fetal rats in monolayer culture, *Exp. Clin. Endocrinol.* **82**:309–319.
81. M. Nechad, 1983, Development of brown fat cells in monolayer culture. II. Ultrastructural characterization of precursors, differentiating adipocytes and their mitochondria, *Exp. Cell Res.* **149**:119–127.
82. P. Kuusela, J. Nedergaard, and B. Cannon, 1986, Beta-adrenergic stimulation of fatty acid release from brown fat cells matured in cell culture, *Life Sci.* **38**:589–599.
83. P. Kuusela and J. Hirvonen, 1984, Marsujen ruskeiden rasvasoluen kasvu ja erilaistuminen primäärisoluviljelmissä, *Finn. Cold Res. Symp.* 1984:21.
84. P. Björntorp, M. Karlsson, H. Pertoft, P. Pettersson, L. Sjostrom, and U. Smith, 1978, Isolation and characterization of cells from rat adipose tissue developing into adipocytes, *J. Lipid Res.* **19**:316–324.
85. G. J. Hausman, J. Novakofski, R. J. Martin, and G. B. Thomas, 1984, The histochemistry of developing adipocytes in primary stromal-vascular cultures of rat adipose tissue, *Histochemistry* **80**:353–358.
86. B. Cannon and O. Lindberg, 1979, Mitochondria from brown adipose tissue: Isolation and properties, *Methods Enzymol.* **55F**:65–78.
87. J. Houstek, B. Cannon, and O. Lindberg, 1975, Glycerol-3-phosphate shuttle and its function in intermediary metabolism of hamster brown-adipose tissue, *Eur. J. Biochem.* **54**:11–18.
88. J. Nedergaard, 1981, Effects of cations on brown adipose tissue in relation to possible metabolic consequences of membrane depolarisation, *Eur. J. Biochem.* **114**:159–167.
89. B. Cannon, V. M. S. Bernson, and J. Nedergaard, 1984, Metabolic consequences of

limited substrate anion permeability in brown fat mitochondria from a hibernator, the golden hamster, *Biochim. Biophys. Acta* **766**:483–491.

90. D. G. Nicholls, 1974, Hamster brown-adipose-tissue mitochondria. The control of respiration and the proton electrochemical potential gradient by possible physiological effectors of the proton conductance of the inner membrane, *Eur. J. Biochem.* **49**:573–583.
91. J. Nedergaard, 1983, The relationship between extra-mitochondrial Ca^{2+} concentration, respiratory rate, and membrane potential in rat brown-adipose-tissue mitochondria, *Eur. J. Biochem.* **133**:185–191.
92. B. Cannon and G. Vogel, 1977, The mitochondrial ATPase of brown adipose tissue. Purification and comparison with the mitochondrial ATPase from beef heart, *FEBS Lett.* **76**:284–289.
93. O. Lindberg, J. de Pierre, E. Rylander, and B. A. Afzelius, 1967, Studies of the mitochondrial energy-transfer system of brown adipose tissue, *J. Cell. Biol.* **34**:293–310.
94. D. G. Nicholls, H. J. Grav, and O. Lindberg, 1972, Mitochondria from hamster brown-adipose tissue. Regulation of respiration in vitro by variations in volume of the matrix compartment, *Eur. J. Biochem.* **31**:526–533.
95. D. G. Nicholls and O. Lindberg, 1973, Brown-adipose-tissue mitochondria. The influence of albumin and nucleotides on passive ion permeabilities, *Eur. J. Biochem.* **37**:523–530.
96. D. G. Nicholls, 1974, Hamster brown-adipose-tissue mitochondria. The chloride permeability of the inner membrane under respiring conditions, the influence of purine nucleotides, *Eur. J. Biochem.* **49**:585–593.
97. R. Locke, E. Real, I. D. Scott, and D. G. Nicholls, 1982, Fatty acids as acute regulators of the proton conductance of hamster brown-fat mitochondria, *Eur. J. Biochem.* **129**:373–380.
98. D. G. Nicholls, 1976, Hamster brown-adipose-tissue mitochondria. Purine nucleotide control of the ion conductance of the inner membrane, the nature of the nucleotide binding site, *Eur. J. Biochem.* **62**:223–228.
99. J. Nedergaard, W. Becker, and B. Cannon, 1983, Effects of dietary essential fatty acids on active thermogenin content in rat brown adipose tissue, *J. Nutr.* **113**:1717–1724.
100. E. Rial and D. G. Nicholls, 1983, The regulation of the proton conductance of brown fat mitochondria. Identification of functional and non-functional nucleotide-binding sites, *FEBS Lett.* **161**:284–288.
101. M. Desautels, G. Zaror-Behrens, and J. Himms-Hagen, 1978, Increased purine nucleotide binding, altered polypeptide composition, and thermogenesis in brown adipose tissue mitochondria of cold-acclimated rats, *Can. J. Biochem.* **56**:378–383.
102. J. Nedergaard and B. Cannon, 1986, Apparent unmasking of GDP-binding in brown-fat mitochondria is due to mitochondrial swelling. Submitted for publication.
103. B. J. Cannon, J. Nedergaard, and U. Sundin, 1981, Thermogenesis, brown fat and thermogenin, In *Survival in Cold,* X. J. Musacchia, and L. Jansky, eds. Elsevier-North, Amsterdam, pp. 99–120.
104. C. S. Lin and M. Klingenberg, 1980, Isolation of the uncoupling protein from brown adipose tissue mitochondria, *FEBS Lett.* **113**:299–303.
105. B. Cannon, A. Hedin, and J. Nedergaard, 1982, Exclusive occurrence of thermogenin antigen in brown adipose tissue, *FEBS Lett.* **150**:129–132.
106. M. E. J. Lean, W. J. Branch, W. P. T. James, G. Jennings, and M. Ashwell, 1983, Measurement of rat brown adipose tissue mitochondrial uncoupling protein by radioimmunoassay: Increased concentration after cold acclimation, *Biosci. Rep.* **3**:61–71.
107. D. Ricquier, J. P. Barlet, J. M. Garel, M. Combes-Georges, and M. P. Dubois, 1983, An immunological study of the uncoupling protein of brown adipose tissue mitochondria, *Biochem. J.* **210**:859–866.

108. J. Nedergaard and B. Cannon, 1985, Parallel increase in GDP binding and thermogenin amount in brown adipose tissue mitochondria from cold-exposed rats, *Am. J. Physiol.*, **248**:C365–C371.
109. M. Klingenberg and E. Winkler, 1985, The reconstituted isolated uncoupling protein is a membrane potential driven H^+ translocator. *EMBO Journal* **4**:3087–3092.
110. O. Lindberg, L. L. Bieber, and J. Houstek, 1976, Brown adipose tissue metabolism: An attempt to apply results from in vitro experiments on tissue in vivo, in *Regulation of Depressed Metabolism and Thermogenesis,* L. Jansky, and W. J. Musacchia, eds., Thomas, Springfield, Ohio, 117–136.

3

The Ultrastructure of Adipocytes

Bernard G. Slavin
University of Southern California

STUDIES ON THE ULTRASTRUCTURE OF ADIPOCYTES

Introduction

The use of the electron microscope to study cellular ultrastructure has enabled investigators to delve into the structure of white adipocytes and brown adipocytes at a level far beyond that possible with the light microscope. Using routine light microscopy to study white adipocytes, one observes a large (up to 200 μm), spherical or ovoid cell with a peripherally placed, flattened nucleus and a thin rim of cytoplasm external to a single large lipid droplet (Fig. 3-1). Brown adipocytes, on the other hand, are generally smaller (up to 60 μm) and polygonal, and they contain multiple lipid droplets and slightly eccentric nuclei (Fig. 3-2). Mitochondria have been demonstrated within the cytoplasm of white and brown adipocytes by the use of a variety of staining techniques (1–3).

In addition, Golgi zones have been described in white adipocytes (4), but other organelles have not been detected with the light microscope. In addition, glycogen accummulation in white and brown adipocytes has been observed by light microscopy in animals injected with insulin, in rats fed after a period of fasting (5, 6), and in insulin-exposed organ culture preparations of adipose tissue (7).

Initial electron microscopic studies of mature adipocytes were of relatively low quality due to inadequate fixation, sectioning, and staining techniques (8–15). However, some important basic morphologic data were obtained from these early studies. For example, the close relationship between collagen fibers, the basal lamina, and the surface of white adipocytes was established (8, 13, 14, 16). In addition, a description of the Golgi apparatus, smooth and rough endoplasmic reticulum, ribonucleoproteins, microtubules, microfilaments, glycogen, and plasmalemmal invaginations (micropinocytotic vesicles) in white adipocytes was reported (10, 14, 15, 16). The fine structure of mitochondria in white and brown adipocytes was also described fairly accurately (2, 8–10, 12, 14–16). The presence of ribonucleoprotein particles, a poorly developed Golgi zone, and a lack of endoplasmic reticulum were observed in brown adipocytes (12).

The use of aldehyde fixation (with osmium tetroxide postfixation) and epoxy resin embedments to achieve superior preservation of cell ultrastructure led to further refinement in the description of adipocyte components.

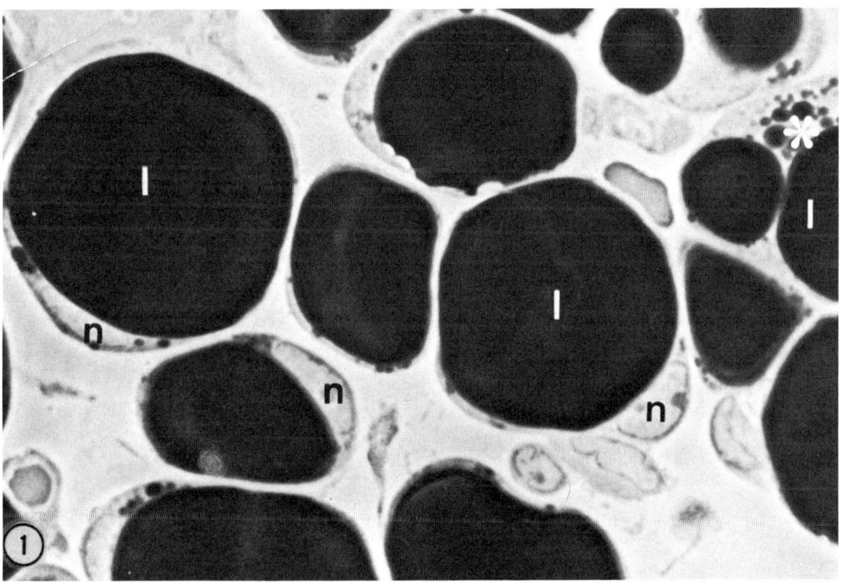

Figure 3-1. Phase contrast micrograph of semithin section of rat epididymal white adipose tissue. Several adipocytes are seen showing the central large lipid droplet (l) and the peripherally located nucleus (n). A cluster of intracytoplasmic small lipid droplets is indicated by the asterisk. × 1950. (*From Heindel et al., 1975; reproduced by permission of Pergamon Press, Ltd., Oxford, England.*)

The following sections describe more recent investigations utilizing the latest electron microscopic techniques to study the ultrastructure of white and brown adipocytes.

Transmission and Scanning Electron Microscopy

White Adipocytes. Modern transmission electron microscopic studies of white adipocytes have focused on (1) a description of normal cell fine structure, (2) ultrastructural changes during development, fasting, diabetes, hormonal stimulation, obesity, and aging, (3) hormone receptor sites, and (4) structure-function relationships with regard to lipid transport.

Normal Ultrastructure. Using modern transmission electron microscopic techniques, a number of investigators observed that the fine structural details of normal mature white adipocytes were essentially similar to those in older studies (17–22). Additional information was gleaned concerning various organelles and inclusions within white adipocytes. For example, in adult

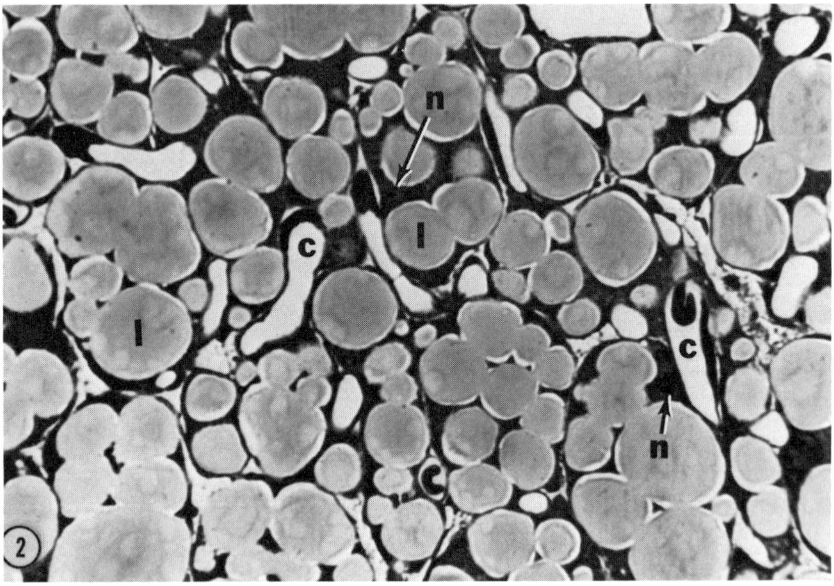

Figure 3-2. Semithin section of rat interscapular brown adipose tissue. Cells contain multiple lipid droplets (l) of varying size and eccentrically placed nuclei (n). c, capillaries. ×1800.

man, developing mammals, chicks, and mature lovebirds, lipid inclusions in white adipocytes were found to be surrounded by a well-developed orthogonal network of filaments (17, 23–27). The filaments measured 75–100 Å in diameter. As the adipocytes matured and lipid droplets enlarged, these filaments gradually disappeared (23). In addition to the findings of filaments surrounding lipid droplets in white adipocytes, randomly distributed cytoplasmic filaments have been observed in developing cells (26, 28). These filaments measured 100 Å in diameter and appeared to be different from tonofilaments, actin, myosin, and neurofilaments of other cells (26).

The plasma membrane of white adipocytes consists of a trilaminar structure typical of all mammalian cells. However, the membrane contains plasmalemmal invaginations similar in morphology to those found in plasma membranes of endothelial cells and smooth muscle cells (Fig. 3-3). Plasmalemmal invaginations occasionally take the form of rosettes or alveoli, which are characterized by several vesicles connected to a single invagination. Freeze-fracture techniques combined with transmission electron microscopy have revealed in white adipocytes the intramembranous portion of the plasma membrane and its associated plasmalemmal invaginations (Fig. 3-4). Also, it has been determined that the number of intramembranous protein particles is higher on the internal leaflet than on the external leaflet of the plasma membrane (18).

Ultrastructural Changes

Development. Transmission electron microscopy has been used by several authors to elucidate the ultrastructure of developing white adipocytes. Basically, the adipocytes develop from fibroblastlike cells (Fig. 3-5) that accummulate lipid in the form of droplets and differentiate into the typical large signet-ring forms (27–33). White adipocytes of bone marrow, on the other hand, appear to develop from perisinusoidal cells that are oval or elongated and lack prominent rough endoplasmic reticulum (34).

Cultured preadipocytes have been used in a model system to study various organelle changes at the electron microscopic level (27). In this system, undifferentiated preadipocytes are characterized by their content of multiple, small lipid droplets and subplasmalemmal actinlike "stress" filaments and bundles of intermediate filaments deeper in the cytoplasm. These filaments are greatly reduced as differentiation proceeds. As with mature white adipocytes, the plasma membrane contains plasmalemmal invaginations that with differentiation assume the structure of rosettes. Using cytochemical techniques at the electron microscopic level, these authors (27) demonstrated an intimate relationship between lipid droplets, mitochondria, microperoxisomes, and endoplasmic reticulum surrounding the lipid droplet.

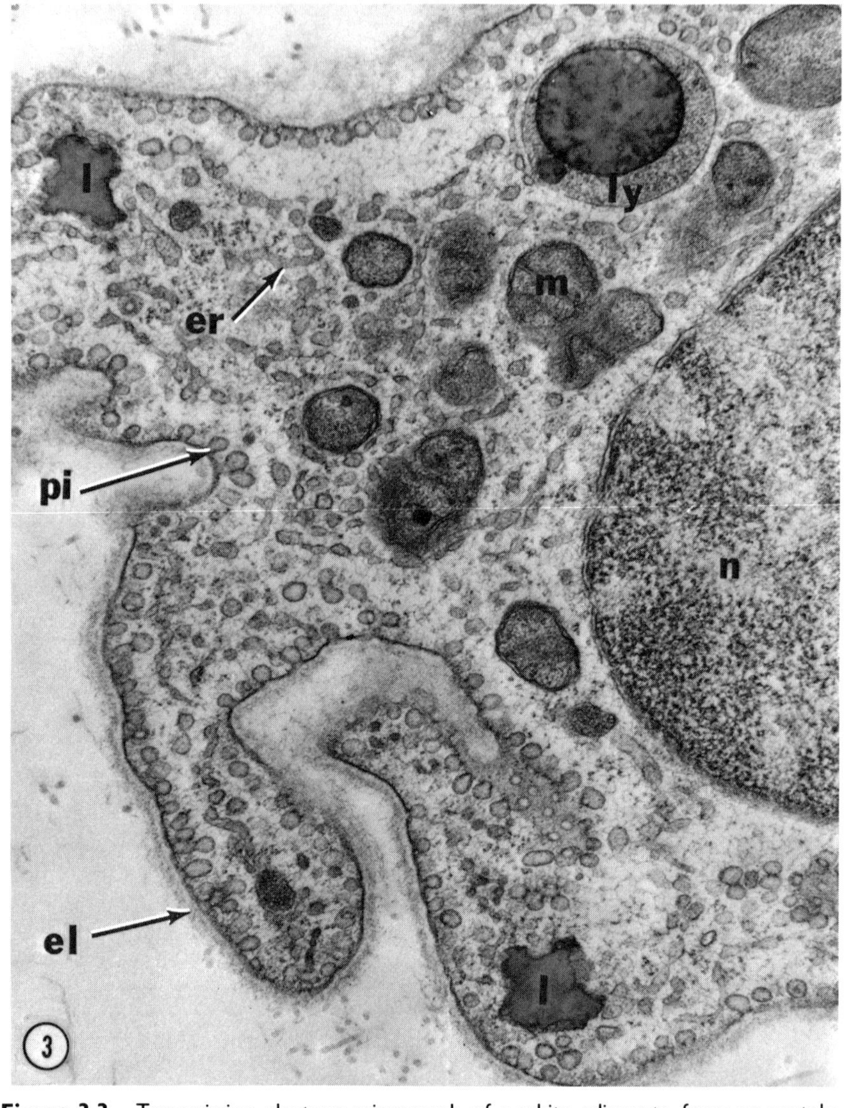

Figure 3-3. Transmission electron micrograph of a white adipocyte from an acutely fasted rat. Note drastic lipid depletion. Two small lipid droplets (l) remain. Note: n, nucleus; m, mitochondrion; ly, lysosome; pi, plasmalemmal invagination; er, endoplasmic reticulum; el, external lamina. ×37,000.

In a recent electron microscopic study, coated vesicles were described as normal constituents of developing white adipocytes (35). Serial section analysis indicated that approximately 50% of these coated vesicles were not connected to the plasma membrane.

Fasting and Diabetes. Ultrastructural studies of white adipocytes in animals exposed to drastic physiological changes such as fasting and diabetes have been reported by several authors (10, 11, 14, 21, 36–41). Such changes involve the transformation of the large unilocular spherical adipocyte

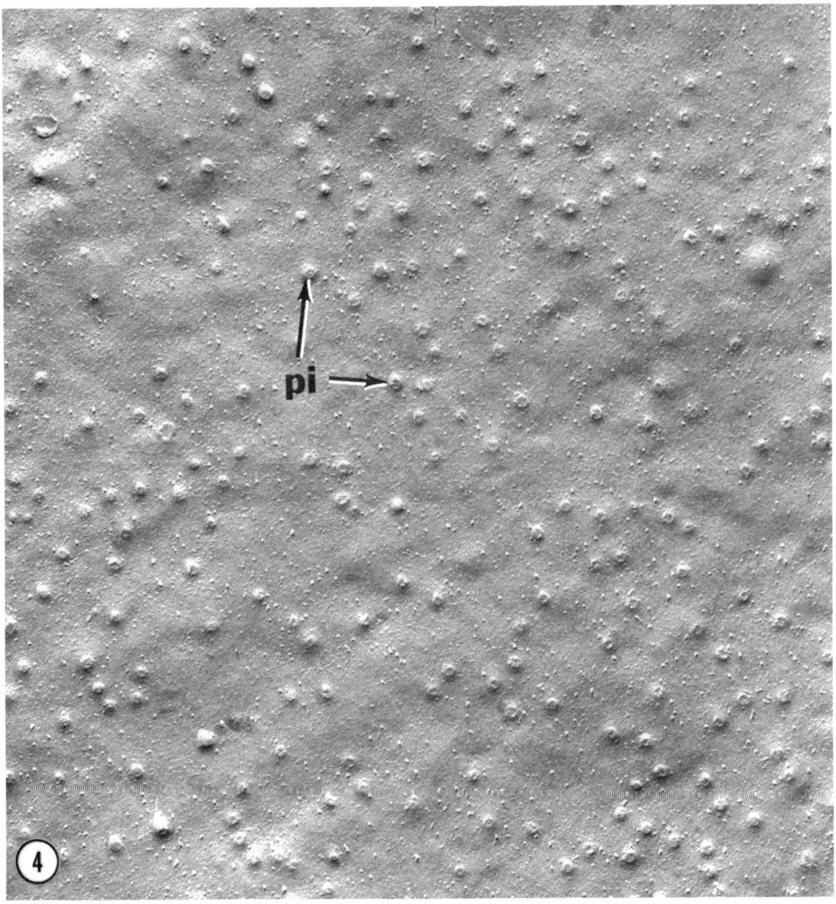

Figure 3-4. Freeze-fracture replica of epididymal white adipocyte from a normal rat. The plane of fracture exhibits the E-face (inner aspect of the outer layer of the plasma membrane). Note randomly scattered plasmalemmal invaginations (pi) surrounded by numerous small intramembranous particles. ×46,500.

through a reduced multilocular stage to that of a lipid-depleted highly stellate form (Fig. 3-6). Mobilization of lipid from white adipocytes in these cases involves hydrolysis of triacylglycerol to fatty acids and glycerol. These hydrolyzed products leave the cell in a theorized route to be discussed in another section of this review.

With the loss of triacylglycerol from white adipocytes, there appears to be an increase in the amount of endoplasmic reticulum, lysosomes, and plasmalemmal invaginations (18, 38). Freeze-fracture replicas of plasma membranes indicate that the total number of plasmalemmal invaginations remains constant during lipid mobilization and only the concentration per unit area of plasma membrane increases (18, Fig. 3-7). The same is true for intramembranous particles within the plasma membrane. However, this increase does represent a basic reorganization of the plasma membrane of white adipocytes during lipid mobilization. An increase in the number of

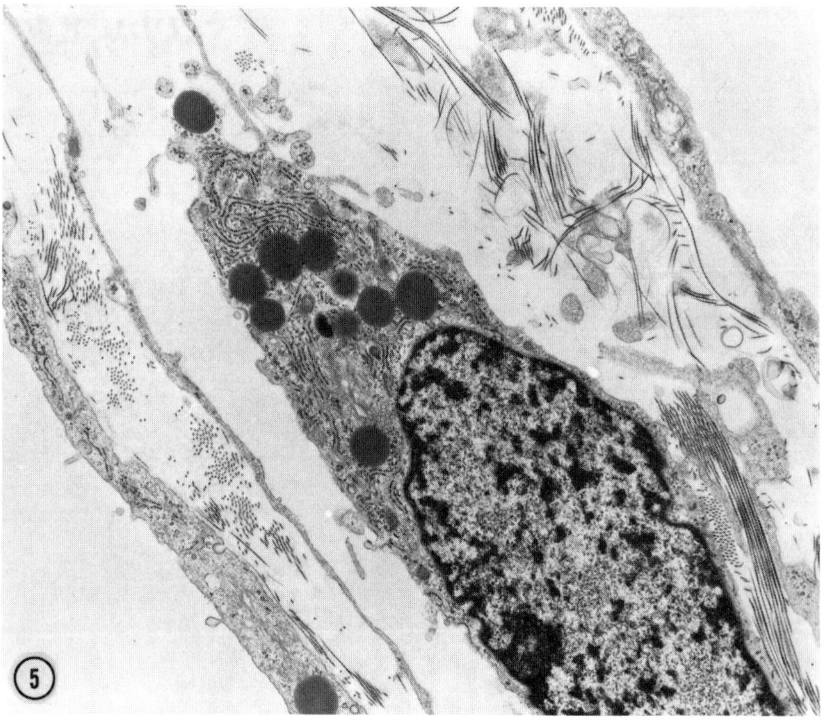

Figure 3-5. Transmission electron micrograph of a developing mesenteric white adipocyte in a newborn rat. Note multiple small lipid droplets and numerous profiles of rough endoplasmic reticulum in this spindle-shaped cell. × 12,000.

lysosomes in the cells with fasting is probably a real phenomenon since only a few lysosomes are noted in the fully lipid-replete cell. Such lysosomes probably function in the lipid degradation process. The increase in profiles of the endoplasmic reticulum of white adipocytes with fasting is most likely because of a concentrating effect of these organelles that is caused by lipid loss and cell shrinkage. This conclusion is based on the knowledge that lipid

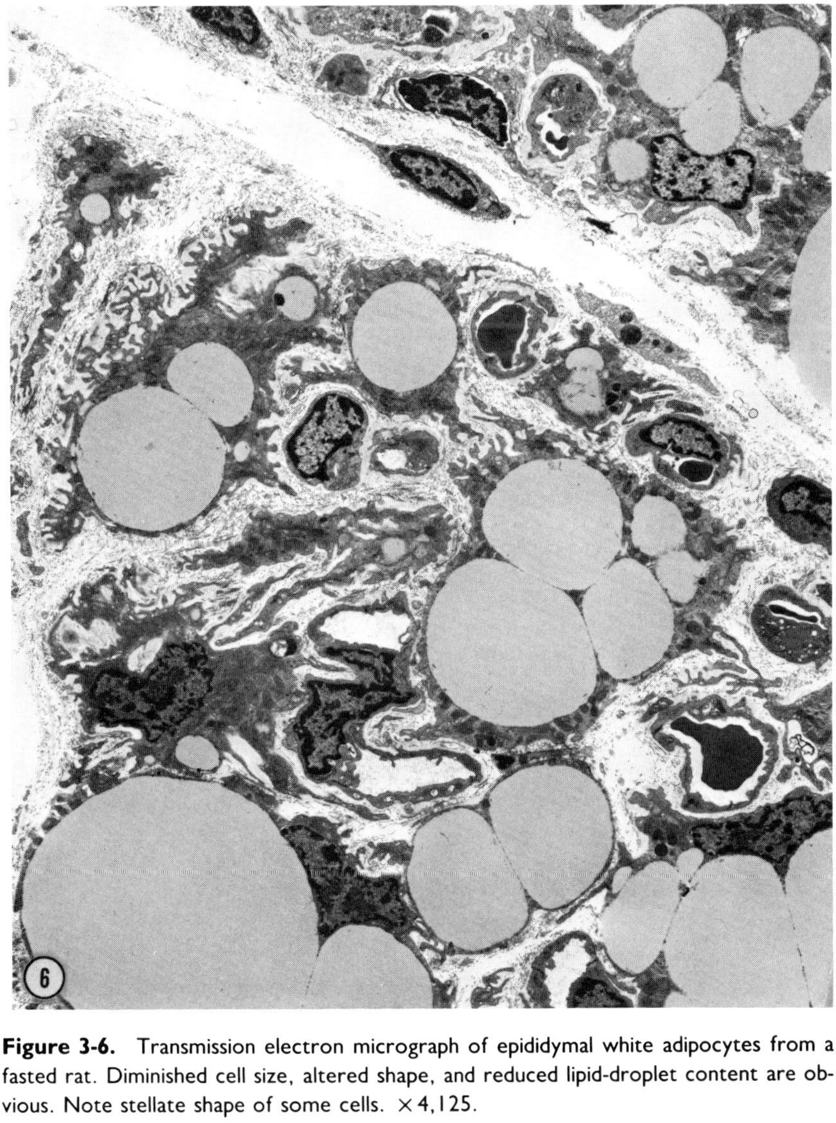

Figure 3-6. Transmission electron micrograph of epididymal white adipocytes from a fasted rat. Diminished cell size, altered shape, and reduced lipid-droplet content are obvious. Note stellate shape of some cells. ×4,125.

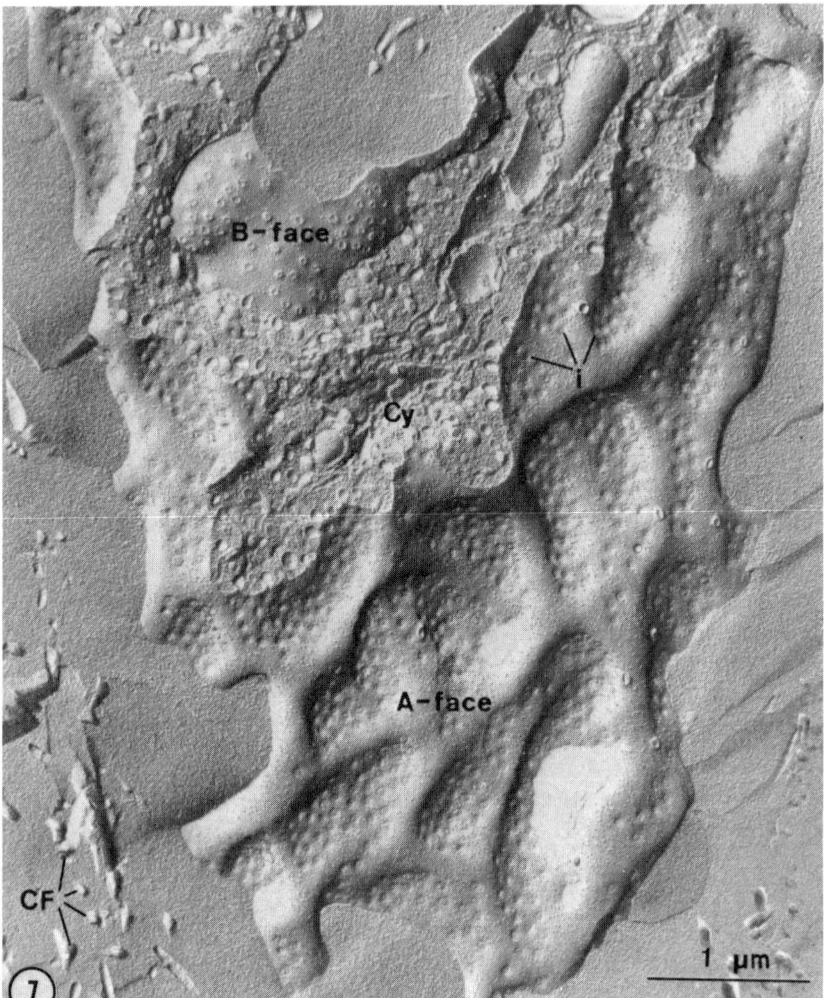

Figure 3-7. Freeze-fracture replica of a portion of an adipocyte after a 9-day fast. The cell surface (A-face) is deeply indented. Invaginations (i) are present in discrete clusters at the bottom of membrane folds. Cy, cytoplasm; CF, collagen fibers. ×27,000. (*From Carpentier et al., 1977; reproduced from* The Journal of Cell Biology, *vol. 72, p. 112 by copyright permission of the Rockefeller University.*)

droplets in normal white adipocytes are surrounded by elements of endoplasmic reticulum. Morphometric analysis would be necessary to clarify this question.

Hormonal Stimulation. Hormonal influences on the ultrastructure of white adipocytes have been reported by several investigators (42–46). Lipolytic hormones (epinephrine and adrenocorticotrophin) and cyclic adenosine 5¹-monophosphate (AMP) phosphodiesterase inhibitors (theophylline and dibutyryl cyclic AMP) have been shown to cause an increase in globular surface protrusions in white adipocytes through combined scanning and transmission electron microscopy (46; Figs. 3-8, 3-9). It was postulated that protrusion formation due to lipolytic agents in white adipocytes may serve as a mechanism by which the cell affects an increased rate of lipid hydrolysis due to exposure of a larger lipid surface area to lipolytic enzymes. Insulin was found to block the actions of the lipolytic hormones. Other investi-

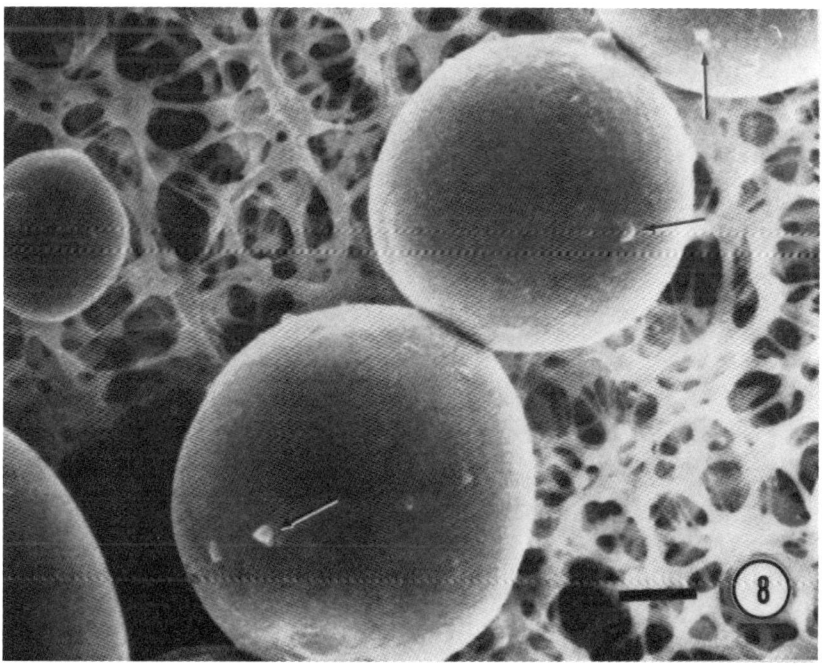

Figure 3-8. A low-magnification scanning electron micrograph of isolated adipocytes. The cells were primarily spherical and ranged from 25 to 125 μm in diameter. The surface of the cell is seen as textured, with several small (< 3 μm in diameter) surface irregularities (arrows). Bar, 10 μm. ×950. (*From Smith and Jarett, 1980; reproduced from* The Journal of Cell Biology, *1980, vol. 84, p. 59 by copyright permission of the Rockefeller University.*)

gators have tested the effects of various hormones on the ultrastructure of the plasma membranes in white adipocytes. In one study, insulin (a lipogenetic hormone) and glucagon and epinephrine (lipolytic hormones) were found to cause an increase in the number of plasmalemmal invaginations (44). The latter lipolytic hormones were shown to cause a decrease in the number of intramembranous particles, while insulin caused an increase in these plasma membrane particles. These findings at the electron microscopic level indicate that hormones can affect structural alterations in the plasma membrane during lipid synthesis and mobilization.

Aging. Only one report of freeze-fracture results has indicated no specific change in the ultrastructure of the plasma membrane of white adipocytes during aging (47). In the latter study, no differences were noted

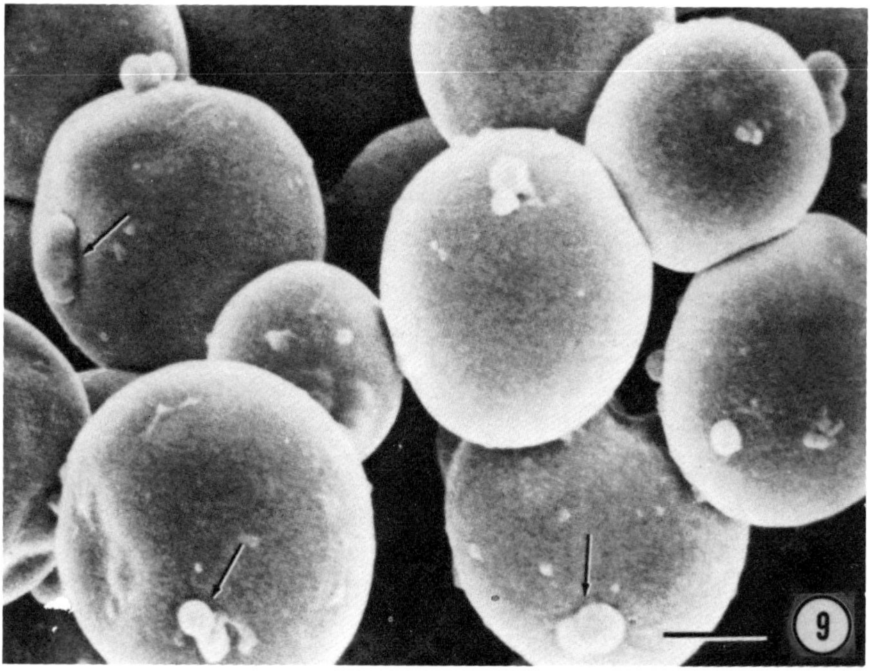

Figure 3-9. A scanning electron micrograph of cells treated with 100 mU/ml ACTH showing increased protrusion formation. Protrusions ranged in size from 8 to 18 μm in diameter. In many instances, multiple closely associated protrusions were seen (arrows). The small surface irregularities were not significantly different from those seen in control adipocytes (Fig. 3-1). Bar, 15 μm ×700. (*From Smith and Jarett, 1980; reproduced from the* Journal of Cell Biology, *1980, vol. 83, p. 63 by copyright permission of the Rockefeller University.*)

between the numbers of plasmalemmal invaginations or intramembranous particles of the adipocytes of 6- and 24-month-old rats.

Obesity. In obese mice, the plasma membranes of white adipocytes contain about half the number of intramembranous particles on both the E- and P-faces compared with control lean mice (48). The authors suggest that this loss of intramembranous particles in obesity results in a change in the lipid-protein ratio of the plasma membrane with unknown effects on the function of the cells.

Hormone Receptor Sites. Electron microscopy has been used to identify hormone receptors on plasma membranes of white adipocytes (49–55). A covalently linked ferritin-insulin complex (Fm-1) has been used as an ultrastructural marker for insulin receptors on isolated and intact white adipocyte plasma membranes (49–51). Results from such studies show that insulin receptors on plasma membranes of the cells appear as irregularly distributed groups or clusters in association with the glycocalyx (Figs. 3-10, 3-11). Other data have confirmed biochemical studies that insulin binding sites are present on the plasma membrane of white adipocytes and within plasmalemmal invaginations and cytosolic vesicles but not on mitochondrial membranes. None of the ferritin-insulin complex was found free within the cytosol or associated with any organelles of the adipocytes, indicating that the site of insulin action is primarily at the level of the plasma membrane.

Most recently, hormone receptors on white adipocytes were made visible

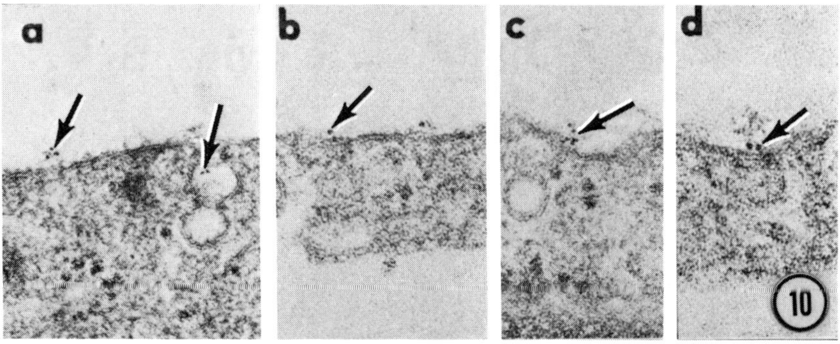

Figure 3-10. Fm-I binding to adipocytes. This composite was prepared to show the various groupings of Fm-I found. The isolated adipocytes were incubated with 500 μUnits Fm-I per ml for 60 min at 24°C and prepared for thin-section analysis. Fm-I (arrows) was found in the glycocalyx on the membrane surfaces as well as in surface-connected vesicles (*a*) in groups of 2 (*d*) to 6 (*c*) as well as single (*b*) molecules. ×95,000. (*From Jarett and Smith, 1977 p. 54; reproduced by permission from Alan R. Liss, Inc., New York.*)

in ultrastructural investigations by the use of insulin covalently bound to amide-modified minibeads of latex (52). This "visual probe" showed hormone receptor sites on the surface of cultured adipocytes through scanning electron microscopic methods (Fig. 3-12).

In other recent ultrastructural studies receptor-mediated internalization (endocytosis) of radioactively labeled insulin and cationic ferritin was described for white adipocytes (35, 53–55). These studies showed that such ligands were localized initially to the plasma membrane, coated pits, coated vesicles, and microvilli of 3T3-L1 adipocytes. With further incubation of white adipocytes with ligands, a significant amount of label was localized to uncoated large vesicles, Golgi-related vesicles, multivesicular bodies, and other lysosomes (Figs. 3-13, 3-14, 3-15). Ligands were least localized to nuclei, mitochondria, lipid droplets, Golgi cisternae, and the cytosol. In

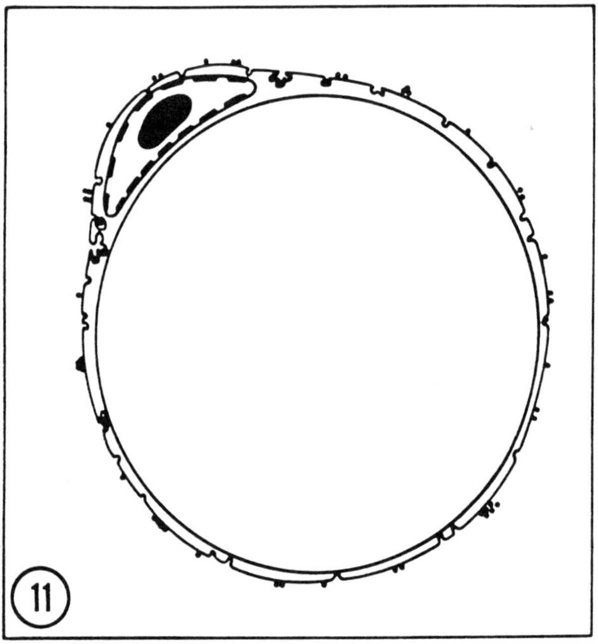

Figure 3-11. Schematic representation of Fm-1 binding to adipocytes. This schematic drawing of a typical adipocyte, with all cytoplasmic structures except the nucleus omitted, is presented to give a representative picture of the Fm-1 distribution on the cell surface and in the surface-connected vesicles. The dark spots representing ferritin are not drawn to scale, but the distribution of the various size groups and spacing of the ferritin molecules around the cell surface were reconstructed from micrographs such as those in Figure 3-10. The magnification of this diagram is approximately × 1,000. (*From Jarett and Smith, 1977 p. 54; reproduced by permission from Allan R. Liss, Inc., New York.*)

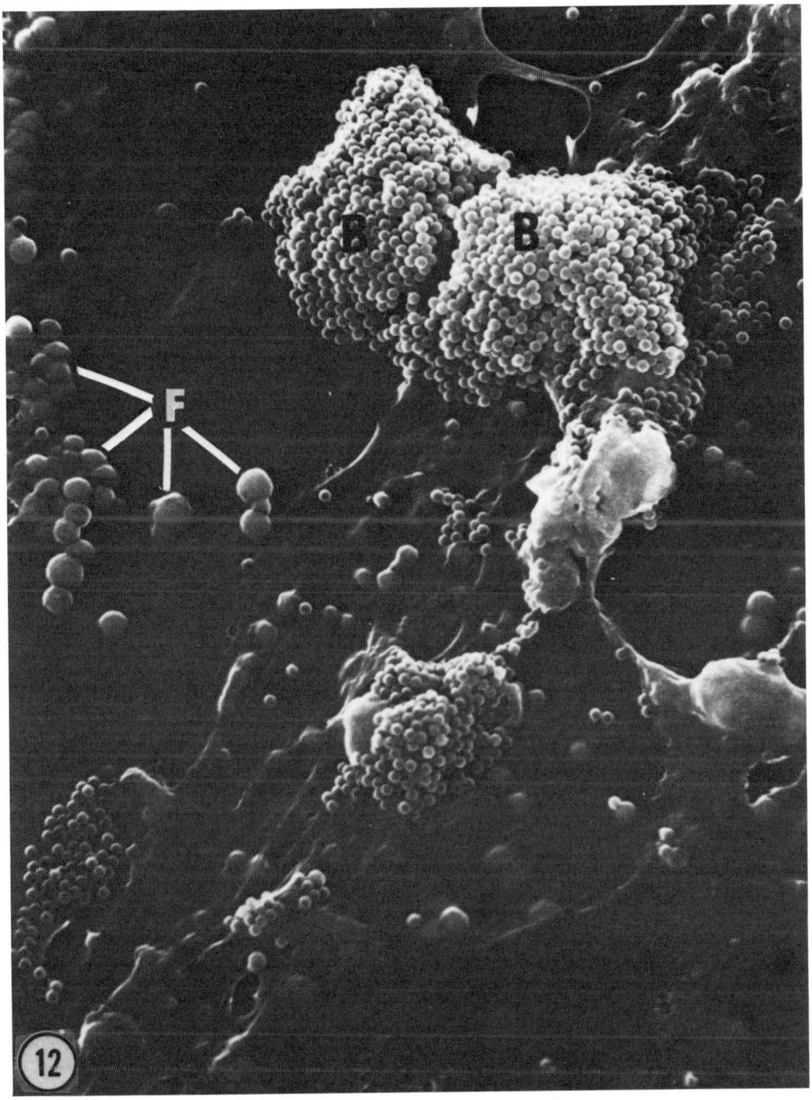

Figure 3-12. An adipocyte-derived fibroblastlike cell, 3 weeks after the initiation of the culture. Note the thin spreading of the cell over the substratum. Small lipid inclusions (F) are still seen in the submembranous location. Large clusters of minibeads (B) are seen on the cell surface. ×5,750. (*From Takahishi and Tavassoli, 1983 p. 238; reproduced by permission from Academic Press, Inc., New York.*)

contrast to previous studies (49–51), binding of ligands to plasmalemmal invaginations was not observed.

Structure-Function Relationships with Regard to Lipid Transport. The correlation between the structure and function of white adipocytes has been investigated through a variety of electron microscopic methods. White adipocytes function at different levels; that is, uptake of substrate (free fatty acids, glucose, amino acids, etc.) from blood capillaries, triacylglycerol synthesis and storage, triacylglycerol hydrolysis, and mobilization and transport of hydrolyzed products from the adipocytes to the capillary lumen. A detailed analysis of these functions related to white adipocyte structure was reviewed several years ago (21, 38) and recently (56) and need not be repeated here. Basically, electron microscopy of various cellular organelles in white adipocytes obtained after cell isolation and ultracentrifugation has been correlated with certain metabolic activities of the cells; for example,

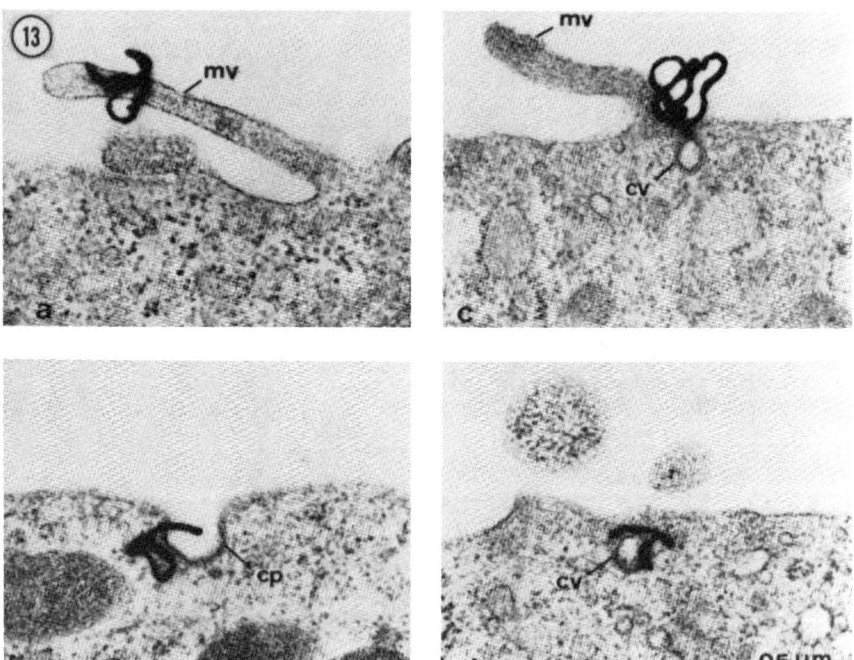

Figure 3-13. Representative examples of cell membrane differentiations with which autoradiographic grains were found preferentially associated after 2 min of incubation at 37°C in the presence of 0.1 nM^{125} I-insulin. (*a*) Microvilli. (*b*) Coated pits. (*c* and *d*) Coated vesicles. As shown in c, coated pits or vesicles, or both, are frequently seen at the base of microvilli. (*From Fan et al., 1982 p. 7789; reproduced by permission from the authors.*)

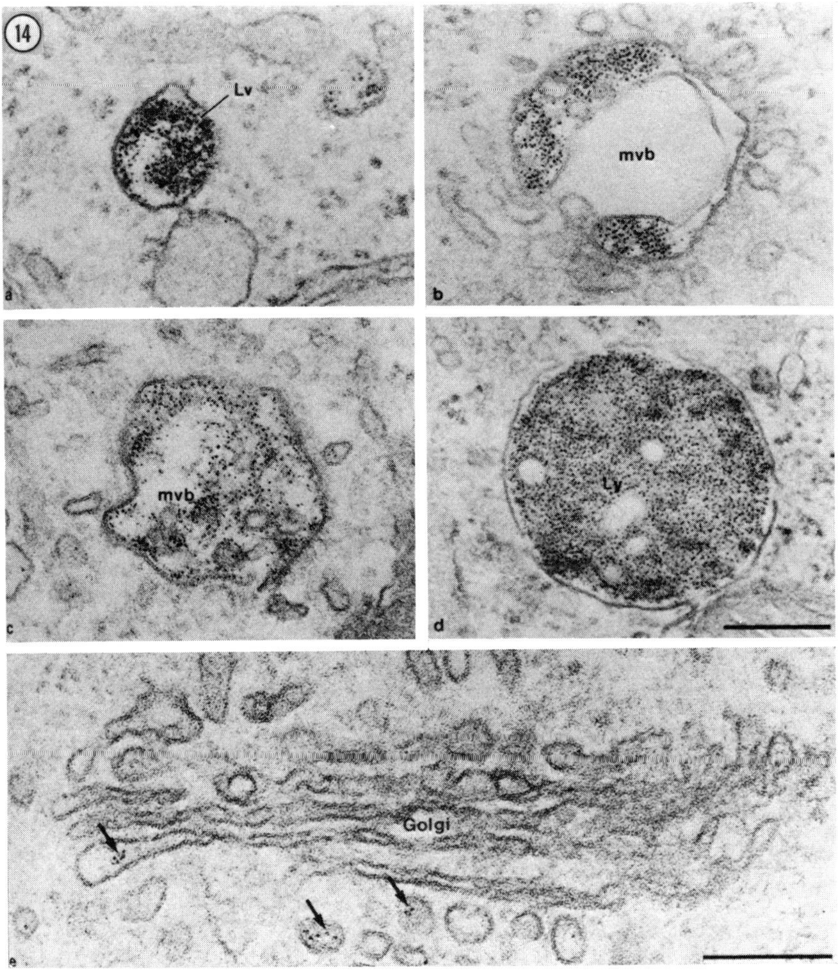

Figure 3-14. Representative examples of intracellular structures with which cationic ferritin was found associated. The 3T3-L1 adipocytes illustrated on these pictures were incubated for 30 or 60 min at 37°C in the presence of 0.75 mg/ml cationic ferritin. (*a*) Large clear vesicles: these structures have a mean size of about 300 nm. (*b, c*) Multivesicular bodies. (*d*) Lysosomelike structure with dense content. (e) Golgi complex: a small percentage of ferritin particles are found within Golgi cisternae and Golgi-associated small vesicles (arrows). Bars 0.2 μm. (*a* to *d*) × 108,000. (e) × 127,000. (*From Fan et al., 1983 p. 129; reproduced by permission from Wissenschaftliche Verlagsgesellschaft Stuttgart.*)

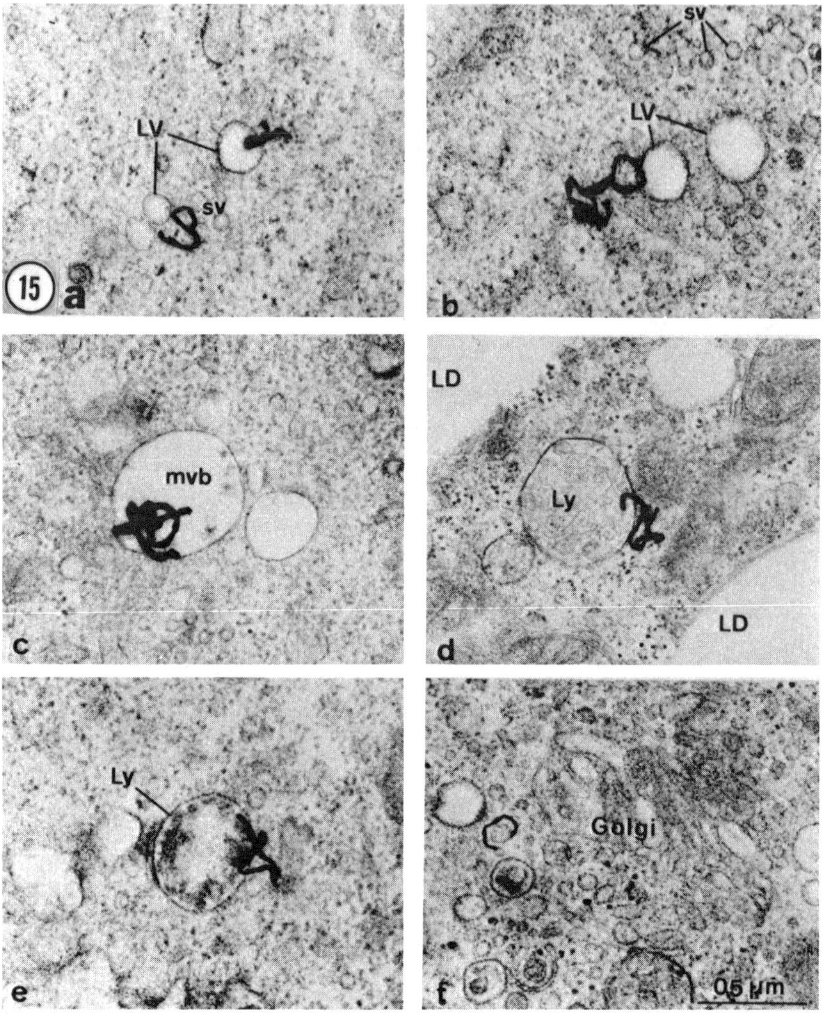

Figure 3-15. Representative examples of intracellular structures with which a preferential association of the labeled material was observed. The 3T3-L1 adipocytes illustrated here were incubated for 60 or 180 min at 37°C in the presence of 10^{10} M ^{125}I-insulin. (*a* and *b*) Large clear vesicles: These structures have a mean size of 291 ± 10 nm (ranging from 200 to 460 nm) and can be easily distinguished from small vesicles (sv), which exhibit a mean size of 64 ± 1 nm (ranging from 50 to 80 nm). (*c*) Multivesicular body (mvb). (*d* and *e*) Lysosomelike structures. As illustrated in *c*, *d*, and *e*, grains are frequently seen overlying the periphery of these structures belonging to the lysosome family. (*f*) Golgi complex: A small number of grains are associated with Golgi complexes only after 180 min of incubation at 37°C. Original magnification ×30,000. (*Reprinted by permission of Elsevier Science Publishing Co., Inc. from "The Interaction of ^{125}I-Insulin with Cultured 3T3-L1 Adipocytes: Quantitative Analysis by the Hypothetical Grain Method," by Fan et al.*, Journal of Histochemistry and Cytochemistry, *Vol. 31, pp. 859–870; copyright 1983 by the Histochemical Society, Inc.*)

glyceride synthesis is associated with mitochondrial and endoplasmic reticulum fractions. In addition, ultrastructural autoradiographic studies of in tact perfused fat pads have shown that labeled fatty acids are taken up from the bloodstream rapidly by white adipocytes (57). The label was found over elements of the endoplasmic reticulum and mitochondria, as well as over lipid droplets, thus implicating the rôle of these organelles in lipid transport.

Plasmalemmal invaginations have been implicated as cellular structures that may participate in the transport of lipid into and out of white adipocytes (21, 38). However, there are conflicting data concerning this view. For example, it was demonstrated in normally fed rats that horseradish peroxidase adhered to the plasma membrane and plasmalemmal invaginations of white adipocytes and there was no migration of the tracer deeper into the cell (58). However, in another study, horseradish peroxidase was found to fill the plasmalemmal invaginations and to migrate within vesicles through the cytosol of the white adipocytes of fed animals (59). This finding is in agreement with findings in fasted rats that horseradish peroxidase particles could be visualized within profiles of endoplasmic reticulum and within lysosomes (58). Also, electron microscopy has indicated that plasma membrane vesicles do not move deeper into the cytosol (50). In fact, the term "surface-connect vesicles" (SVC) was coined to describe this system of plasmalemmal invaginations (50).

A proposed model for lipid transport between blood capillaries and white adipocytes has been studied through the use of primary glutaraldehyde fixation followed by tannic acid treatment for electron microscopy. In this case, intracellular channels resembling endoplasmic reticulum were noted to extend from the plasma membrane to the intracellular lipid droplets of white adipocytes (22, 60, 61; Figs. 3-16, 3-17). It was suggested therefore that these intracellular channels acted to transmit and esterify fatty acids from the extracellular space. In addition, intracellular channels were observed across the endothelial cell, and the outer leaflet of the capillary plasma membrane was shown to be continuous with the plasma membrane of the adipocyte (22, 60–64). In this way, a route of lipid transport was observed from the luminal surface of the capillary to the intracellular lipid droplet of the adjacent adipocyte in a process involving an "interfacial continuum" of plasma membranes and lateral movement of lipid in the "continuum."

Brown Adipocytes

Normal Ultrastructure. Fine structural details of brown adipocytes have been reported by a number of investigators (3, 12, 36, 65–68). Like that of white adipocytes, the plasma membrane of brown adipocytes displays plas-

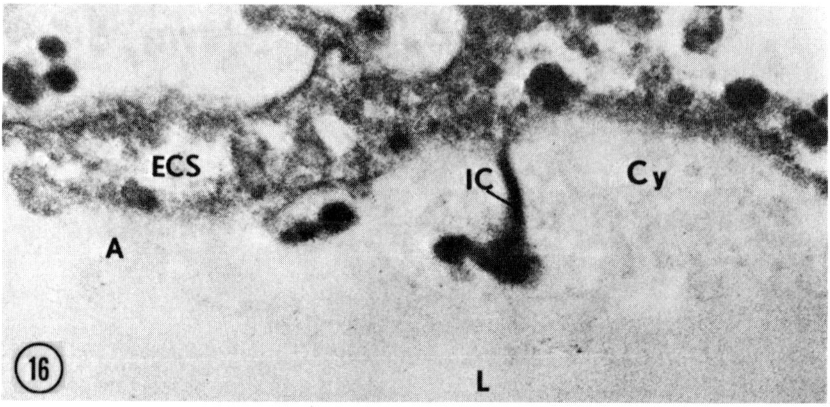

Figure 3-16. A section of adipose tissue from a 9-day-old rat. The micrograph shows a portion of cytoplasm (Cy) and central lipid droplet (L) of an adipocyte (A). The electron-opaque material in the intracellular channel (IC) indicates that this channel was open to the extracellular space (ECS) after fixation with glutaraldehyde and during treatment with tannic acid. Preparation: glut(4°C)/TA(4°C)/Os(4°C). × 106,000. (*From Blanchette-Mackie and Scow, 1981a p. 283; reproduced by permission from Academic Press, Inc., New York.*)

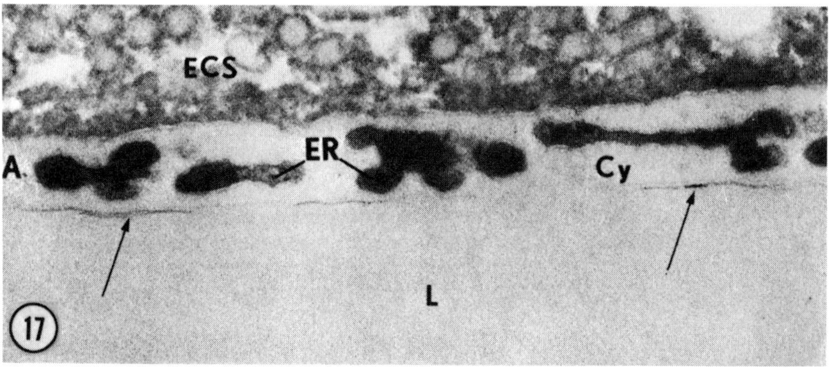

Figure 3-17. A section of adipose tissue from a 9-day-old rat showing a portion of cytoplasm (Cy) and central lipid droplet (L) of an adipocyte (A). The presence of electron-opaque material in lumina of the endoplasmic reticulum (ER) indicates that they were also open to the extracellular space (ECS) when the tissue was treated with tannic acid. The arrows indicate the presence of electron-opaque material along the border of the lipid droplet. Fixation: glut (4°C)/TA(4°C)/Os(4°C). × 120,000. (*From Blanchette-Mackie and Scow, 1981 p. 283; reproduced by permission from Academic Press, Inc., New York.*)

malemmal invaginations and is surrounded by a complete basal lamina (Figs. 3-18, 3-19). Gap junctions and electrical coupling between brown adipocytes have been observed (69). Multiple spherical intracellular lipid droplets varying in size are a salient feature of brown adipocytes. A close association of smooth endoplasmic reticulum with lipid droplets has been reported by several authors (3, 22, 69, 70). Another conspicuous feature of brown adipocytes is the presence of numerous, large, pleomorphic mitochondria (>0.5 μm diameter) with cristae densely arranged in parallel fashion (Fig. 3-20). The wide variation in mitochondrial size has been found to be correlated with certain physiological states; for example, development, hibernation, and cold exposure (66). The mitochondrial matrix of brown adipocytes appears to vary from medium to high electron density and, at times, contains dense inclusion granules of varying shape and size (Figs. 3-21, 3-22). Golgi zones, elements of rough endoplasmic reticulum, glycogen, and microfilaments or microtubules are not well developed in brown adipocytes. However, some authors have described abundant smooth endoplasmic reticulum in these cells (3, 71). Ribosomes (non-membrane bound) are often found in brown adipocytes. Lysosomal bodies are not normally

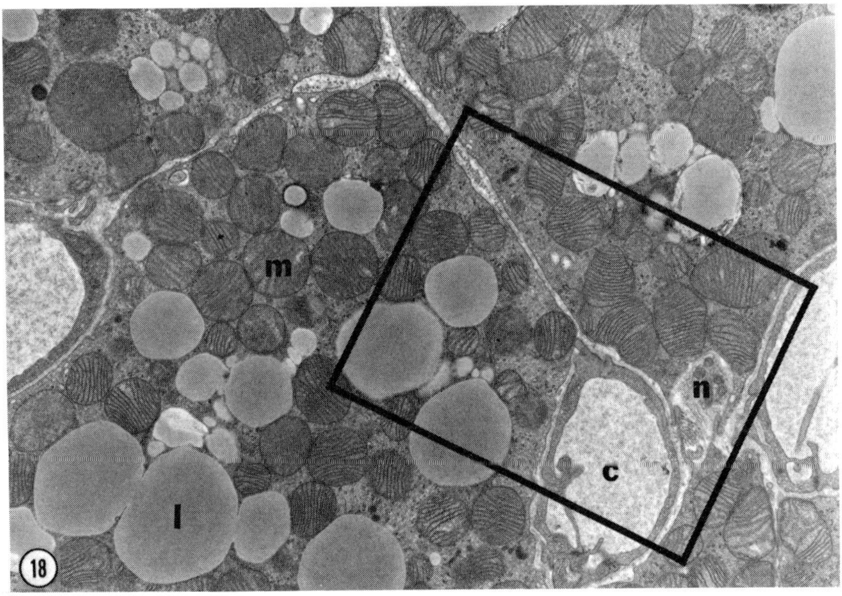

Figure 3-18. Transmission electron micrograph of interscapular brown fat of a mouse. Note multiple lipid droplets (l) and numerous mitochondria (m) in brown adipocyte. c, capillary; n, nerve terminal. $\times$ 10,000.

found in these cells, but their numbers can increase considerably during metabolically active states (72, 73). Brown adipocytes are richly supplied by sympathetic nerve endings (74, 75).

Ultrastructural Changes With Thermogenesis and Lipid Transport. Brown adipocytes function to generate heat in newborn animals, in mammals exposed to cold stress, and in hibernating mammals during arousal (76). Nonshivering thermogenesis in these adipocytes is induced by a number of biochemical events; that is, adenyl cyclase activation via norepinephrine stimulation, lipolysis of stored triglycerides, and transduction of energy stored as fatty acids. Mitochondria are the organelles responsible for the oxidation of fatty acids during this process (77).

Changes in the fine structural morphology of brown adipocyte mitochondria have been reported, reflecting their role in heightened metabolic activity during thermogenesis. During cold adaptation in rats or during arousal from hibernation in squirrels, for example, the number and size of

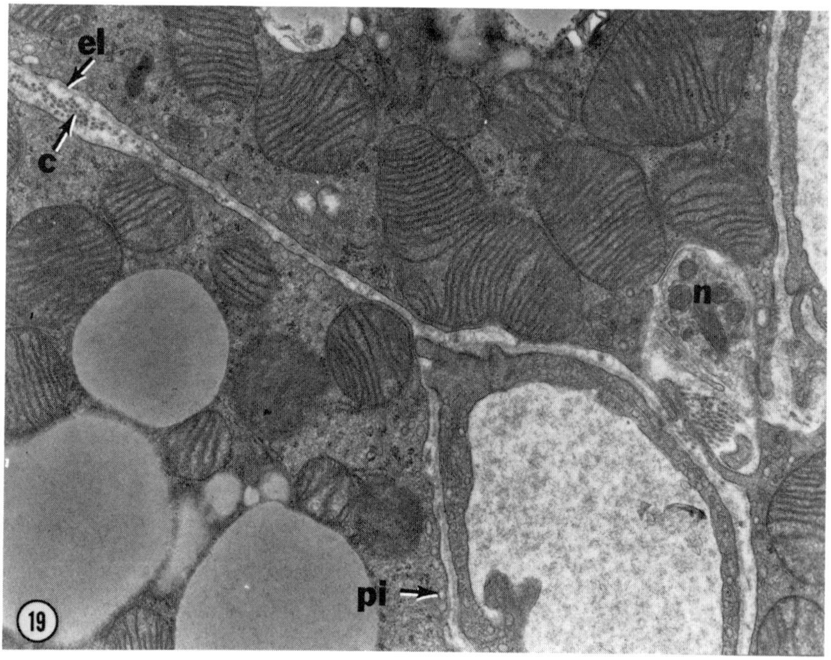

Figure 3-19. Higher power of boxed area in Figure 3-18. Note the basal or external lamina (el) enveloping the brown adipocytes. (c) collagen fibers; (pi) plasmalemmal invaginations; (n) nerve terminal. ×20,000.

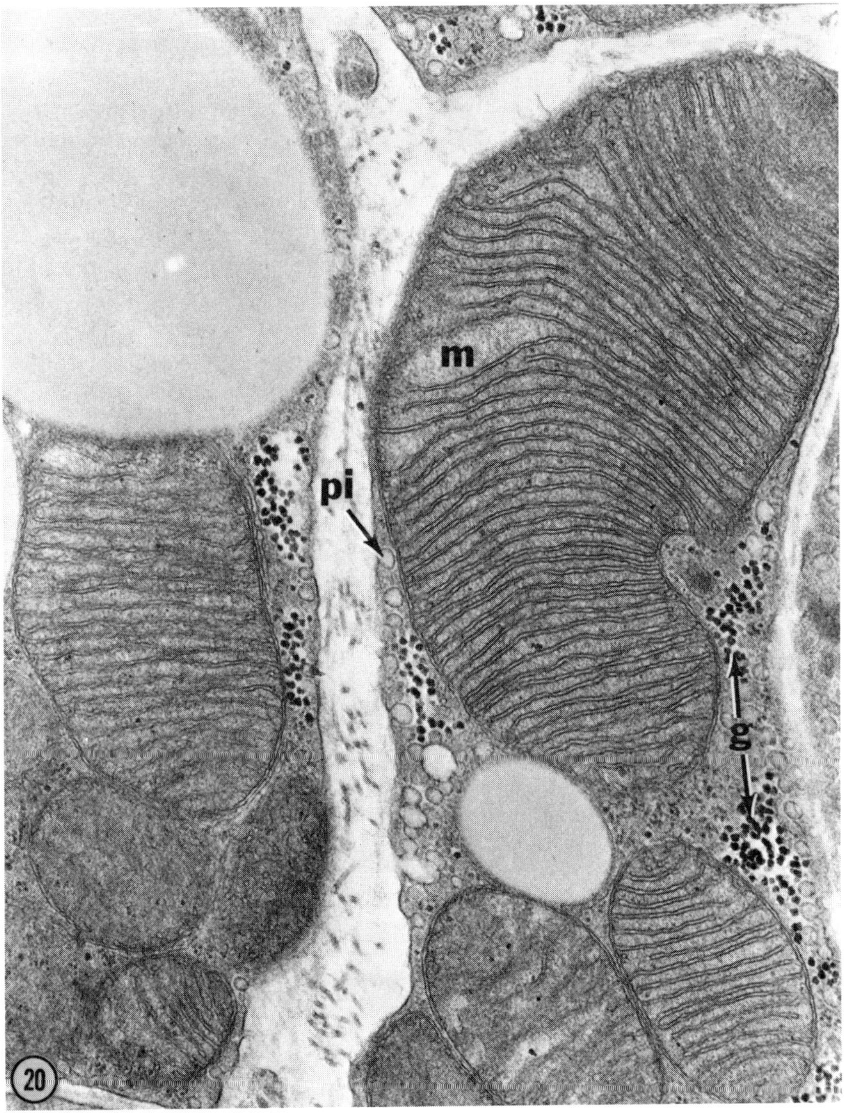

Figure 3-20. Transmission electron micrograph of brown adipocytes in cold-exposed mice. Note the huge mitochondrion (m), plasmalemmal invaginations (pi), and glycogen granules (g). ×38,000.

mitochondria reportedly increased, as did the number of cristae per mitochondrion (68, 78–80, Fig. 3-20). These morphologic findings have been correlated with reports of increased respiratory enzymes in brown adipocytes in biochemical studies (81). Dramatic changes in mitochondrial morphology have been described at the electron microscopic level during normal activity, hibernation, and arousal in golden-mantled squirrels (80). Not only did mitochondrial size increase during hibernation and arousal, but the cristae assumed a tubular appearance and formed whorls, loops, and numerous interconnections (Fig. 3-23). In the arousal state, some mito-

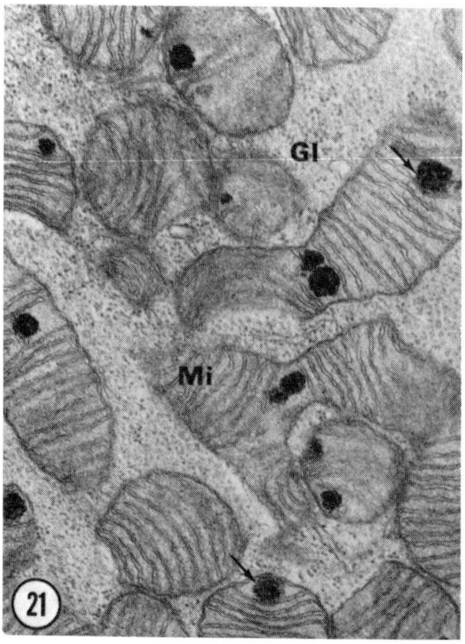

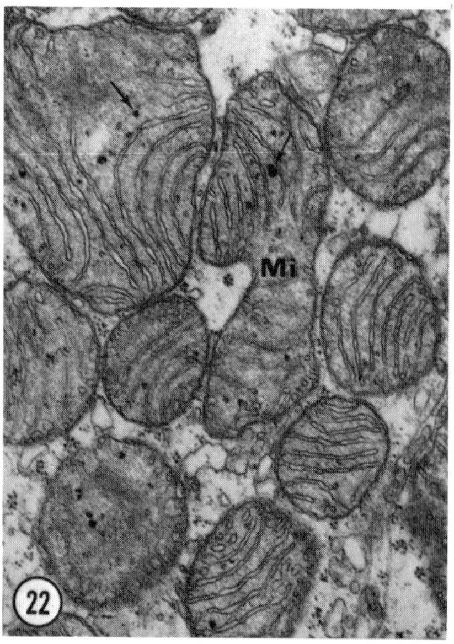

Figure 3-21. Portion of brown adipocyte of rat, 1 day before birth. At this stage, the intramitochondrial dense granules have reached their maximum diameter. They are very numerous and frequently show an obvious substructure (arrow). There appears to have been a further increase in the number of cristae per mitochondrial profile. Mi, mitochondrion; Gl, glycogen. ×30,000. (*From Barnard and Lindberg, 1969 p. 303; reproduced by permission from Academic Press, Inc., New York.*)

Figure 3-22. Portion of brown adipocyte of rat, day of birth. The intramitochondrial dense granules (arrow) in this field are somewhat smaller than the average for this stage. The mitochondria have increased considerably in volume and they are well filled with cristae. Notice the absence of glycogen particles. Mi, mitochondrion. ×30,000. (*From Barnard and Lindberg, 1969 p. 303; reproduced by permission from Academic Press, Inc., New York.*)

chondria were cup-shaped while others of giant size appeared to be "exploding" and giving rise to new, smaller, mitochondria (Fig. 3-23).

Intramitochondrial inclusions of varying size and configuration have been reported in brown adipocytes during the perinatal period (78, 82, 83; Figs. 3-21, 3-22). These inclusions are composed mainly of lipids, according to the results of a study in which inclusions were absent following lipid extraction procedures (84). Such inclusions decrease drastically within 1 to 2 days after birth (78, 83). The fact that mitochondria increase in size concomitant with a decrease in mitochondrial inclusions suggests that these inclusions may give rise to new mitochondrial membranes (84).

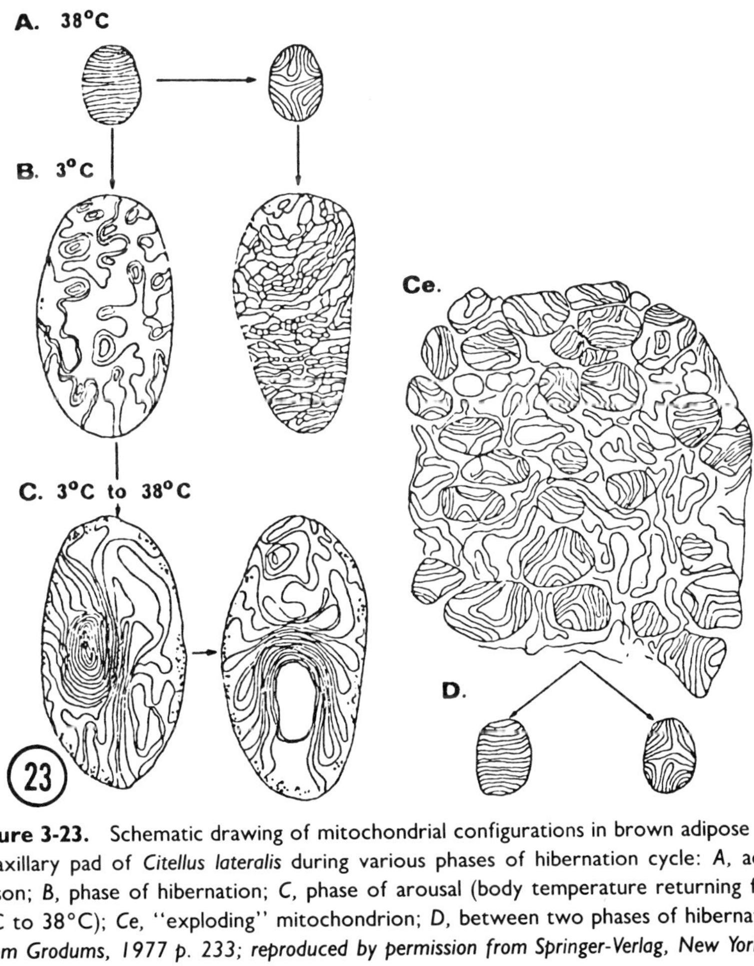

Figure 3-23. Schematic drawing of mitochondrial configurations in brown adipose cells of axillary pad of *Citellus lateralis* during various phases of hibernation cycle: *A*, active season; *B*, phase of hibernation; *C*, phase of arousal (body temperature returning from 3°C to 38°C); *Ce*, "exploding" mitochondrion; *D*, between two phases of hibernation. (*From Grodums, 1977 p. 233; reproduced by permission from Springer-Verlag, New York.*)

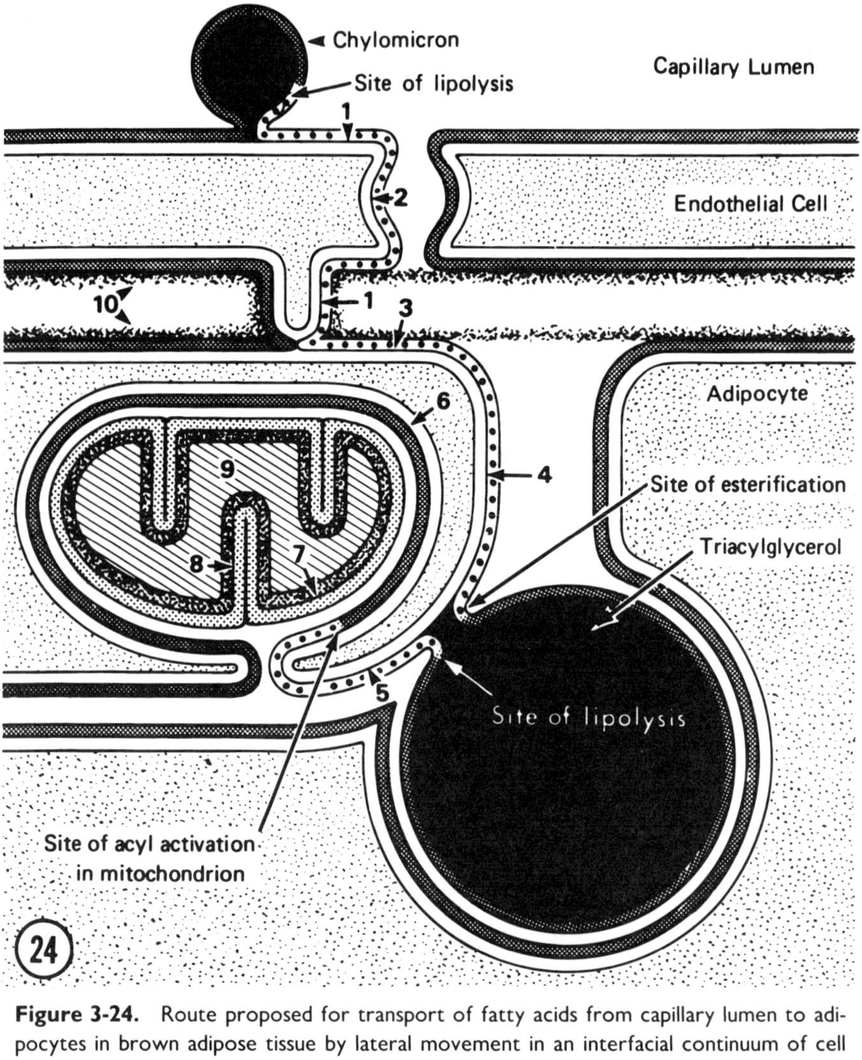

Figure 3-24. Route proposed for transport of fatty acids from capillary lumen to adipocytes in brown adipose tissue by lateral movement in an interfacial continuum of cell membranes. The interfacial continuum, represented by the broad stippled line, is composed of the external leaflets of the plasma membrane of endothelium (1), the membrane of transcellular channels in endothelium (2), the plasma membrane of adipocytes (3), and the membrane of surface-connected channels (4), the membrane of endoplasmic reticulum (5), and the outer mitochondrial membrane (6) in adipocytes. The external leaflet of the plasma membrane of the endothelial cell (1) and that of the adipocyte (3) are continuous at the site of contact between the cells. The interfacial continuum would include the chylomicron surface film in fed animals when the surface film is fused temporarily with the external leaflet of the endothelial plasma membrane (1). Fatty acids (black dots) formed

A model for lipid transport in white adipocytes has recently been proposed for brown adipocytes (70). In this model, fatty acids are transported from capillary lumen to lipid droplets of brown adipocytes via "lateral movement in an interfacial continuum of cell membranes." The process by which lipid is transported to mitochondria for oxidation in brown adipocytes has been demonstrated by the use of fine structural analysis (70). Incubating fixed brown adipose tissue at 25° C before postfixation with OsO_4 (osmium tetraoxide) revealed fatty acids in the form of lamellar whorls. It was suggested that such fatty acid whorls were formed by lipolysis in the fixed brown adipose tissue, and that these fatty acids were transported to mitochondria via continuous membrane connections (Fig. 3-24).

METHODOLOGY FOR STUDYING THE ULTRASTRUCTURE OF ADIPOCYTES

Advantages and Disadvantages

Obviously, transmission electron microscopy of adipocytes has the distinct advantage of resolving subcellular details not visible with the light microscope. However, due to the large size of mature white adipocytes and their central lipid droplet, the attenuated peripherally placed cytosol exhibits little in the way of organelles. However, rarely obtained subsurface sections of white adipocytes have demonstrated various organelles in clear view (15). Examination of white adipocytes from fasted or diabetic animals or those derived from the 3T3-L1 cell line where the intracellular lipid content is reduced shows a plethora of organelles not normally seen in lipid-filled adi-

---

by the action of lipoprotein lipase on chylomicron triacylglycerol, as in fed animals, would enter the continuum in the capillary lumen, move in the continuum to the endoplasmic reticulum in the adipocyte, and leave the continuum when they are reesterified to triacylglycerol and accumulate as lipid droplets between leaflets of the reticulum. Fatty acids formed by the action of tissue (hormone-sensitive) lipase on intracellular triacylglycerol, as in cold-stressed animals, would reenter the continuum at the site of lipolysis in the endoplasmic reticulum, move in the continuum to mitochondria, and leave the continuum when fatty acids are activated in the outer mitochondrial membrane for transfer to the inner mitochondrial membrane (7). Fatty acids produced by intracapillary lipolysis could bypass the pedicel of the lipid droplet and leave the continuum in mitochondria. Other components of mitochondria labeled in the drawing are cristae (8) and the inner mitochondrial matrix (9). Note that the cellular projection of the endothelial cell penetrates the basement membrane (10) to make contact with the other cell. Cytoplasmic leaflets of membranes are represented by broad white lines. (*From Blanchette-Mackie and Scow, 1983 p. 241; reproduced by permission from Lipid Research, Inc., Bethesda, Md.*)

pocytes. Another advantage of using transmission electron microscopy is that one can localize radioactive grains or other electron-dense probes to specific organelles (35, 54).

Transmission electron microscopy of white or brown adipocytes does not reveal en face views of plasma membranes to best advantage in sectioned adipose tissue. For this, freeze-fracture replicas of white or brown adipocytes observed with the transmission electron microscope allows for the en face demonstration of plasma membrane changes, plasmalemmal invaginations, intramembranous particles, hormone receptor sites, and other such features.

Any disadvantages of studying the electron microscopy of adipocytes lie in the increased time and effort expended to obtain the final results. One difficulty with this method is that the lipid droplets of white and brown adipocytes are not easily preserved by fixatives currently employed. In addition, embedding this lipid-rich tissue is often unsuccessful. As a result, ultramicrotome sectioning of blocks of adipose tissue embedded in epoxy resin may result in unwanted artifacts.

In addition, osmication of the lipid droplets often causes "chatter" artifacts that are probably due to uneven hardening of the tissue block. This author has been unsuccessful in avoiding the chatter effect when sectioning adipose tissue. A thorough study that compares the use of various fixatives, plastic embedments, sectioning techniques, and other factors might be useful in determining the cause of such artifacts.

Materials and Equipment

No special or exclusive materials or items of equipment are required to study the ultrastructure of adipocytes. A routinely equipped electron microscopy laboratory will suffice for all phases of preparation of thin sections, freeze-fracture replicas, or scanning electron microscopic specimens. Current textbooks dealing with the various methods of electron microscopy are readily available. The following section briefly summarizes some of the usual methods of studying the ultrastructure of adipocytes.

Methods

Routine Transmission Electron Microscopy Using Thin Sections

Fixation, Dehydration, and Embedding Techniques. Although some investigators have successfully preserved white and brown adipose tissue by immersion fixation in aldehyde solutions followed by postfixation in OsO_4

(26, 80, 85, 86), this author prefers to fix adipose tissue primarily with aldehydes by vascular perfusion. Following gravity drip perfusion of the entire animal via the heart for approximately 30 minutes, adipose tissue samples (mesentery, epididymis, etc.) are removed carefully, diced with a sharp razor blade, and placed in fresh fixative for an additional 30 minutes at room temperature. Secondarily, tissue blocks are postfixed in 1% OsO_4 for 1–2 hours at 4° C. Careful handling of adipose tissue with surgical instruments is of utmost importance since the plasma membranes and peripheral cytosol of adipocytes are quite fragile and can be easily broken by mechanical pressure.

In earlier studies (14, 36), white adipocytes were fixed by immersion fixation in cold buffered OsO_4, followed by rapid dehydration in ascending concentrations of cold ethanol, and embedded in Vestopal W, Maraglas, or Epon. While the preservation of fine structural detail was attained, extraction of cytosolic material was noted (36). Currently accepted methods for the preparation of adipose tissue blocks begin with the use of primary glutaraldehyde fixation (2–5%) or a combination of glutaraldehyde (2%) and paraformaldehyde (2%) buffered with sodium cacodylate or sodium phosphate at pH 7.4. Postfixation with osmium tetroxide (1–3%) buffered with the same solutions as before is required to preserve lipid droplets; otherwise, the droplets will dissolve with the use of dehydration agents. Good results have been obtained with isolated white adipocytes with the use of OsO_4 at a concentration of 3% and postfixing for 3 hours (20). Dehydration of adipose tissue blocks with increasing concentrations of ethanol is entirely acceptable, as is embedding of blocks in currently available epoxy resins (Araldite, Epon, etc.).

Sectioning and Staining Techniques. Producing thin sections of mature adipocytes (white or brown) on an ultramicrotome is a most difficult task because of the huge amount of stored lipid. The thought of a glass or diamond knife cutting through a mass of osmicated spheres is enough to frighten off any experienced microscopist! This author has found that success in obtaining adequate thin sections depends largely on the size of the block face; that is, the smaller the face the better the sectioning. Successful sectioning of adipocytes from fasting, diabetic, or developing animals is possible because of reduced intracellular lipid content.

Staining of thin sections with uranium and lead is no different for adipocytes than it is for other tissues.

Scanning Electron Microscopy of Adipocytes. A few scanning electron microscopic experiments performed by this author entailed the use of lipid-depleted epididymal or mesenteric adipose tissue from acutely starved rats. Adipose tissue was fixed by vascular perfusion with 3% glutaraldehyde in

0.1 M cacodylate buffer, at pH 7.4, at room temperature for 1 hour. Diced segments of fat were dehydrated with ethanol and critical point dried with carbon dioxide as the intermediate fluid. The diced specimens were attached to aluminum stubs with double-stick tape, coated with gold-palladium in a Hummer IV sputter coater, and viewed with the scanning electron microscope.

In a more recent scanning electron microscopic study, isolated white adipocytes were obtained by collagenase digestion (46) and fixed for 30 minutes in phosphate-buffered 2% glutaraldehyde at pH 7.4. The fixative was added slowly to the isolated cell suspension with constant gentle mixing to prevent the clumping of cells. The isolated adipocytes were postfixed with 2% OsO_4 in 0.1 M cacodylate buffer for 30 minutes with constant stirring. The latter investigators (46) discovered that dehydration with polar solvents (alcohol, acetone) caused extraction of lipid and collapse of the cells with attendant poor results in the scanning electron microscope. However, if nonpolar solvents (graded series of ethylene glycol and ethylene glycol monomethyl ether) were used, the collapse of the adipocytes was prevented due to lipid droplet retention.

Freeze-Fracture Preparations. Methods for obtaining freeze-fracture replicas of brown or white adipose tissue are basically similar to those for most other tissues (18). Glutaraldehyde-fixed blocks of adipose tissue were placed in 30% glycerol (phosphate buffered, 0.1 M, pH 7.4) for 30 min and rapidly frozen in Freon 22 (cooled with liquid nitrogen). Blocks of tissue were then transferred to a Balzers BAF301 apparatus, fractured with a razor blade, and shadowed with carbon and platinum (87). The resulting replicas were cleaned with sodium hypochlorite for 2 hours, placed in freshly prepared chloroform and methanol (2:1) twice for 20 minutes each, and soaked three times for 20 minutes each in dimethylformamid. The cleaned replicas were rinsed in distilled water, placed on copper grids, and viewed with the transmission electron microscope.

INTERPRETATION

Ultrastructural examination of mature white and brown adipocytes in routine thin sections reveals two distinct cell types differing in size and content of intracellular organelles and inclusions as was indicated in the early part of this chapter. Difficulty in interpreting the differences between white and brown adipocytes arises when the morphological appearances of mature brown adipocytes and developing white adipocytes are considered. During the differentiation of white adipocytes a multilocular stage occurs when the differences between mature brown adipocytes and developing white adipocytes are insignificant. At the electron microscopic level although the multilocularity of lipid droplets is very similar in the two cell types, the morphology of mitochondria and other organelles is distinctly different.

Electron microscopy, to date, has not allowed for the distinction to be made between the connective tissue fibroblast and the adipocyte progenitor (preadipocyte). It has been known for some time that white and brown adipocytes are derived from fusiform fibroblastlike cells that accumulate lipid in the form of droplets. As the lipid droplets increase in size the cell shape becomes oval and finally spherical as maturity is attained. However, no morphological marker at the electron microscopic level can help distinguish between cells programmed to act as fibroblasts (collagen producing only) and fibroblastlike cells destined for adipocyte differentiation.

ACKNOWLEDGMENTS

The author wishes to acknowledge the technical assistance of Stephen Harris and Aileen Kuda.

REFERENCES

1. D. W. Fawcett, 1952, A comparison of the histological organization and cytochemical reactions of brown and white adipose tissues, *J. Morphol.* **90**:363–405.
2. J. D. Lever, 1975, The fine structure of brown adipose tissue in the rat with observations on the cytological changes following starvation and adrenalectomy, *Anat. Rec.* **128**:361–372.
3. Y. Umahara, 1968, Light and electron microscopic studies on the brown adipose tissue in the bat, *Arch. Histol. Jpn.* **29**:459–509.
4. J. E. Miller and L. H. Eastlick, 1952, Studies on the transplanted embryonic limbs of the chick. IV. The cytology of the "adipose tissue," *Trans. Am. Microsc. Soc.* **71**:1–19.
5. E. Wertheimer and B. Shapiro, 1948, The physiology of adipose tissue, *Physiol. Rev.* **28**:451–463.
6. D. W. Fawcett, 1948, Histological observations on the relation of insulin to the deposition of glycogen in adipose tissue, *Endocrinology* **42**:545–567.
7. R. L. Sidman, 1956, The direct effect of insulin on organ cultures of brown fat, *Anat. Rec.* **124**:723–740.
8. W. H. Chase, 1959, Structure of basement membranes of fat cells, *AMA. Arch. Pathol.* **67**:550–557.
9. W. H. Chase, 1959, Fine structure of rat adipose tissue, *J. Ultrastruct. Res.* **2**:283–287.
10. F. Wasserman and T. F. McDonald, 1960, Electron microscopic investigation of the surface membrane structures of the fat-cell and of their changes during depletion of the cell, *Z. Zellforsch.* **52**:778–800.
11. F. Wasserman and T. F. McDonald, 1963, Electron microscopic study of adipose tissue (fat organs) with special reference to the transport of lipids between blood and fat cells, *Z. Zellforsch.* **59**:326–357.

12. L. Napolitano and D. Fawcett, 1958, The fine structure of brown adipose tissue in the newborn mouse and rat, *J. Biophys. Biochem. Cytol.* **4:**685–692.
13. T. Imaeda, 1959, The fine structure of human subcutaneous fat cells, *Arch. Histol. Jpn.* **18:**57–68.
14. H. Sheldon, C. H. Hollenberg, and A. I. Winegrad, 1962, Observations on the morphology of adipose tissue. 1. The fine structure of cells from fasted and diabetic rats, *Diabetes* **11:**378–387.
15. M. C. Schotz, J. E. Stewart, A. S. Garfinkel, C. F. Whelen, M. Baker, M. Cohen, T. J. Hensley, and M. Jacobson, 1969, Isolated fat cells: Morphology and possible role of released lipoprotein lipase in deposition of lipoprotein fatty acid, in *Drugs Affecting Lipid Metabolism,* W. L. Holmes, A. Carlson, and R. Paoletti, eds., Plenum Press, New York, pp. 161–183.
16. L. Napolitano, 1963, The differentiation of white adipose cells. An electron microscopic study, *J. Cell Biol.* **18:**663–679.
17. M. Rupec, 1981, Zur ultrastruktur der Fettzellen, *Dermatologica* **163:**385–392.
18. J. -L. Carpentier, A. Perrelet, and L. Orci, 1977, Morphological changes of the adipose cell plasma membrane during lipolysis, *J. Cell Biol.* **72:**104–117.
19. S. W. Cushman, 1970, Structure-function relationships in the adipose cell. I. Ultrastructure of isolated adipose cell, *J. Cell Biol.* **46:**326–341.
20. R. Pictet, B. Jeanrenaud, L. Orci, A. Renold, and C. Rouiller, 1968, Cellules adipeuses in situ et isolées. Essai de fixation pour la microscopie electronique, *Z. Ges. Exp. Med.* **148:**255–274.
21. J. J. Heindel, L. Orci, and B. Jeanrenaud, 1975, Fat mobilization and its regulation by hormones and drugs in white adipose tissue, in *International Encyclopedia of Pharmacology and Therapeutics,* C. Peters, ed. Pergamen Press, Oxford, Scotland, Section 24, vol. 1, pp. 175–373.
22. E. J. Blanchette-Mackie and R. O. Scow, 1982, Continuity of intracellular channels with extracellular space in adipose tissue and liver: Demonstrated with tannic acid and lanthanum, *Anat. Rec.* **203:**205–219.
23. M. Imaizumi, 1969, On the fine structure of the surface of lipid droplets in adipose cells, *Arch. Histol. Jpn.* **30:**353–365.
24. L. M. Luckenbill and A. S. Cohen, 1966, The association of lipid droplets with cytoplasmic filaments in avian subsynovial adipose cells, *J. Cell Biol.* **31:**195–199.
25. E. M. Wood, 1967, An ordered complex of filaments surrounding the lipid droplets in developing adipose cells, *Anat. Rec.* **157:**437–448.
26. H. Sheldon and C. C. Ferguson, 1970, On the occurrence of cytoplasmic filaments in adipose cells, Soc. Fransiase Microscop. Elect. 6th International Congress of Electron Microscopy, vol. 3, p. 534 Grenoble. Société française de microscopie électronique.
27. A. B. Novikoff, P. M. Novikoff, O. M. Rosen, and C. S. Rubin, 1980, Organelle relationship in cultured 3T3-L1 preadipocytes, *J. Cell Biol.* **87:**180–196.
28. B. G. Slavin, 1979, Fine structural studies on white adipocyte differentiation, *Anat. Rec.* **195:**63–72.
29. H. J. Mersmann, J. R. Goodman, and L. J. Brown, 1975, Development of swine adipose tissue: Morphology and composition, *J. Lipid Res.* **16:**269–279.
30. I. Dardick, M. J. Poznanski, I. Waheed, and G. Setterfield, 1976, Ultrastructural observations on differentiating human preadipocytes cultured in vitro, *Tissue and Cell* **8:**561–571.
31. F. Desnoyers and N. Vodovar, 1977, Etude histocytologique comparee chez le Porc et le Rat due tissue adipeux ṕerirénal au stade de son apparition, *Biol. Cell* **29:**177–182.

32. K. Iyama, K. Ohzono, and G. Usuku, 1979, Electron microscopical studies on the genesis of white adipocytes: Differentiation of immature pericytes into adipocytes in transplanted preadipose tissue, *Virchows Arch. B. Cell Pathol.* **31**:143–155.
33. G. J. Hausman, D. R. Campion, R. L. Richardson, and R. J. Martin, 1981, Adipocyte development in the rat hypodermis, *Am. J. Anat.* **161**:85–100.
34. M. Tavassoli, 1976, Ultrastructural development of bone marrow adipose cell, *Acta Anat.* **94**:65–77.
35. J. Y. Fan, J. L. Carpentier, P. Gordon, E. Van Obberghen, N. M. Blackett, C. Grunfeld, and L. Orci, 1982, Receptor-mediated endocytosis of insulin: Role of microvilli, coated pits and coated vesicles, *Proc. Natl. Acad. Sci. (USA)* **79**:7788–7791.
36. L. M. Napolitano, 1965, Observations on the fine structure of adipose cells, *N. Y. Acad. Sci.* **131**:34–42.
37. L. Napolitano, 1965, The fine structure of adipose tissues, in *Handbook of Physiology,* Sect. 5., Adipose Tissue, A. E. Renold, and G. F. Cahill, Jr., eds. American Physiological Society, Washington, D.C., pp. 109–123.
38. B. G. Slavin, 1972, The cytophysiology of mammalian adipose cells, *Int. Rev. Cytol.* **33**:297–334.
39. L. M. Napolitano and H. T. Gagne, 1963, Lipid-depleted white adipose cells: An electron microscopic study, *Anat. Rec.* **147**:273–293.
40. J. R. Williamson, 1964, Adipose tissue. Morphological changes associated with lipid mobilization, *J. Cell Biol.* **20**:57–74.
41. F. Desnoyers, N. Vodovar, and G. Durand, 1975, Observations ultrastructurales, chez le rat, de cellules adipeuses totalement dépourvues de lipide d'accumulation, puis de leur remplissage de novo, *C. R. Acad. Sci. Paris* **280**:1717–1720.
42. R. J. Barrnett and E. G. Ball, 1960, Metabolic and ultrastructural changes induced in adipose tissue by insulin, *J. Biophys. Biochem. Cytol.* **8**:83–101.
43. S. W. Cushman, 1970, Structure-function relationships in the adipose cell. II. Pinocytosis and factors influencing its activity in the isolated adipose cell, *J. Cell. Biol.* **46**:342–353.
44. J. -L. Carpentier, A. Perrelet, and L. Orci, 1976, Effects of insulin, glucagon, and epinephrine on the plasma membrane of the white adipose cell: A freeze-fracture study, *J. Lipid Res.* **17**:335–342.
45. D. Soifer, T. Braun, and O. Hechter, 1971, Insulin and microtubules in rat adipocytes, *Science* **172**:269–271.
46. R. M. Smith and L. Jarett, 1980, Surface structure changes of rat adipocytes during lipolysis stimulated by various lipolytic agents, *J. Cell Biol.* **84**:57–65.
47. A. Jacobs, C. Remacle, and M. Jeanjean, 1979, Lack of morphological changes of the adipose cell plasma membrane during aging: Application of the freeze-fracture technique, *Age* **2**:70–72.
48. J. -L. Carpentier, A. Perrelet, and L. Orci, 1976, Modification ultrastructure de l'organization des membranes plasmiques des adipocytes chez les Souris obeѐses (ob/ob), *C. R. Acad. Sci. Paris* **283**:1417–1419.
49. L. Jarett and R. M. Smith, 1974, Electron microscopic demonstration of insulin receptors on adipocyte plasma membranes utilizing a ferritin-insulin conjugate, *J. Biol. Chem.* **249**:7024–7031.
50. L. Jarett and R. M. Smith, 1975, Ultrastructural localization of insulin receptors on adipocytes, *Proc. Natl. Acad. Sci. (USA)* **72**:3526–3530.
51. L. Jarett and R. M. Smith, 1977, The natural occurrence of insulin receptors in groups on adipocyte plasma membranes as demonstrated with monomeric ferritin-insulin, *J. Supramol. Struct.* **6**:-45–59.

52. K. Takahashi and M. Tavasoll, 1983, Modulation of insulin receptors in cultured adipocytes as studied by a latex mini-bead probe, *J. Ultrastruct. Res.* **83**:233–241.
53. J. Y. Fan, J. L. Carpentier, E. Van Obberghen, N. M. Blackett, C. Grunfeld, P. Gordon, and L. Orci, 1983, The interaction of ^{125}I-insulin with cultured 3T3-L1 adipocytes: Quantitative analysis by the hypothetical grain method, *J. Histochem. Cytochem.* **31**:859–870.
54. J. Y. Fan, J. -L. Carpentier, E. Van Obberghen, C. Grunfield, P. Gordon, and L. Orci, 1983, Surface interactions and intracellular localization of cationic ferritin: Similarity to ^{125}I-insulin in 3T3-L1 adipocytes, *Eur. J. Cell Biol.* **31**:125–131.
55. E. A. Seiss, M. L. Nestorescu, and O. H. Wieland, 1974, Localization of insulin on fat cell ghosts by ferritin labelled insulin antibodies, *Diabetologia* **10**:439–447.
56. B. G. Slavin, 1985, The morphology of adipose tissue, in *New Perspectives in Adipose Tissue: Structure, Function and Development,* A. Cryer, and R. L. R. Van, eds., Butterworth and Co., Ltd., Butterworths, London. pp. 23–43.
57. O. Stein, R. O. Scow, and Y. Stein, 1970, FFA-^{3}H uptake by perfused adipose tissue: Electron microscopic autoradiographic study, *Am. J. Physiol.* **219**:510–518.
58. B. G. Slavin, and C. Hodel, 1969, Uptake of horseradish peroxidase by adipose cells in organ culture, *Anat. Rec.* **157**:389.
59. B. Jeanrenaud, 1971, Adipocytes, available energy and endocrine pancreas, *Diabetologia* **7**:209–222.
60. E. J. Blanchette-Mackie and R. O. Scow, 1981a, Membrane continuities within cells and intercellular contacts in white adipose tissue of young rats, *J. Ultrastruct. Res.* **77**:277–294.
61. E. J. Blanchette-Mackie and R. O. Scow, 1981b, Lipolysis and lamellar structures in white adipose tissue of young rats: Lipid movement in membranes, *J. Ultrastruct. Res.* **77**:295–318.
62. E. J. Blanchette-Mackie and R. O. Scow, 1971, Sites of lipoprotein lipase activity in adipose tissue perfused with chylomicrons. Electron microscopic cytochemical study, *J. Cell Biol.* **51**:1–25.
63. R. O. Scow, E. J. Blanchette-Mackie, and L. C. Smith, 1976, Role of capillary endothelium in the clearance of chylomicrons. A model for lipid transport from blood by lateral diffusion in cell membranes, *Circ. Res.* **39**:149–162.
64. R. O. Scow, E. J. Blanchette-Mackie, and L. C. Smith, 1980, Transport of lipid across capillary endothelium, *Fed. Proc.* **39**:2610–2617.
65. E. R. Suter, 1969, The fine structure of brown adipose tissue. III. The effect of cold exposure and its mediation in newborn rats, *Lab. Invest.* **21**:259–268.
66. E. R. Suter, 1969, The fine structure of brown adipose tissue. II. Perinatal development in the rat, *Lab. Invest.* **21**:246–258.
67. R. J. Barrnett, 1962, The morphology of adipose tissue, with particular reference to its histochemistry and ultrastructure, in *Adipose Tissue as an Organ,* L. W. Kinsell, ed. Charles C. Thomas, Springfield, Illinois, pp. 3–78.
68. J. F. Thomson, D. A. Habeck, S. L. Nance, and K. L. Beetham, 1969, Ultrastructural and biochemical changes in brown fat in cold-exposed rats, *J. Cell Biol.* **41**:312–334.
69. D. Picard, F. Tasso, and G. Cotte, 1966, Aspect ultrastructural de la lipogénèse dans le tissu adipeux brun, *Z. Zellforsch. Mikrosk. Anat.* **69**:260–273.
70. E. J. Blanchette-Mackie and R. O. Scow, 1983, Movement of lipolytic products to mitochondria in brown adipose tissue of young rats: An electron microscopic study, *J. Lipid Res.* **24**:229–244.
71. R. F. Dyer, 1968, Morphological features of brown adipose cell maturation in vivo and vitro, *Am. J. Anat.* **123**:25–281.

72. L. Napolitano, 1963, Cytolysosomes in metabolically active cells, *J. Cell Biol.* **18**:663–679.
73. D. Hull, 1966, The structure and function of brown adipose tissue, *Br. Med. Bull.* **22**:92–96.
74. W. Bargmann, G. V. Hehn, and E. Lindner, 1968, Über die Zellen des braunen Fettqewebes und ihre Innervation, *Z. Zellforsch.* **85**:601–613.
75. J. Ochi, M. Konishi, and H. Yoshikawa, 1969, Morphologischer Nachweis der sympathischen Innervation des braunen Fettgewebes in der Ratte, *Z. Anat. Entwick-Gesch.* **129**:159–267.
76. O. Lindberg, 1970, *Brown Adipose Tissue,* Am. Elsevier Publishing, New York.
77. K. J. Hittleman and O. Lindberg, 1970, Fatty acid uncoupling in brown fat mitochondria, in *Brown Adipose Tissue,* O. Lindberg, ed. Elsevier, New York, pp. 245–262.
78. T. Barnard and O. Lindberg, 1966, Ultrastructural changes in the chondriome during perinatal development in brown adipose tissue of rats, *J. Ultrastruct. Res.* **29**:293–310.
79. E. R. Suter, 1969, The fine structure of brown adipose tissue. I. Cold-induced changes in the rat, *J. Ultrastruct. Res.* **26**:216–241.
80. E. I. Grodums, 1977, Ultrastructural changes in the mitochondria of brown adipose cells during the hibernation cycle of *Citellus lateralis, Cell Tissue Res.* **185**:231–237.
81. T. Barnard, J. Skala, and O. Lindley, 1969, Changes in interscapular brown adipose tissue of the rat during perinatal and early postnatal development and after cold acclimation. I. Activities of some respiratory enzymes in the tissue, *Comp. Biochem. Physiol.* **33**:499–508.
82. J. Skala, T. Barnard, and O. Lindberg, 1970, Changes in interscapular brown adipose tissue of the rat during perinatal and early postnatal development and after cold exposure. II. Mitochondrial changes, *Comp. Biochem. Physiol.* **33**:509–528.
83. E. R. Suter, 1969, The ultrastructure of brown adipose tissue in perinatal rats, *Experientia* **25**:286–287.
84. E. R. Suter and W. Staubli, 1970, An ultrastructural histochemical study of brown adipose tissue from neonatal rats, *J. Histochem. Cytochem.* **18**:100–106.
85. T. Hasegawa and K. Ueda, 1971, The ultrastructure of human subcutaneous fat cells, *Jpn. J. Dermatol.,* Series B. **81**:467–469.
86. G. J. Hausman and L. R. Richardson, 1982, Histochemical and ultrastructural analysis of developing adipocytes in the fetal pig, *Acta Anat.* **114**:228–247.
87. H. Moore and K. Muhlethaler, 1963, Fine structure in frozen-etched yeast cells, *J. Cell Biol.* **17**:609–628.

4

Enzymology of Adipocytes

Roy J. Martin and Alayne Makula
University of Georgia

Adipose cell enzymes have been studied extensively. Measurements of V_{max} (maximum velocity) and K_m (Michaelis Constant) allow the researcher to make certain conclusions about the quantitative and qualitative importance of various metabolic pathways. Researchers have emphasized the importance of measuring the key or rate-limiting enzymes in a metabolic pathway. Using such measurements is usually a powerful technique for examining cellular adaptation to a change in developmental or metabolic status. For example, glycerol phosphate dehydrogenase has been used as a marker for the development of adipose cells in culture (1). Fatty acid synthetase and glucose-6-phosphate dehydrogenase have been closely linked to insulin status (2, 3).

At the same time it is essential to point out that enzyme changes do not always reflect changes in substrate influx through a pathway. Leveille and Chakrabarty (4, 5) found that adaptation to a change in feeding status or pattern was initially keyed to substrate flux rather than the enzyme V_{max} changes. Therefore, many investigators will measure metabolic flux by using tracer studies in addition to studying enzyme activity. In the later stages of the response to dietary manipulation, metabolic shifts were found to be reflected in V_{max} changes in many enzymes involved in carbohydrate and fat metabolism (6).

It is likely that the allosteric activation of rate-limiting enzyme activity is an initial or early response to pathway stimulation by hormones or diet.

Measuring the active form of the enzyme is usually rather difficult because of the rapid intracellular changes that occur after an animal is killed for tissue analysis. Even in cell culture it may be difficult to "fix" the enzyme in its activated state so that the percent activation can be determined. An example of the fixing of an allosterically activated enzyme is the technique developed by Clark and Clark (7) for acetyl coenzyme A (acetyl-CoA) carboxylase. This technique is presented later in this chapter.

This chapter provides information on enzymes involved in carbohydrate, lipid, and amino acid metabolism of adipose cells. A brief review of adipose enzyme function and regulation and a description of the methodology used for assays is presented.

CARBOHYDRATE METABOLISM

Conversion of carbohydrate to lipid is a major mechanism by which the body stores excess calories when on a carbohydrate diet (8). Adipose tissue has the primary responsibility for storing this energy. The metabolic pathways and some of the key enzymatic steps are outlined in Figure 4-1. Hyperglycemia and hyperinsulinemia are stimulated when glucose is taken in excess of energy demands. Insulin is thought to enhance the uptake of glucose by adipose cells by stimulating the translocation of intracellular glucose transport systems to the plasma membrane (9).

Glucose Transport Systems

The number of glucose transport systems in the adipose cell membrane is assayed by measuring specific D-glucose-inhibitable cytochalasin B binding (10). On a high-carbohydrate diet the insulin-stimulated transport of glucose into adipose cells was nearly twice that of rats on a high-fat diet (9). Cytochalasin B binding sites were also doubled in the adipose cell plasma membranes of rats on the high-fat diet. These observations illustrate the importance of the glucose transport systems in determining glucose uptake by isolated adipose cells. This important step in the uptake of glucose by the adipose cell has not been extensively investigated in the various animal models of obesity. It is not known if there is an increase in the plasma membrane glucose transport systems during the preobese phase.

Cytochalasin B Binding Assay. Adipose cells are homogenized at 24°C in a buffer containing 20 mM Tris-HCl, 1 mM ethylenediaminetetraacetate (EDTA), and 255 mM sucrose. All steps following homogenization are carried out at 4°C with this same buffer. Each homogenate is centrifuged at 16,000 × *g* for 15 min. The supernatant is aspirated and saved for preparation of the microsomal membrane fraction, the fat cake is discarded, and the pellet (containing plasma membranes, mitochondria, nuclei, and broken cells) is washed once and resuspended for preparation of the plasma membrane fraction. The plasma membrane fraction is obtained by centrif-

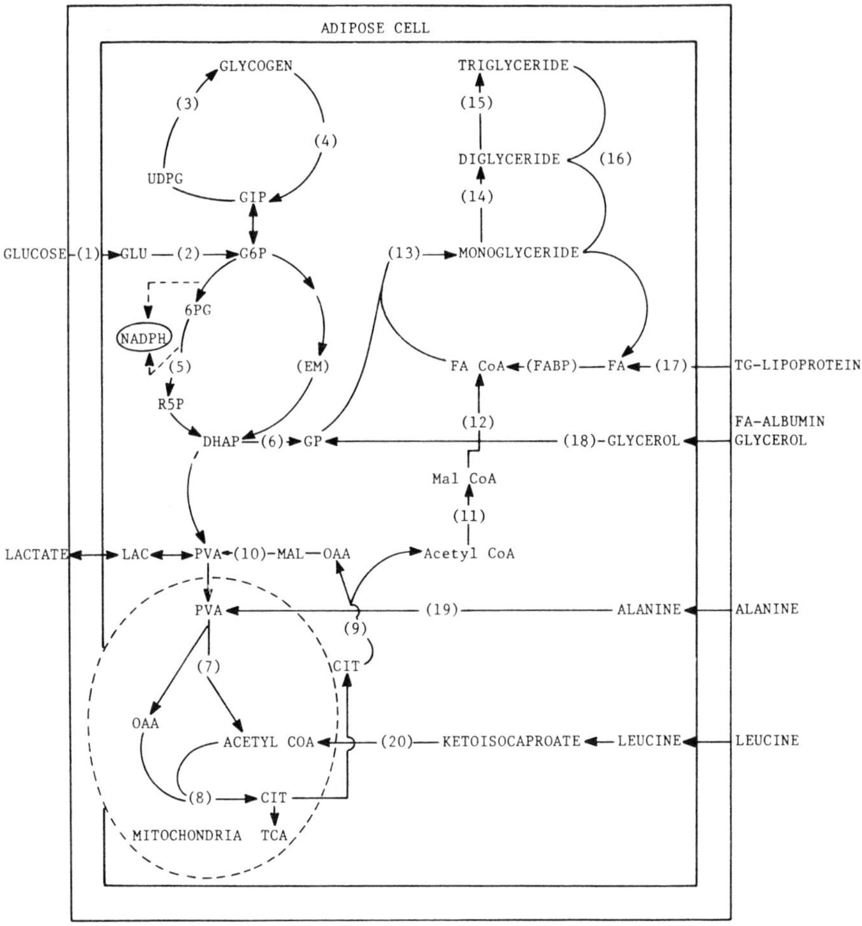

Figure 4-1. Intermediary metabolism of adipose cells—important enzymatic reactions: (1) Glucose transport system, (2) hexokinase, (3) glycogen synthetase, (4) glycogen phosphorylase, (5) pentose phosphate pathway dehydrogenase, (6) glycerophosphate dehydrogenase, (7) pyruvate dehydrogenase, (8) citrate synthetase, (9) citrate cleavage enzyme, (10) malic enzyme, (11) acetyl-CoA carboxylase, (12) fatty acid synthetase complex, (13) glycerol-3-P acyltransferase, (14) monoacylglycerol acyltransferase, (15) diacylglycerol acyltransferase, (16) hormone-sensitive lipase, (17) lipoprotein lipase, (18) glycerokinase, (19) alanine amino transferase, (20) branch chain amino acid dehydrogenase.

ugation at 23,000 × *g* for 60 min on a discontinuous 1.12 M sucrose gradient in 20 mM Tris-HCl and 1 mM EDTA. The plasma membrane band is washed twice by resuspension in buffer and centrifugation at 48,000 × *g*, and resuspended to a final concentration of 3–5 mg protein/ml. The microsomal membrane fraction is obtained from the initial supernatant by centrifugation at 160,000 × *g* for 70 min. The pellet is washed once and resuspended to a final concentration of 5–7 mg protein/ml. Equilibrium D-glucose-inhibitable cytochalasin B binding to each fraction is then measured.

The assay of cytochalasin B binding has been made more specific by the use of cytochalasin E. Cytochalasin E does not inhibit either D-glucose-inhibitable cytochalasin B binding or glucose transport itself but suppresses cytochalasin B binding to other sites present in purified adipose cell membranes, that account for up to 95% of the cytochalasin B binding. In the presence of cytochalasin E, cytochalasin B binds almost exclusively to the D-glucose-inhibitable site.

The cytochalasin B binding assay is carried out as follows. Initially, 40 μl of cytochalasin E (Aldrich Chemical Co.) solution in buffer containing 25% ethyl alcohol is added to 500 μl of membrane suspension and mixed. Two 250-μl samples are removed from the mixture, added to 20 μl of buffer either containing or not containing D-glucose, and mixed again. Four 100-μl samples are removed from each of the two resulting suspensions and mixed with 20 μl of buffer containing increasing concentrations of [^{3}H]cytochalasin B (New England Nuclear and Aldrich Chemical Co.) and tracer [U-^{14}C]sucrose (New England Nuclear). The final concentrations of ethyl alcohol, cytochalasin E, and D-glucose are, respectively, less than 1%, 2,000 nM, and 500 nM. The four final concentrations of cytochalasin B are approximately 40, 170, 420, and 20,000 nM. [U-^{14}C]Sucrose is used to aid in determining trapped, unbound [^{3}H]cytochalasin B in the membrane pellets obtained during the binding assay and is, therefore, added in a constant proportion to the concentration of [^{3}H]cytochalasin B such that the final [^{14}C] and [3H] counts per min in the mixture will be approximately the same.

Duplicate 50-μl samples of each of the eight final suspensions are then transferred to 175-μl centrifuge tubes, the membranes in each sample are pelleted by centrifugation, 25 μl of each supernatant is removed for measurement of the free [^{3}H]cytochalasin B concentration, and the remaining supernatant in each sample is aspirated and discarded. Each centrifuge tube with its pellet is transferred directly to a scintillation vial for measurement of bound [^{3}H]cytochalasin B. The plasma membrane fractions are centrifuged at 48,000 × *g* for 20 min. The microsomal membrane fractions are centrifuged at 220,000 × *g* for 60 min. All scintillation counting is carried out in 5-ml polyethylene scintillation vials with 3.0 ml of scintillation fluid (Aquasol, New England Nuclear) containing 1.7% H_2O (11).

Hexokinase

The hexokinase isoenzyme system is unique in that it is the only enzymatic step required for all pathways of glucose utilization. Two adipose isoenzymes of hexokinase have been characterized. Adipose tissue hexokinases I and II have been shown to behave like adaptive enzymes in vivo, decreasing with fasting and increasing to normal again with refeeding (12, 13).

The ratio of type II to type I hexokinase was high in insulin-sensitive tissues of the rat (12). The low activity of adipose tissue hexokinase II in metabolic states of diminished rates of glucose utilization (14, 15) suggests a key role for this isoenzyme in determining the rate of glucose utilization by adipose cells. Dietary carbohydrate content also influenced the activity of this enzyme. The ultimate activity of hexokinase II in fasted rats refed seven different diets was dependent mainly on the carbohydrate content of the diet (16).

Hexokinase Assay (HK; EC 2.7.1.1). The basic reaction in the hexokinase assay is as follows:

$$\text{Gucose} + \text{ATP} \rightarrow \text{G–6–P} + \text{ADP},$$

where ATP = adenosine 5^1-triphosphate; G-6-P = glucose-6-phosphate; and ADP = adensine 5^1-diphosphate. See Table 4-1 for information on the hexokinase assay.

Table 4-1. Spectrophotometric Analysis of Adipose Tissue Enzymes

Enzyme	*Reference*
Pyruvate kinase (PK) (EC 2.7.1.40) (phosphoenolpyruvate kinase, phosphoenol transporylase) Tris-HCl, 0.44 mmol, pH 7.6; KCl, 0.61 mmol; $MgCl_2$, 66 μmol (use a premix: 50 ml 0.1 M Tris; 50 ml 0.14 M KCl; 2.5 ml 0.3 $MgCl_2$; 0.9 ml/assay); phosphoenol pyruvate, 0.475 mol; ADP, 1.3 mol; NADH, 0.11 mol (in 0.001 M·KOH); lactate dehydrogenase, 3.5 units.	Bucher and Pfleiderer, (17)
Hexokinase (HK) (EC 2.7.1.1) (heterophosphatase) Glycylglycine, 0.038 mmol, pH 7.5; $MgCl_2$, 4.6 μmol; NADP, 1.0 μmol; G-6-Pd, 37 units (check for HK contamination by omitting sample). (Use a premix: 45 ml 0.05 M glycylglycine; 1.5 ml 0.18 M $MgCl_2$; 1.5 ml 0.04 M NADP; 1.5 ml 1466 units/ml G-6-PD; 3–6 ml, H_2O depending on extract volume. 0.9–0.95 ml/assay). ATP^2,	Katzen et al. (18)

Table 4-1. Continued

Enzyme	*Reference*
2.7 μmol (pH 7.5 with NADH); glucose, 0.0125 μmol. (A lag time of 10–15 min may be necessary.)	
Glycerol phosphate dehydrogenase (GPD) (EC 1.1.1.8) (sn-glycerol-3-phosphate dehydrogenase) Triethanolamine-HC1, 0.1 mmol, pH 7.5, (EDTA) 2.5 μmol; NADH, 0.12 μmol; dihydroxyacetone phosphate, 0.2 μmol; B-mercaptoethanol, 0.1 μmol.	Kozak and Jensen (19) Modification Wise and Green (20)
Lactate dehydrogenase (LDH, lactic acid dehydrogenase) (EC 1.1.1.27) Tris buffer 0.95 mmol, pH 9.0; NAD, 0.5 μmol; lactate[2], 5.0 μmol pH 9.0 (make fresh daily).	Neilands (21)
NADP-isocitrate dehydrogenase (ICDH-NADP) (EC 1.1.1.42) Tris buffer, 34 μmol, pH 7.4; EDTA, 0.34 μmol; $MnCl_2$, 1.5 μmol; $NADP^+$, 0.1 μmol; gelatin 0.03%; DL-isocitrate, 1.5 μmol.	Plaut (22) Modified by Bauman et al. (23)
Malate dehydrogenase (MDH) (EC 1.1.1.37) Tris buffer, 95 μmol, PH 9.0; NAD, 0.5 μmol; pH 9.0 (make fresh daily)	Baldwin and Milligan (24)
Citrate cleavage enzyme (CCE) (EC 1.1.49) Tris-HCl, 0.1 mmol, pH 8.7; citric acid, 20 μmol pH 8.0 (with KOH); $MgCl_2$, 10 μmol; dithiothreitol 10 μmol (make fresh); NADH, 0.14 μmol; CoA (SH) 0.33 μmol; MDH, 2.5 units (more if needed to ensure excess); ATP, 50 μmol, pH 7–7.5 (with KOH) (a 5–10 min lag time may be needed before ATP addition).	Cottam and Srere (25)
Aspartate amino transferase (GOT) (EC 2.6.1.1) (Glutamic-oxaloacetic transaminase, glutamic-aspartic transaminase, transaminase A); potassium phosphate (monabasic), 0.18 mmol, pH 7.5; α-ketoglutarate[2], 5.0 μmol; L-aspartate, 0.013 μmol (pH 7.0); NADH, 0.085 μmol (in 0.001 N KOH); dehydrogenase, 3.33 units.	Sizer and Jenkins (26)
L-alanine amino transferase (GPT) (EC 2.6.1.2) (glutamic-pyruvic transaminase, glutamic alanine transaminase) potassium phosphate (monobasic), 0.17 mmol, pH 7.3; α-ketoglutarate[2], 5 μmol; L-alanine, 25 μmol; NADH, 8.5 μmol (in 0.00I N KOH); lactate dehydrogenase, 1.25 units (omit if slope is not increased when present; dilute if activity is present in the absence of sample).	Segal and Matsuzaiva (27)
Malic enzyme (ME) (EC 1.1.1.40) (malate dehydrogenase) Glycine, 0.045 mmol, pH 7.5; $MnCl_2$, 3 μmol; NADP, 0.25 μmol; malate, 5 μmol (make fresh daily, pH 7.0).	Ochoa (28)
G-6-PD (EC 1.1.49) Glycine, 0.09 mmol, pH 7.6; $MgCl_2$, 25 μmol; NADP, 0.08 μmol; G-6-P, 1.25 μmol.	Glock and McLean (29)
6-phosphogluconate dehydrogenase (6-PDG) (EC 1.1.1.44) (6-phosphogluconate carboxylase). Add 6-phosphogluconate[2], 1.25 μmol to G-6-PD assay. Subtract G-6-PD activity from combined.	Glock and McLean (29)

Glycogen Synthetase

Meal feeding has been used to demonstrate adaptive enzymes in adipose tissue. The influx of nutrients after a large meal stimulated the enzymes and metabolic pathways that were important in storing excess energy in the form of either glycogen or fat. Glycogen accumulation in adipose tissue proceeded at a rate more than 50 times greater in meal-fed rats (4). Adipose tissue glycogen synthetase activity also increased in meal-fed rats (30) and in fasted refed rats (31). Glycogen synthesis in adipose tissue was catalyzed by multiple glucose-6-phosphate-dependent forms of glycogen synthetase. There was no evidence for a glucose-6-phosphate-independent form of glycogen synthetase in adipose cells (32). Hormones regulate glycogen metabolism by promoting reversible interconversions between these forms.

Glycogen Synthetase Assay (EC2.4.1.11). The basic reaction in the glucose synthetase assay is as follows (UDP = uridine 5[1]-diphosphate):

$$\text{UDP-Glucose} + [\text{Glucose}]_n \rightarrow \text{UDP} + [\text{Glucose}]_{n+1}$$

Epididymal fat pads are homogenized in cold 50 mM Tris-HCl and 1 mM Sodium-EDTA (Na-EDTA) at pH 7.8 by using a glass Potter-Elvehjem tube and Teflon pestle. Samples are centrifuged at 12,000 × g for 20 min, and the infranatant layer is removed and used for the assays.

The reaction mixture contains 25 μl of 200 mM Tris-HCl, 10 mM Na-EDTA, glycogen (40 mg/ml), 2 mM UDP (U = ^{14}C) glucose (5 microcuries/ml (μCi/ml)) pH 7.8 and 25 μl of 10–25 mM glucose-6-phosphate warmed to 30°C. The addition of 50 μl of the cold sample, in a final dilution of 1:10 based on wet weight, starts the reaction (32). The reaction is terminated after 10–20 min by the addition of 2 ml of 70% ethyl alcohol with 0.1 M ammonium acetate. After centrifugation, the pellets are rinsed in resuspended water and centrifuged four times. The final suspension is added to a scintillation vial containing 10 ml Omnifluor and counted (33).

An alternate method is to spot 75 μl of the incubation reaction on Whatman No. 31 filter paper. The papers are dropped into cold 66% ethyl alcohol (7–8 ml/paper) and stirred for 20–25 min after the last paper is added, allowing no contact between the papers and the Teflon bar. A second 10-min ethyl alcohol wash takes place at room temperature and a 5-min acetone wash removes the remains of the ethyl alcohol and water. The papers are dried and placed in glass scintillation vials. Ten milliliters of 0.5% 2, 5-diphenyloxazole in toluene (weight/volume, w/v) is added and activity is counted (34).

Glycogen synthase activity ratios are calculated by dividing the activity found in the absence of added glucose-6-phosphate by the maximal activity found in the presence of glucose-6-phosphate.

Pentose Phosphate Pathway

The pentose phosphate pathway is closely associated with changes in the rate of lipogenesis (29, 35, 36). Plots of both pentose phosphate activity and fatty acid synthesis versus glucose uptake by adipose cells revealed that both pathways were stimulated in proportion to the amount of glucose uptake (37). The reducing equivalents required for fatty acid synthesis were supplied primarily by two enzymes of the pentose phosphate pathway, glucose 6-phosphate dehydrogenase and 6-phosphogluconate dehydrogenase. Both enzymes were induced by fasting-refeeding (38, 39), meal feeding (5), and genetic obesity (3, 40), and were depressed by feeding a high-fat diet that suppressed de novo lipogenesis (4, 41, 42). Insulin stimulation of adipose tissue lipogenesis resulted in a twofold increase in the rate of synthesis of glucose-6-phosphate dehydrogenase enzyme and a decrease in the degradation of the enzyme (2). In addition to changes in enzyme protein an allosteric control of glucose-6-phosphate dehydrogenase has been suggested (43, 44).

Pentose Phosphate Assay. The following reactions are studied in the pentose phosphate Pathway Enzyme assay (refer to Table 4-1).

$$\text{G-6-P} + \text{NADP} \rightarrow \text{6-phosphogluconate} + \text{NADPH}$$
$$\text{6-Phosphogluconate} + \text{NADP} \rightarrow \text{R-5-P} + \text{NADPH},$$

where NADP = nicotinamide dinucleotide phosphate; NADPH = reduced nicotinamide dinucleotide phosphate; and R-5-P = ribulose-5-phosphate.

Malate Cycle

The activity of the pentose phosphate pathway does not provide all of the reducing equivalents for lipogenesis (35, 45). The pentose cycle was found to provide about 60% of the NADPH required for fatty acid synthesis (37). The malate cycle (Fig. 4-1) illustrates another pathway that supplied reducing equivalents for lipogenesis. Key enzymes in this pathway are citrate cleavage enzyme and malic enzyme; the latter enzyme provided the reducing equivalents by oxidation and decarboxylation of malate. Both enzymes responded to food intake (6, 46) and obesity (Fig. 4-2; 2, 47–50).

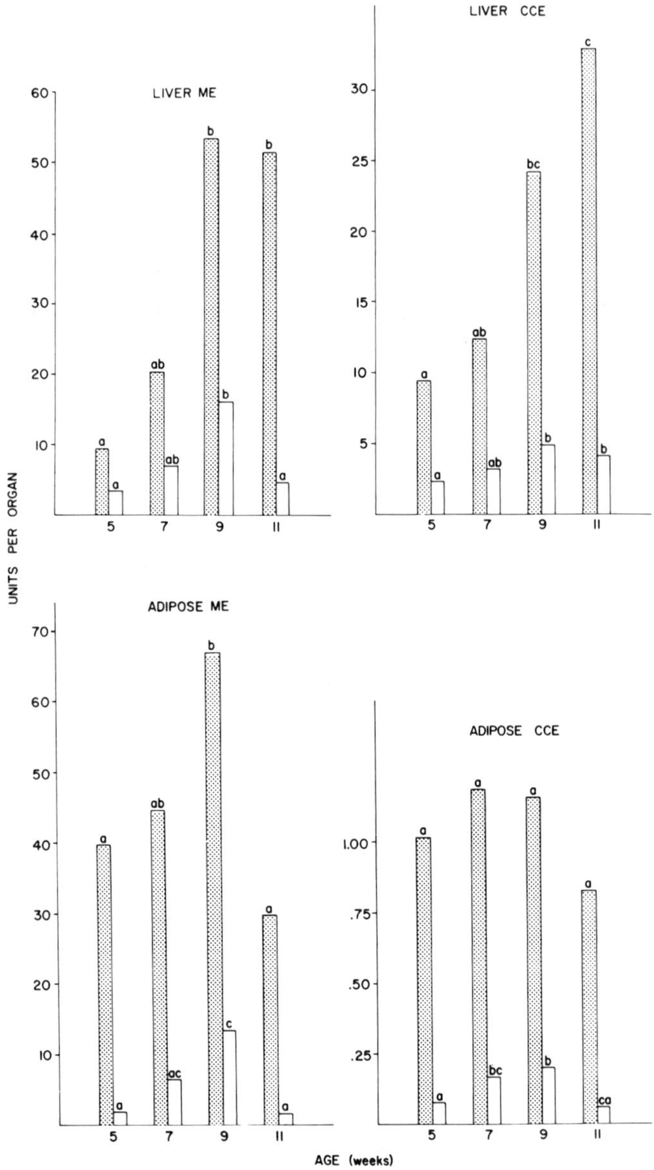

Figure 4-2. Liver and adipose ME and CCE in the lean (open bars) and obese (stippled bars) Zucker rat at different ages. A unit is the amount of enzyme required to catalyze 1 nmol/min. Means in the same genotype with different superscripts are significantly different ($P < 0.05$). N.S.—Means between lean and obese in the same age group are not significantly different ($P < 0.05$); all not so labeled are significantly different ($P < 0.05$).

Malate Cycle Assay. The following reactions are studied in the malate cycle assay (refer to Table 4-1).

Citrate + CoASH-(Citrate Cleavage Enzyme) → OAA + Acetyl-CoA
OaA + NADPH-(Malate Dehydrogenase) → Malate + NAD
Malate + NADP-(Malic Enzyme) → Pyruvate + NADPH

Sources of Glycerophosphate

The capacity of adipose cells to esterify fatty acids is dependent on the supply of glycerophosphate (Fig. 4-3). In the fed state glucose was the primary source of the glycerophosphate used in esterification of fatty acids. The enzymatic reaction that diverts some of the carbons from glycolysis for this purpose was glycerophosphate dehydrogenase. The activity of this enzyme in adipose tissue was usually enhanced during the hyperlipogenic state (51) and in animal models of obesity (40). In addition, it has been utilized as an early marker of adipose cell differentiation in cell culture (1, 52).

In food-restricted rats the rates of adipose cell lipolysis increased as well as the rates of reesterification (53). This elevated rate of reesterification was maintained in the absence of glucose. Therefore, another source of precursors for glycerophosphate formation must exist. It has been shown that pyruvate decreased the rate of adipose cell lipolysis by increasing fatty acid esterification (54). Adipose tissue phosphoenolpyruvate carboxykinase and pyruvate carboxylase may play an important role in the conversion of pyruvate to glycerophosphate (55). Glucocorticoids have been shown to bind to adipose cell receptors and cause a decrease in phosphoenolpyruvate carboxykinase (PEPCK) (56).

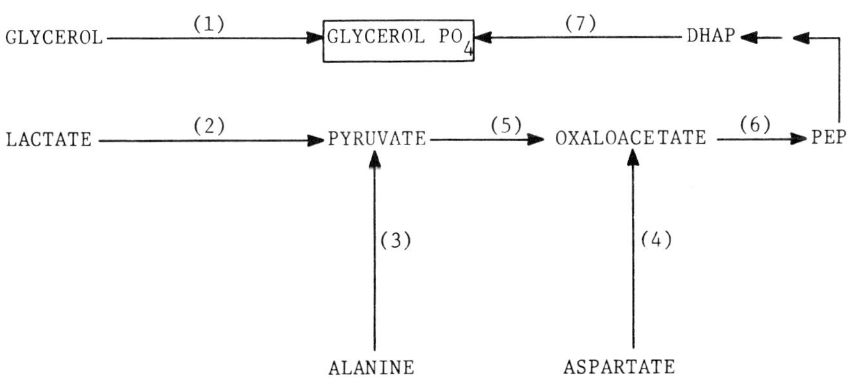

Figure 4-3. Pathways for glycerophosphate generation in adipose cells.

Adipose cell glycerokinase (EC 2.7.1.30) is also a potential mechanism by which glycerophosphate is regenerated. Interest in this mechanism was enhanced by Mayer (57), who suggested that glycerokinase was present in adipose cells and was the primary genetic defect in obese hyperglycemic mice. Since then numerous studies of this enzyme in adipose tissue have been reported. The enzyme activity was increased in obese mice (58, 59, 60), obese rats (48, 61), and dietary obesity (62).

Glycerokinase Assay (EC 2.7.1.30). The following reaction takes place in the glycerokinase assay:

$$\text{Glycerol} + \text{ATP} \rightarrow \text{L-glycerol-3-phosphate} + \text{ADP}$$

Samples of adipose tissue are homogenized in cold 0.25 M sucrose and centrifuged at 27,000 × *g* for 20 minutes at 0°C. The infranatant portion is removed for protein precipitation by ammonium sulfate (final concentration 0.5 M). The precipitate is resuspended in 1 ml 0.25 M sucrose.

The incubation medium (6 mM ATP, 5 mM $MgCl_2$, 25 mM NaF, 1 mM EDTA, 20 mM mercaptoethanol, 100 mM Tris, and 1 mM glycerol-2-^{14}C (specific activity 10 μC/ml of medium) is placed in stoppered 2-ml autoanalyzer tubes in a 37°C waterbath. Addition of the enzyme sample initiates the reaction. After a specific time (1–10 min) the addition of 100 μl of 98% ethyl alcohol stops the reaction. Aliquots are pipetted onto diethylaminoethyl-cellulose (DEAE-cellulose) filter discs, dried, and washed with 100 to 200 ml deionized distilled water. The discs are dried in a 95°C oven and placed in a scintillation vial for counting (63, 64).

Phosphoenolpyruvate Carboxykinase Assay (PEP CK, EC 4.1.1.32). The reaction studied in the PEPCK assay is as follows:

$$\text{Oxaloacetate} + \text{GTP} \rightarrow \text{Phosphoenolpyruvate} + \text{GDP} + CO_2,$$

where GTP = guanosine 5^1-triphosphate; GDP = guanosine 5^1-diphosphate.

The epididymal fat pads are homogenized in 4 volumes of cold 0.2 M sucrose, pH 7.5, containing 0.02 M triethanolamine, 1 mM glutathione, and 1 mM EDTA, with a glass homogenizer and a Teflon pestle. The homogenate is centrifuged for 30 min at 100,000 × *g*. The supernatant fraction is used for assays.

The assay medium contains imidazole buffer, 100 mM, pH 6.6; PEP, 1.25 mM; Mn Cl_2, 2 mM; IDP, 1.25 mM; $KH^{14}CO_3$, 50 mM (10^5 cpm/

mol); GSH, 2 mM; NADH, 2.5 mM; malate dehydrogenase, 2 units; and tissue extract in a final volume of 1 ml with pH 7.0. The addition of PEP initiates the reaction. After a 15 min incubation at 30°C, the reaction is stopped by the addition of 1 ml of 2 N HCl. The reaction tubes are centrifuged and CO_2 is vigorously bubbled through the aqueous portion for 5 min to remove the excess $^{14}CO_2$. Aliquots of 0.5 ml are transferred to scintillation vials and taken to dryness in an 85°C oven. One milliliter of water and 10 ml of scintillator are added before radioactivity is counted (65, 66).

Pyruvate Carboxylase Assay (EC 6.4.1.1). The following reaction takes place in the pyrovate carboxylase assay:

$$\text{Pyruvate} + CO_2 + \text{ATP} + H_2O \rightarrow$$
$$\text{oxaloacetate} + \text{ADP} + \text{Phosphate} + 2H^+$$

The tissue preparation is the same as for PEPCK. The supernatant fraction is kept at 25°C. The reaction medium contains Tris-HCl, 25 mM, pH 7.4; sodium pyruvate, 10 mM; ATP, 2.5 mM, acetyl-CoA, 0.75 mM; $KH^{14}CO_3$, 50 mM (2 μc) $MnCl_2$, 5 mM; and tissue extract to a 1 ml total volume. The reaction is stopped by the addition of 0.5 ml Trichloroacetic acid (TCA) (10% w/v). The precipitate is centrifuged out and unlabeled CO_2 is bubbled through the aqueous solution for 3 min. Aliquots are removed and counted with Diotol as the solvent (67).

Pyruvate Dehydrogenase Complex

The hormonal and dietary control of carbohydrate conversion to fatty acids has been associated with changes in adipose cell pyruvate dehydrogenase (68, 69, 70). The activity of this enzyme complex in adipose tissues of obese mice was twofold greater than in lean mice (71).

Three separate enzymes are involved in the conversion of pyruvate into acetyl-CoA (Fig. 4-4). Control of this enzyme complex has been shown to involve a phosphorylation-dephosphorylation mechanism (72) and end product inhibition (68).

Pyruvate Dehydrogenase Assay (EC 1.2.4.1). The reaction taking place in the pyruvate dehydrogenase (also known as pyruvate:lipoate oxidoreductase) assay is as follows:

$$\text{Pyruvate} + \text{CoA} + \text{NAD}^+ \rightarrow \text{Acetyl-CoA} + CO^2 + \text{NADH} + H^+$$

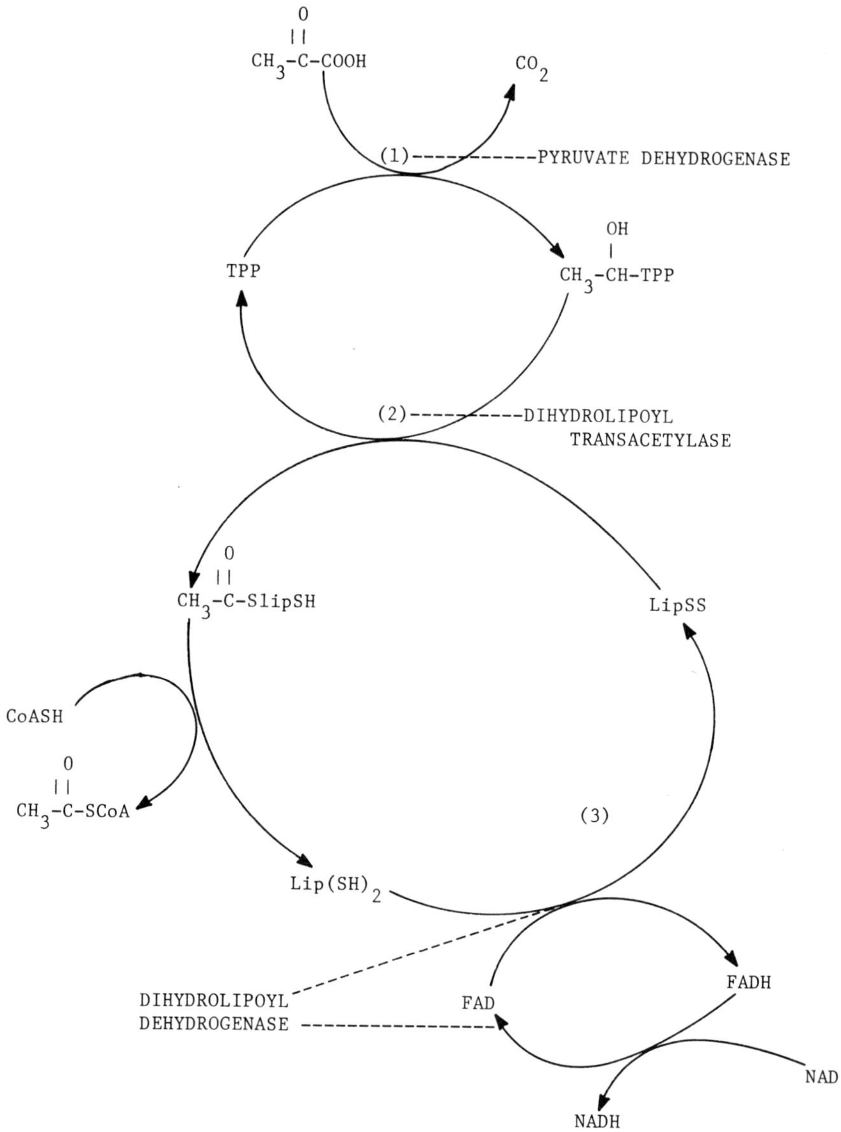

Figure 4-4. Pyruvate dehydrogenase complex reactions. The abbreviations used are: TPP, thiamine pyrophosphate; LipSS and Lip$(SH)_2$, lipoyl moiety and its reduced form; CoASH, coenzyme A; FAS, flavin adenine dinucleotide; NAD^+ and NADH, nicotinamide adenine dinucleotide and its reduced form.

Approximately 200 mg frozen adipose tissue or fat cells (prepared according to Rodbell, 73) are homogenized in 0.5 ml potassium phosphate buffer, 15 mM, pH 7.0, with 40 mg BSA/ml and 3 ml silicone oil (density 0.95) with a high-speed tissue disintegrator. The tissue is homogenized at 0°C, two times for 30 sec each with a 30-sec interval, then centrifuged at 10,000 × *g* for 2 min. The aqueous portion is used for the assay.

The assay medium contains: 25 mM potassium phosphate buffer, pH 8.0; 0.15 mM p-nitroaniline; 1 mM NAD^+; 1 mM thiamine pyrophosphate; 10 mM mercaptoethanol; 20 mM sodium pyruvate; 0.4 mM $MgCl_2$; 10 μg crystalline lactate dehydrogenase; 50 mU milliunits arylamine acetyl-transferase and 50 μl of adipose tissue extract. The addition of 0.1 mM coenzyme A (CoA) initiates the reaction. The final volume is 250 μl. The reaction is recorded at 405 nm at 25°C. Total pyruvate dehydrogenase activity is measured as above after dephosphorylation of the inactive form by incubating the adipose tissue extracts with 10 mM $MgCl_2$, 0.05 mM $CaCl_2$ and partially purified pyruvate dehydrogenase phosphatase, 12 μg protein in 120 μl final volume. The reaction is stopped after 10 min (74).

LIPID METABOLISM

Adipose cells function primarily as a dynamic energy reservoir storing excess energy as lipid during positive energy balance and mobilizing this stored energy when in negative energy balance. The major enzymes involved in these metabolic processes are covered in this section.

Acetyl CoA Carboxylase

De novo lipogenesis is primarily regulated by two enzyme systems, acetyl-CoA carboxylase and fatty acid synthetase. The former has frequently been cited as the rate-limiting enzyme in fatty acid synthesis (75, 76, 77). Evidence for coordinated control of both enzymes has been reported (78, 79). It has also been suggested that acetyl-CoA carboxylase may be the rate-limiting enzyme in short-term control (80).

Allosteric regulation of acetyl-CoA carboxylase is of prime importance in the control of enzyme activity (81). Citrate activation of the enzyme was achieved by polymerization of inactive protomers to active polymeric filaments (Fig. 4-5). The major effect of the activator was on the maximum velocity of the enzyme rather than on the K_m values for substrates. A lower polymer/protomer ratio for acetyl-CoA carboxylase was associated with reduced rates of fatty acid synthesis (7, 82). The most potent inhibitors were CoA thioesters of C_{16}–C_{18} fatty acids that were competitive with respect to citrate (81).

A second model for the activation of acetyl-CoA carboxylase is the phosphorylation-dephosphorylation model (77, 83; Fig. 4-4). Adipose tissue acetyl-CoA carboxylase was phosphorylated and deactivated when exposed to epinephrine or dibutyryl cyclic adenosine 5[1]–monophosphate (CAMP) (83).

Acetyl-CoA Carboxylase Assay (EC 6.4.1.2). The reaction taking place in the acteyl-CoA carboxylase (ACC) assay is as follows (malonyl-CoA = malonyl coenzyme A):

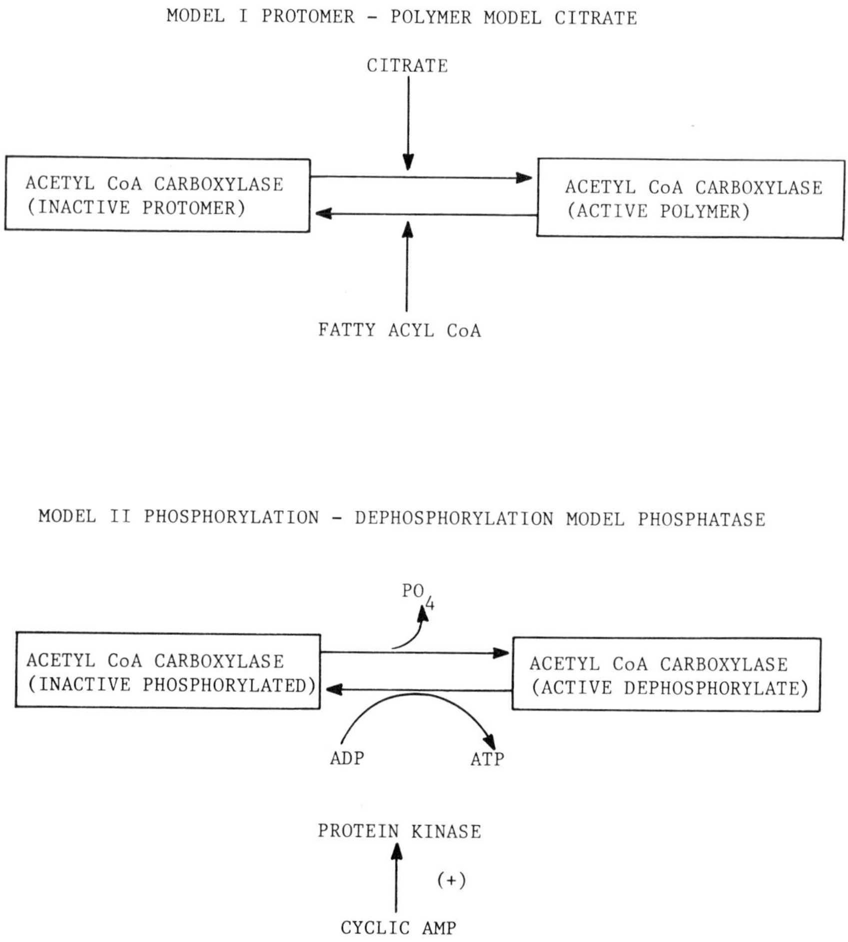

Figure 4-5. Mechanisms of activation for acetyl-CoA carboxylase. Two models have been proposed for the activation of acetyl-CoA carboxylase.

$$\text{Acetyl-CoA} + \text{ATP} + \text{HCO}_3\ \text{ADP} + \text{malonyl-CoA} + \text{Pi}$$

Epididymal fat pads are homogenized in 3 volumes of medium containing 0.1 M potassium phosphate, 0.15 M potassium chloride, and 0.1 mM EDTA, at pH 7.0 and 25°C, in a blender. The homogenate is centrifuged at 20,000 × *g* for 30 min. The fat cake and precipitate are discarded and the supernatant solution is filtered through a double layer of cheesecloth. The supernatant solution is saturated to 45% with ammonium sulfate and centrifuged at 13,000 × *g* for 30 min. The pellet is dissolved in 0.01 M potassium phosphate, 0.1 mM EDTA, and 10 mM 2-mercaptoethanol, at pH 7.0 and 25°C. This solution is dialyzed against the same buffer for 3 hr and centrifuged at 105,000 × *g* for 1 hr. The enzyme preparation (1–2 mg protein/ml) is preincubated for 30 min at 37°C in a final volume of 0.25 ml. The medium contains 60 mM Tris-HCl, pH 7.5, 3 mM reduced glutathione, 8 mM $MgCl_2$, 0.1 mM EDTA, 5 mM potassium citrate, and 0.15 mg BSA. The 2-min incubation assay medium is the same as before with 0.6 mg bovine serum albumin BSA/ml used and the addition of 2 mM ATP, 0.2 mM acetyl-CoA, and 10 mM NaH $^{14}CO_3$ (0.2 μCi/μmole). The addition of the preincubated enzyme begins the reactions. The final volume is 0.5 ml. The reaction is terminated by the addition of 0.1 ml 6 N HCl. Aliquots are pipetted into scintillation vials and dried under an infrared lamp at 85°C for 40 min. Water, 200 μl, and 15 ml scintillation fluid are added. Radioactivity is counted in a liquid scintillation counter (84).

Fatty Acid Synthetase Complex

Fatty acid synthetase is a multienzyme complex arranged as two polyfunctional polypeptides (81, 85). This multienzyme complex affects the synthesis of a fatty acid by catalyzing the sequential addition of two carbon fragments derived from malonyl-CoA to an acyl group. The acyl chain remains covalently attached to the multienzyme via a thioester linkage during synthesis. Growth in the chain length of the fatty acid is terminated when the thioester bond is hydrolyzed.

Adaptation of this enzyme complex occurs under a number of conditions. Adipose tissue fatty acid synthetase decreased during the initial phases of adaptation to meal feeding (6) but was elevated over twofold in the final phase (86). Diurnal changes in adipose tissue fatty acid synthetase reflected the diurnal pattern of food intake (87). Genetically induced obesities had high adipose tissue lipogenesis during the dynamic phase of rapid weight gain. During in vitro differentiation of adipocytes, the specific activity of fatty acid synthetase increased 13-fold (88).

Fatty Acid Synthetase Assay. The reaction taking place in the fatty acid synthetase assay is as follows:

$$\text{Acetyl-CoA} + 7\ \text{malonyl-CoA} + 14\ \text{NADPH} + 14\ H^+ \rightarrow \text{palmitic acid} + 7CO_2 + 8\ \text{CoA} + 14\ \text{NADP} + 6\ H_2O$$

The epididymal fat pads are rinsed and homogenized in cold 0.25 M sucrose containing 0.5 mM dithiothreitol and 2 mM EDTA (5 ml/g tissue) in a glass homogenizer with a Teflon pestle, cooled in ice. The homogenate is centrifuged at 105,000 × *g* for 30 min. The infranatant fraction is used for analysis (88).

The following 0.145 ml of assay medium is preincubated for 2-1/2 hr at 0°C: 0.275 M potassium phosphate, pH 6.6 containing 8.3 mM EDTA, 2.75 mM dithiothreitol, 1.24 mM NADPH, 34.5 μM acetyl-CoA 90 μg albumin, and the enzyme preparation. The medium is warmed to 37° and the reaction is started by the addition of 55 μl of 0.5 mM [2-^{14}C] malonyl-CoA (1300–2200 cpm/nmol). The reaction is terminated after 3 min by adding 50 μl of 2 N NaOH. Samples are heated in boiling water for 20 min, cooled, acidified with 40 μl 5 N HCl, and extracted five times with 1.5 ml of n-pentane. The pentane extracts are pooled, evaporated, and counted for radioactivity (88).

Triacylglycerol Synthesis (TGS)

Biosynthesis of triacylglycerol occurs by a series of stepwise acylations of glycerol-3-phosphate with long-chain acyl-CoAs catalyzed by a system of enzymes associated with the endoplasmic reticulum (Fig. 4-1; 89). The last enzyme of this pathway, diacylglycerol acyltransferase (DA EC 2.3.1.20) was thought to be the rate-limiting step in the formation of triacylglycerols (89, 90). This enzyme appeared to be inactivated by a AMP-dependent proteinkinase and activated by a phosphatase enzyme (91).

Sn-Glycerol-3-Phosphate Assay (EC 2.3.1.15) (Glycerol-P-acyl-transferase). The reaction taking place in the Sn-glycerol-3-phosphate assay as as follows:

$$\text{Glycerol phosphate} + \text{acyl-CoA} \rightarrow \text{Acyl-glycerol-P} + \text{CoA}$$

The incubation mixture (75 mM Tris-HCl, pH 7.4, 4 mM $MgCl_2$, 2 mg/ml BSA, 8 mM NaF, 50 μM palmitoyl coenzyme A (palmitoyl-CoA), and 20 μM [^{14}C]glycerol-phosphate (7.6 mCi/mmol) is placed in screw-capped culture (13 x 100 mM) at 23°C. The addition of up to 12 μg sample protein begins the reaction. The final volume is 200 μl.

After 10 min the reaction is stopped by adding 3 ml of chloroform:methanol (1:2, v:v) and 0.6 of 1% perchloric acid (a modification of Bligh and Dyer's procedure, 92). After 5 min, 1 ml of chloroform and 1 ml of 1% perchloric acid are added, vigorously mixed, and centrifuged. The upper layer is discarded. The lower chloroform layer is washed three times with 2 ml of 1% perchloric acid. One milliliter of the last chloroform layer is transferred to scintillation vials and evaporated to dryness under nitrogen. Four milliliters of aquasol are added, and the samples are counted in a liquid scintillation counter (93).

Diacylglycerol Acyltransferase Assay (EC 2.3.1.20).

Acyl-CoA + 1,2-diacyl-sn-glycerol → Triacylglycerol + CoASH

The assay is performed in a medium of 175 mM Tris-HCl (pH 8.0), 8 mM $MgCl_2$, 1 mg/ml BSA, and 30 μM palmityl-CoA, containing about 4 μg microsomal sample. The addition and vortexing of 20 μl of 2 mM 1,2-[^{3}H]diacyl-sn-glycerol (56 μCi/μmol) in ethyl alcohol begins the reaction. The final volume is 200 μl. The assay is performed at 23°C. The termination mixture (1.5 ml isopropanol:heptane:water, 80:20:2 [v:v:v) containing [^{14}C]tripalmitin is added after 10 min (94, 95).

A 20 μl solution containing 20 μg triolein and 5,550 dpm of [^{14}C] tripalmitin is added and mixed, 1 ml heptane and 0.5 ml water are added, and the tubes are capped and vortexed. The heptane phase is transferred to disposable tubes and evaporated to dryness under nitrogen.

The material is dissolved in 100 μl diethylether and spotted on a silica gel plate activated by acetone. The plates are run in petroleum ether:diethylether:acetic acid, 8:20:1 (v:v:v), and dried, and components are visualized by an iodine in methanol spray. The triolein-marked region is scraped into a scintillation vial, 8 ml of aquasol is added, and the solution is counted in a liquid scintillation counter (96).

Lipoprotein Lipase

At the luminal surface of capillaries, the enzyme lipoprotein lipase (LPL) catalyzes the breakdown of lipoprotein triglycerides, allowing fatty acids from triglycerides to enter the adipose cell. The kinetics of LPL are less well understood than those of most mammalian enzyme systems. The action of LPL at a lipid-water interface is the greatest methodological barrier to discovering key facts about the enzyme reaction.

Phospholipids and apolipoprotein C-II were required for maximum LPL activity in vitro (97, 98). Apolipoprotein C-II contains 79 amino acids (99) and was found in association with all major classes of plasma lipoproteins. Under normal conditions, this cofactor was not a rate-limiting factor in the catabolism of chylomicron triglyceride (97, 98).

Feedback inhibition of LPL by fatty acids has been investigated. Scow and Olivecrona (100) showed that when LPL acted on chylomicrons in vitro, hydrolysis slowed down as soon as the binding sites on albumin were filled. The strong inhibition of LPL and the loss of lipolysis-stimulating effects of apolipoprotein C-II provided a feedback regulation of the action of the lipase on lipoproteins at the capillary endothelium (101).

Adipose LPL activity is under nutritional and hormonal control. Increasing food intake by rat pups by reducing litter size resulted in increased adipose tissue LPL (102). High-fat diets stimulated adipose tissue LPL (103, 104). Fructose diets lowered adipose tissue LPL in female rats (105). The mechanisms by which diet influenced adipose tissue levels of LPL are not clearly understood.

Evidence for hormonal control of adipose tissue LPL activity has been provided. Studies in vivo have revealed a strong positive correlation between adipose tissue LPL and plasma insulin (106). Diurnal variations in plasma insulin concentrations parallel the pattern of LPL activity in meal-fed rats (107). Stimulatory effects of insulin on 3T3-L1 adipocyte LPL activity has provided more direct evidence for a role of this hormone in regulation (108).

Lipoprotein lipase has been measured in fetal pig adipose tissue and found to be correlated with adipocyte lipid filling (109). A genetic influence on pig adipose tissue LPL has also been identified during fetal development (110). In genetically obese Zucker rats, adipose LPL has been suggested to be a primary cause of the obesity (11). As early as 7 days of age, adipose LPL activity has been found to be elevated in obese Zucker rats (112). More recent observations have challenged the role of adipose LPL as a primary cause of the obesity in Zucker rats (113, 114).

Lipoprotein Lipase Assay (LPL EC 3.1.1.34). The reaction taking place in the lipoprotein lipase assay is as follows:

$$\text{Triolein } (^3\text{H}) \rightarrow (^3\text{H}) \text{ oleic acid} + \text{glycerol}$$

Adipose tissue (1 g) is homogenized in 3 ml of cold buffer (0.15 M NaCl, 20% glycerol, 0.005 M Barbital, pH 7.0) in a glass homogenizer with a Teflon pestle. The pestle is rinsed with 1 ml of the buffer, and the rinse is added to the sample. The homogenate is centrifuged at 10,000 rpm for 10 min. The fat layer is discarded. (If samples are from older animals, the supernatant solution and the pellet are rehomogenized and centrifuged as before. If the supernatant solution is jellied, 1–2 ml of buffer is mixed in with a spatula, but not homogenized, and centrifuged. (It is necessary to keep track of the total volume.) The supernatant solution is transferred to a glass tube and is the enzyme homogenate.

The stock solution of triolein is prepared by mixing in a 50-ml beaker 234 Ci glycerol tri-(^{3}H) oleate and 20 mg egg yolk phosphatidyl choline (1 g/10 ml heptane) and drying under nitrogen. Pure triolein, 206.7 mg, is poured down the side of the beaker, and 30 ml 0.156 M Tris base, pH 8.6, is added. This mixture is sonicated three times for 2 min each at full power with a Fisher sonic dismembrator having a standard tip. This stock solution will be stable up to 1 month if foil wrapped and refrigerated. Prior to use, 4.4 mg fatty acid-free bovine serum albumin is added and solubilized at 30°C.

To prepare the assay substrate, 1/4 volume of cofactor, serum or apoprotein (13.3 mg/3 ml Tris, pH 8.2) is mixed gently with 3/4 volume of triolein stock solution in a small glass tube, plugged with a rubber stopper. It is warmed to 30°C and the pH is adjusted if necessary to 8.5–8.6. The substrate, 200 μl, is preincubated in screw-capped glass tubes in a 30°C metabolic shaker for 30 min. The addition of 50 μl of undiluted enzyme homogenate begins the reaction. The tubes are recapped and the incubation continues for 30 min. The blanks contain 50 μl of 0.156 M Tris base in place of the enzyme homogenate. The reaction is terminated by adding 4.05 ml chloroform:methanol:heptane (1.41:1.25:1.00, v:v:v), immediately followed by 1.4 ml buffer (0.05 M KOH, 0.04 M K_2CO_2, 0.014 M potassium tetraborate). This mixture is vortexed for 5–10 sec and then centrifuged at 1,600 rpm at room temperature for 10–20 min. One milliliter of the upper, aqueous phase is pipetted into a scintillation vial, scintillation fluid is added, and the mixture is counted in a liquid scintillation counter. The upper phase is 54% of the total in this system (114).

Hormone-Sensitive Lipase

The mobilization of stored lipid is controlled by the activity of hormone sensitive lipase (HSL) which catalyzes the conversion of triacylglycerol to 1,2-diacylglycerol plus FFA and to 2-monoacylglycerol plus FFA. Hydrolysis of triacylglycerol may be the rate-limiting step because of the lower enzyme activity toward this substrate (115). Very few studies have been reported on the kinetics of this enzyme. The apparent subunit molecular weight of 84,000 daltons was reported (115). The activity of this enzyme was controlled by phosphorylation and dephosphorylation (116, 117, 118).

The lipase was activated by a cAMP-dependent protein kinase. A schematic outline of the key steps regulating the interconversion of hormone sensitive lipase a and b is shown in Figure 4-6.

A second mechanism for altering the rate of intracellular lipolysis and HSL activity has been suggested (117). This mechanism involves a modification in the nature of HSL and substrate interaction and can be demonstrated only in the presence of endogenous substrate.

The HSL system developed in in vitro culture of 3T3-L1 cells (which develops into adipose-like cells); this in vitro system was similar to that found in vivo (119). In this in vitro system insulin stimulated the protein phosphatase activity. Therefore, insulin may reduce lipolysis by other mechanisms not directly associated with its effects on cAMP levels in the cell (120, 121).

Although very few studies have been done on the nutritional control of

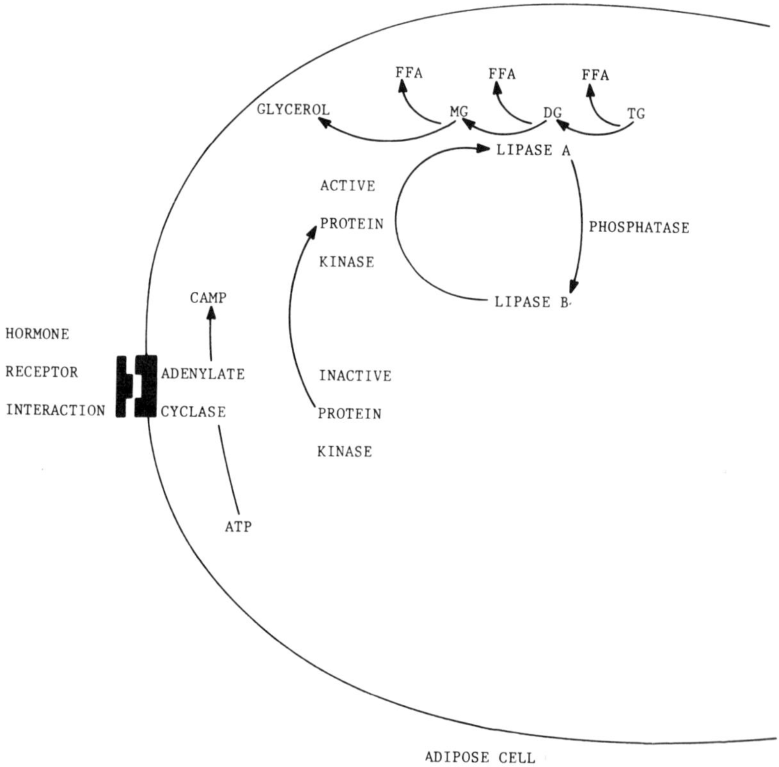

Figure 4-6. Regulation of hormone-sensitive lipase.

HSL, some evidence indicates that nutrition has an influence. Meal feeding caused an increase in adipose tissue HSL (104).

Hormone-Sensitive Lipase Assay. The hormone-sensitive lipase assay involves the following general reaction:

$$\text{Triglyceride} \rightarrow 3\ \text{Fatty acids} + \text{glycerol}$$

Fat pads are incubated for 2 hr in 4 ml of medium (Krebs Ringer Bicarbonate buffer, pH 7.4 with 4% albumin) under 95% O_2–5% CO_2 to reduce lipoprotein lipase activity. Fat pads are homogenized in 10 mol cold 0.25 M sucrose with 10 mM Tris, pH 7.11, and 10 mM EDTA. The homogenate is first centrifuged at low speed to remove the fat layer. The infranatant portion can be further purified by centrifugation at 78,000 × *g* for 1 hr. The infranatant fraction is used for analysis.

The assay medium consists of 1 μmol [^{14}C]triolein (emulsified by sonication and stored at −20°C, resonicated prior to use), 30 μmol EDTA, 20 mg albumin, and 40 μmol phosphate buffer at pH 6.8. To 0.6 ml of this is added 0.2 ml of the infranatant fraction to start the reaction. The incubation is carried out for 1 hr at 30°C.

The reaction is stopped by adding 3 ml chloroform:methanol:benzene 2:2.4:1 (v:v:v) containing 0.3 μmol nonradioactive oleic acid. This is followed by the addition of 0.1 ml of 1 N NaOH to give a final pH of 11–11.5. The samples are vigorously vortexed for 15 sec, then centrifuged at room temperature for 10 min at 2,500 rpm. Aliquots (1.8 ml) of the upper phase are transferred to scintillation vials and 10 ml scintillation fluid:Triton X-100 (2:1; v:v) is added. Counting is done in a liquid scintillation counter (122).

Fatty Acid Binding Protein

A small cytoplasmic fatty acid binding protein (FABP) with a molecular weight of approximately 12,000 daltons was first reported by Ockner and co-workers (123). It was proposed that a FABP might participate in the translocation of fatty acids from the cellular membrane to the endoplasmic reticulum through the aqueous medium of the cytoplasm, in which the long-chain fatty acids were only sparingly soluble. Physiological functions proposed for fatty acid binding protein include:

1. Cellular uptake of blood-borne fatty acids
2. Presentation of fatty acids and acyl-CoAs to intracellular organelles that metabolize these lipids

3. Modulation of the free and protein-bound fatty acid and acyl-CoA pools
4. Allosteric activation of certain enzymes of triacylglycerol synthesis

Recent reports of a self-aggregation property of FABP provide evidence for a potential mechanism by which membrane-bound enzyme activity was regulated (124). The allosteric short-term regulation of lipid metabolizing enzyme has been proposed for acetyl-CoA carboxylase (125), triacylglycerol synthesis (126, 127), and palmitoyl-CoA hydrolase (128).

Hepatic levels of FABP were influenced by diet (129, 130) and endocrine status (131). Adipose tissue levels of the protein have not been extensively investigated.

Fatty Acid Binding Protein Assay. A new procedure (132) for the detection and quantitation of low-molecular-weight cytosolic FABPs in chromatographic fractions is described. Aliquots of the fractionated cytosol are incubated with radiolabeled palmitate, and the unbound fatty acid ligand is quickly removed by the addition of a dextran gelatin-coated charcoal suspension. The FABP is quantitated by counting the protein-bound radioactivity in the supernatant fraction after a brief centrifugation step. Validation studies have shown the assay to be linear over a range of 10–40 *g*.

The composition of the assay solution is shown in Table 4-2. The charcoal suspension is prepared immediately before use by mixing 200 ml of ice-cold 20 mM Tris-HCl, pH 7.7, with 1 g of dextran, 1 g of gelatin, 1 g of the alkaline Norit-A powdered carbon, and 5.8 g of NaCl in a beaker. The suspension is kept at 0°C. The contents of the inner beaker are stirred at all times by use of a magnetic stirring apparatus and stirred at least 30 min prior to use in order to allow the dextran to swell and adequately coat the charcoal particles. Once prepared, the dextran-gelatin-charcoal suspension is used within 2–3 hr in order to avoid unacceptably high blank values.

One milliliter aliquots of the labeled palmitate working solution (40,000 dpm/ml) are pipetted into disposable 12 $\times$ 75 mm glass culture tubes in tube racks commonly used for radioimmunoassay. The racks are partially immersed in a 45°C water bath set. Fraction aliquots containing 20–80 μg of FABP (10–40 μg of FABP for the high-specific-activity preparation) are then added in a total volume of 500 μl. Each sample is vortexed briefly and returned to the water bath.

To minimize loss of palmitate label from binding sites on the proteins during subsequent steps, the tubes are immersed in an ice bath for 10 min to stabilize the bound ligands. One-milliliter aliquots of the dextran gelatin-coated charcoal mixture are removed from the stirring suspension and quickly added to each tube. The latter are then vortexed briefly (approximately 1 sec) and immediately centrifuged (5 min, 3,200 $\times$ *g*, 4°C) with a Dupont-Sorvall radioimmunoassay swinging bucket rotor (Model HS-4).

Table 4-2. Composition of Assay Solutions

[^{14}C]Palmitate Working Solution	*Charcoal Suspension*
20 mM Tris-HCl, 200 ml	20 mM Tris-HCl, 200 ml
0.5 M NaCl	0.5 M NaCl
Propylene glycol (10%)	pH adjusted to 7.7 (25°C)
pH adjusted to 8.4 (25°C)	1 g Dextran
3.6μCi [^{14}C]Palmitate[a]	1 g Gelatin
(40,000 dpm/ml)	1 g Powdered carbon

[a]Radiolabeled [^{14}C]palmitate with a specific activity of 7.6 mCi/mmol is recommended for FABP concentrations of 30–120 μg/ml, or 42.6 mCi/mmol for FABP concentrations of 10–40 μg/ml.

Exposure of the binding proteins to the unsedimented charcoal suspension for periods greater than 3–4 min reduces net palmitate uptake.

Aliquots (500 μl) are removed from each tube and counted with an aqueous immiscible scintillation fluid. Blank values for evaluating the efficiency of charcoal absorption of unbound fatty acid are determined by substituting the fraction samples with an elution buffer containing glycerol.

AMINO ACID METABOLISM

Amino acids are required for the induction of adipose cell lipogenesis (133). Fasted rats were refed a balanced diet or a diet devoid of protein. Adipose cell fatty acid synthesis was increased 10-fold in the rats receiving the balanced diet but was not induced in the rats fed the protein-devoid diet. Apparently the stimulation of lipogenesis required dietary amino acids for the synthesis of new enzyme protein.

Amino acids may also be utilized by adipose cells for energy or as a carbon source for de novo lipogenesis. The disposal of the amino group occurs by transamination to form either alanine or glutamate. Alanine amino transferase and glutamate-oxaloacetate transaminase are two key enzymes involved in this process. The carbon residues of amino acids are catabolized by metabolic pathways analogous to those found in other mammalian tissues, into short chain di- or monocarboxylic acids. These carboxylic acids, which may be straight or branch-chained acids, are potential fatty acid precursors. Specific pathways of leucine and glutamate metabolism are presented below.

Leucine is metabolized in adipose tissue to form ketone bodies, fatty acids, and glutamine. In vitro release of ketones (134) and glutamine (135) has been demonstrated in adipose tissue. Production of glutamine in adipose tissue appeared to be a major route for the disposal of amino groups released in the transamination of leucine and other branch-chain amino acids. Leucine oxidation to carbon dioxide was decreased in protein deficiency, and the formation of alpha-ketoisocaproic acid was increased (135).

Branch-chain alpha-keto acid dehydrogenase activity is thought to limit the rate of leucine oxidation in adipose tissue. This enzyme was activated in vitro by insulin and by leucine (136). The mechanism of regulation of this enzyme appeared to be similar to that proposed for pyruvate dehydrogenase.

Leucine utilization for fatty acid synthesis was a minor pathway in adipose cells (40). Less than 10% of the fatty acids synthesized in porcine adipose cells in vitro could be attributed to leucine carbon. While lipogenesis was increased nearly threefold in adipocytes from obese pigs, there was no significant difference in leucine incorporation into lipids.

Glutamate is incorporated into fatty acid by adipose cell fatty acids by two different pathways (see Fig. 4-7). One of these pathways, the forward pathway, involves the entry of alpha-ketoglutarate into mitochondria and metabolism via the Krebs cycle; the second pathway involves the backward

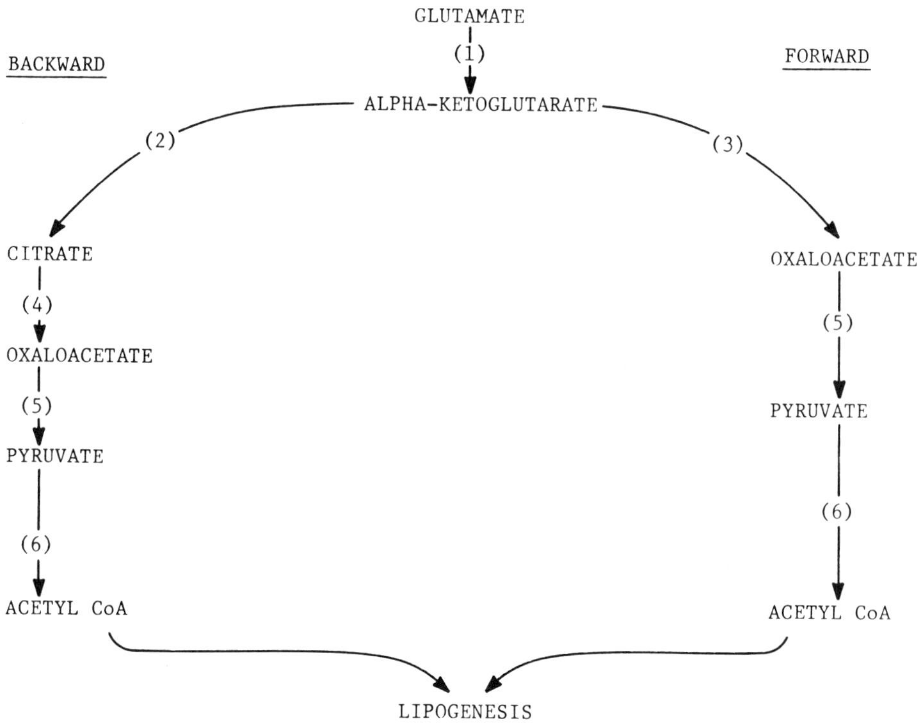

Figure 4-7. Glutamate pathways for lipogenesis in adipose tissue. 1) Glutamate-aspartate transaminase, 2) isocitrate dehydrogenase (NADPH), 3) alphaketoglutarate dehydrogenase, 4) citrate cleavage enzyme, 5) malate Dehydrogenase and malic enzymes, 6) pyruvate dehydrogenase.

flow of alpha-ketoglutarate to citrate in the cytoplasm, and cleavage of citrate to supply acetyl-CoA for lipogenesis. The occurrence of the backward pathway is made possible by the reversibility of the cytoplasmic NADP-dependent isocitrate dehydrogenase. (EC 1.1.1.42). Both pathways in adipose tissue were stimulated in meal feeding-induced hyperlipogenesis (137).

Glutamate is utilized by adipose cells for synthesis into glyceride-glycerol and fatty acid and for energy. Approximately 60% was utilized for energy, 25% for glyceride synthesis, and 15% for de novo fatty acid synthesis (137, 138). These estimates excluded the conversion of glutamate to glutamine (glutamine synthetase), which could be a very active process. During differentiation of 3T3-L1 adipose cells in culture, glutamine synthetase increased more than 100-fold (139). It was concluded that glutamine synthetase activity may be required for adipocyte differentiation when adipose cell cultures are maintained in media lacking glutamine.

Comparative aspects of amino acid utilization by adipose cells and other cell types have not been extensively investigated. Further study of amino acid metabolism in adipose cells is needed.

For information on assays for alanine amino transaminase, glutamate-oxaloacetate transaminase, and isocitrate dehydrogenase see Table 4-1.

APPENDIX

Preparation of Adipose Tissue Extracts for Enzyme Analysis

Adipose samples are removed and homogenized on ice in cold sucrose buffer (0.25 M sucrose, 1 mM EDTA, 5 mM Tris, 1 mM dithiothreitol, pH 7.4) using a Brinkman polytron to make a 6–10% solution. The resulting homogenates are centrifuged at 27,000 × *g* for 20 min at 0–4°C. The supernatant solutions are aliquoted to vials (1/assay) and frozen for future analysis.

Assays are run on a recording spectrophotometer at 25°C. Cuvette volumes are 1 or 2 ml. Assays listed in Table 4-1 are calculated for 1-ml-volume cuvettes. Homogenate volumes are usually 0.02 0.05 ml. Where dilutions of samples are necessary, the sucrose buffer is used (138, 139).

Preparation of Isolated Adipose Cells and Subcellular Fractions

For those assays using subcellular fractions, the following procedures are used:

Epididymal fat pads are removed, rinsed in 0.85% NaCl, cut into 1-g pieces, and incubated for 20 min at 37°C in a metabolic shaker. The medium consists of a Krebs Ringer bicarbonate buffer with 10 mg collagenase and 3 mol glucose/ml (73, 140).

After centrifugation for 1 minute at 400 × *g* at room temperature, the fat cells on the surface are removed and the isolated cells are washed in 10 volumes of 10 mM Tris-HCl, pH 7.4, 1 mM EDTA, and 0.25 M sucrose. After centrifugation at 400 × *g*, the cells are suspended in 3.5 volumes of the Tris-HCl medium, transferred to a glass homogenizer utilizing a Teflon pestle with a clearance of 0.006–0.009 in. Ten strokes regulated by a variable-speed motor driven at 2,600 rpm homogenizes the cells. The cells are chilled on ice. A supernatant solution is obtained by centrifugation of the homogenate at 16,000 × *g* for 15 min. A needle and syringe is used to remove the supernatant solution, avoiding the lipid and the fat cake. This supernatant is resuspended in the Tris-HCl buffer by using a glass homogenized and Teflon pestle with a clearance of 0.004–0.006 in.

The pellet resulting from centrifugation at 160,000 × *g* for 70 min is rinsed gently with the medium. The tube is wiped to further eliminate any lipid, and the pellet is resuspended in 1 ml of medium and transferred to a clean tube for a second centrifugation at 160,000 × *g* for 70 min. The pellet is resuspended for assay use (141).

References

1. R. A. Coleman and R. M. Bell, 1980, Selective changes in enzymes of the sn-glycerol-3-phosphate and dihydroxyacetone-phosphate pathways of triacylglycerol biosynthesis during differentiation of 3T3-L1 preadipocytes, *J. Biol. Chem.* **255**:7681–7687.
2. R. W. Geisler, A. E. Roggeveen, and R. J. Hansen, 1978, The effect of insulin on the turnover of glucose 6 "P" dehydrogenase in epididymal adipose tissue of the rat, *Biochem. Biophys. Acta* **544**:284–293.
3. R. J. Martin, R. F. Welton, and B. R. Baumgart, 1973, Adipose and liver tissue enzyme profiles in obese hyperglycemic mice, *Proc. Soc. Exp. Biol. Med.* **142**:241–245.
4. G. A. Leveille and K. Chakrabarty, 1967, Diurnal variations in tissue glycogen and liver weight of meal fed rats, *J. Nutr.* **93**:546–550.
5. G. A. Leveille, 1970, Adipose tissue metabolism: Influence of periodicity of eating and diet composition, *Fed. Proc.* **29**:1294–1301.
6. M. K. Armstrong, D. R. Romsos, and G. A. Leveille, 1976, Time sequence of lipogenic changes in adipose tissue of rats when converted from ad libitum feeding to meal eating, *J. Nutr.* **106**:884–891.
7. B. A. Clark and S. D. Clarke, 1982, Polymer-protomer transition of acetyl-CoA carboxylase as a regulator of lipogenesis, *Arch. Biochem. Biophys.* **218**:92–100.
8. R. L. Baldwin, 1970, Metabolic functions affecting the contribution of adipose tissue to total energy expenditure, *Fed. Proc.* **29**:1277–1283.
9. S. W. Cushman, L. J. Wardzala, P. J. Hissin, E. Karnieli, I. A. Simpson, and L. B. Solans, 1983, Mechanism of insulin resistant glucose transport in isolated rat adipose

cell, in *The Adipocyte and Obesity,* A. Angel, C. H. Hollenberg, and D. A. K. Roncari, eds., Raven Press, New York.

10. L. J. Wardzala, S. W. Cushman, and L. B. Salans, 1978, Mechanism of insulin action on glucose transport in isolated rat adipocytes, *J. Biol. Chem.* **253**:8002–8005.
11. S. W. Cushman and L. J. Wardzala, 1980, Potential mechanism of insulin action on glucose transport in the isolated rat adipose cell. Apparent translocation of intracellular transport systems to the plasma membrane, *J. Biol. Chem.* **255**:4758–4762.
12. H. M Katzen, 1967, The multiple forms of mammalian hexokinase and their significance to the action of insulin, *Adv. Enzym. Regul.* **5**:335–348.
13. R. Hansen, S. Pilkis, and M. E. Krahl, 1967, Properties of adaptive hexokinase isoenzymes of the rat, *Endocrinology* **81**:1397–1404.
14. B. Borreback, 1976, Adaptable hexokinase activity in epididymal adipose tissue studied in vivo and in vitro, *Biochem. Biophys. Acta* **141**:221–226.
15. R. J. Hansen, S. J. Pilkis, and M. E. Krahl, 1970, Effect of insulin on the synthesis in vitro of hexokinase in rat epididymal adipose tissue, *Endocrinology* **86**:57–65.
16. R. S. Bernstein, M. C. Marshall, and A. L. Carney, 1977, Effects of dietary composition on adipose tissue hexokinase. II. And glucose utilization in normal and streptozotocin diabetic rats, *Diabetes* **26**:770–779.
17. T. Bucher and G. Pfleider, 1955, Pyruvate kinase from muscle, in *Methods in Enzymology,* vol. 1, S. P. Colowich and N. O. Kaplan, eds., Academic Press, New York, p. 435.
18. H. M. Katzen, D. D. Soderman, and C. E. Wiley, 1970, Multiple forms of hexokinase, *J. Biol. Chem.* **245**:4081–4096.
19. L. P. Kozak and J. T. Jensen, 1974, Genetic developmental control of multiple forms of L-glycerol-3-phosphate dehydrogenase, *J. Biol. Chem.* **249**:7775–7781.
20. L. S. Wise and H. Green, 1979, Participation of one isozyme of cytosolic glycerophosphate dehydrogenase in the adipose conversion of 3T3 cells, *J. Biol. Chem.* **254**:273–275.
21. J. B. Neilands, 1955, *Methods in Enzymology,* vol. 1, S. P. Colowick and N. O. Kaplan, eds., Academic Press, New York, pp. 449–454.
22. G. W. E. Plaut, 1962, in *Methods in Enzymology,* vol. 5, S. P. Colowick and N. O. Kaplan, eds., Academic Press, New York, pp. 645–651.
23. D. E. Bauman, R. E. Brown, and C. L. Davis, 1970, Pathways of fatty acid synthesis and reducing equivalent generation in mammary gland of rat, sow, and cow, *Biochem. Biophys. Acta,* **140**:237–244.
24. R. L. Baldwin and L. P. Milligan, 1966, Enzymatic changes associated with the initiation and maintenance of lactation in the rat, *J. Biol. Chem.* **241**:2058–2066.
25. G. L. Cottam and P. A. Srere, 1969, The sulfhydryl groups of citrate cleavage enzyme, *Arch. Biochem. Biophys.* **130**:304–309.
26. W. Sizer and W. T. Jenkins, 1962, Glutamic aspartic transaminase from pig ventricles, *Methods Enzymol.* **5**:677–684.
27. H. L. Segal and T. Matsuzavia, 1970, *Methods in Enzymology* vol. XVII, Academic Press, p. 153–159.
28. S. Ochoa, 1955, Malic enzyme from pigeon liver and wheat germ, in *Methods in Enzymology,* vol. I, S. P. Calarvick, and N. O. Kaplan, eds., Academic Press, New York, pp. 739–753.
29. G. E. Glock and P. McLean, 1953, Further studies on the properties and assay of glucose 6-phosphate dehydrogenase and 6-phosphogluconate dehydrogenase of rat liver, *Biochemistry* **55**:400–408.
30. J. H. Wiley and G. A. Leveille, 1970, Influence of periodicity of eating on the activities

of adipose tissue and muscle glycogen synthesizing enzymes in the rat, *J. Nutr.* **100**:85–93.

31. H. R. Kaslow and S. E. Mayer, 1979, Adaptations of glycogen metabolism in rat epididymal adipose tissue during fasting and refeeding, *J. Biol. Chem.* **254**:4678–4683.
32. H. R. Kaslow, R. D. Eichner, and S. E. Mayer, 1979, Interconversions between multiple glucose-6-P dependent forms of glycogen synthetase in adipose tissue, *J. Biol. Chem.* **254**:4674–4677.
33. R. D. Eichner, 1976, Rat adipose tissue glycogen synthase, *J. Biol. Chem.* **251**:2316–2322.
34. J. A. Thomas, K. K. Schlender, and J. Larner, 1968, A rapid filter paper assay for UDP glucose-glycogen glucosyltransferase, inducing an improved biosynthesis of UDP-^{14}C-glucose, *Anal. Biochem.* **25**:486–499.
35. J. Katz, B. R. Landau, and G. E. Bartsch, 1966, The pentose cycle, triose phosphate isomerization and lipogenesis in rat adipose tissue, *J. Biol. Chem.* **241**:727–731.
36. A. L. Greenbaum, K. A. Gumaa, and P. McLean, 1971, The distribution of metabolites and the control of pathways of carbohydrate metabolism in animals of different dietary and hormonal status, *Arch. Biochem. Biophys.* **143**:617–622.
37. H. Kather, M. Rivera, and K. Brand, 1972, Interrelationship and control of glucose metabolism and lipogenesis in isolated fat cells, *Biochem. J.* **128**:1089–1095.
38. R. A. Freedland and B. Szepesi, 1971, Control of enzyme activity: Nutritional factors, in *Enzyme Synthesis and Degradation in Mammalian Systems,* Karger, Basel, p. 103–140.
39. M. S. Gimenez and B. C. Johnson, 1981, Pair-feeding in dietary control of glucose-6-P dehydrogenase, *J. Nutr.* **111**:260–265.
40. R. J. Martin and J. H. Herbein, 1976, A comparison of the enzyme levels and the in vitro utilization of various substrates for lipogenesis in obese pigs, *Proc. Soc. Exp. Biol. Med.* **151**:231.
41. G. L. Allee, E. K. O'Hea, G. A. Leveille, and D. H. Baker, 1971, Influence of dietary protein and fat on lipogenesis and enzymatic activity in pig adipose tissue, *J. Nutr.* **101**:869–878.
42. B. M. Berke and M. L. Kaplan, 1983, Effect of high fat and high carbohydrate diets on development of hepatic and adipose lipogenesis in fa/fa and non fa/fa rats, *J. Nutr.* **113**:820–834.
43. B. D. Sanwal, 1970, Regulatory mechanisms involving nicotinamide adenine nucleotide as allosteric effectors, *J. Biol. Chem.* **245**:1626–1631.
44. A. Kawaguchi and K. Bloch, 1974, Inhibition of glucose 6 P dehydrogenase by palmitoyl CoA, *J. Biol. Chem.* **249**:5793–5800.
45. J. P. Flatt and E. G. Ball, 1964, Studies on the metabolism of adipose tissue, *J. Biol. Chem.* **239**:675–680.
46. R. J. Martin, M. Ezekewe, J. H. Herbein, and J. H. Ziegler, 1974, Effect of neonatal nutritional experience on growth and development of the pig, *J. Anim. Sci.* **39**:521–527.
47. M. L. Kaplan and G. H. Fried, 1973, Adaptive enzyme responses in adipose tissue of obese hyperglycemic mice, *Biochem. Biophys. Acta* **158**:711–719.
48. R. J. Martin and P. Lamprey, 1975, Early development of adipose cell lipogenesis and glycerol utilization in Zucker obese rats, *Proc. Soc. Exp. Biol. Med.* **149**:35–39.
49. S. Deb and R. J. Martin, 1975, Effect of exercise and of food restriction on the development of spontaneous obesity in rats, *J. Nutr.* **105**:1543–1549.
50. R. Bazin and M. Lavau, 1982, Development of hepatic and adipose tissue lipogenic enzymes and insulinemia during suckling and weaning in Zucker rats, *J. Lipid Res.* **23**:839–849.

51. K. Chakrabarty and G. A. Leveille, 1968, Influence of periodicity of eating on activity of various enzymes in adipose tissue, *J. Nutr.* **96:**76–81.
52. F. Pairault and H. Green, 1979, A study of the adipose conversion of suspended 3T3 cells by using glycerophosphate dehydrogenase as a differentiation marker, *Proc. Natl. Acad. Sci.* (USA) **76:**5138–5142.
53. B. Jeanrenaud, 1967, Effect of glucocorticoid hormones on fatty acid mobilization and reesterification in rat adipose tissue, *Biochem. J.* **103:**627–631.
54. L. Reshef, R. W. Hanson, and F. J. Ballard, 1970, A possible physiological role for glyceroneogenesis in rat adipose tissue, *J. Biol. Chem.* **245:**5979–5984.
55. E. Gorin, Z. Talyor, and E. Shafrir, 1969, Glyceroneogenesis in adipose tissue of fasted, diabetic and triamcinolone treated rats, *Eur. J. Biochem.* **8:**370–375.
56. D. Feldman, 1977, Glucocorticoid receptors and regulation of phosphoenolpyruvate carboxikinase activity in rat kidney and adipose tissue, *Am. J. Physiol.* **233:**E147–E152.
57. J. Mayer, 1968, *Overweight: Causes, Cost and Control,* Prentice Hall, Inc., Englewood Cliffs, New Jersey.
58. S. W. Thenen and J. Mayer, 1975, Hyperinsulinemia and fat cell glycerokinase activity in obese (ob/ob) and diabetic mice, *Horm. Metab. Res.* **8:**80–83.
59. R. F. Welton, R. J. Martin, R. W. Scholz, and B. R. Baumgardt, 1973, Effects of dietary manipulation on adipose tissue glycerokinase activity and plasma metabolite levels in the chicken (*Gallus domesticus*), *J. Nutr.* **103:**890–898.
60. R. F. Welton, R. J. Martin, and B. R. Baumgardt, 1973, Effect of feeding and exercise regimens on adipose tissue glycerokinase activity and body composition of lean and obese mice, *J. Nutr.* **103:**1212–1219.
61. S. W. Thenen and J. Mayer, 1975, Adipose tissue glycerokinase activity in genetic and acquired obesity in rats and mice, *Proc. Soc. Exp. Biol. Med.* **148:**953–959.
62. J. S. Stern, J. Hirsch, A. Drewnowski, A. C. Sullivan, P. R. Johnson, and C. K. Cohn, 1983, Glycerokinase activity in adipose tissue of obese rats and mice: Effects of diet composition, *J. Nutr.* **113:**714–720.
63. E. Newsholme, J. Robinson, and K. Taylor, 1967, A radiochemical enzyme activity assay for glycerokinase and hexokinase, *Biochem. Biophys. Acta* **132:**338–346.
64. R. W. Welton, 1972, The relationship of adipose tissue glycerokinase activity to obesity in some laboratory and farm animals, Ph.D. Thesis, Pennsylvania State University, University Park, Pennsylvania.
65. Hitoshi Maruyama, Richard L. Easterday, Huei-Che Chang, and M. Daniel Lane, 1966, The enzymatic carboxylation of phosphoenolpyruvate. I. Purification and properties of phosphenolpyruvate carboxylase, *J. Biol. Chem.* **241:**2405–2412.
66. H. Chang and M. Daniel Lane, 1966, The enzymatic carboxylation of phosphoenolpyruvate. II. Purification and properties of liver mitochondrial phosphoenolpyruvate carboxykinase, *J. Biol. Chem.* **241:**2413–2420.
67. F. J. Ballard and Richard W. Nanson, 1967, The citrate cleavage pathway and lipogenesis in rat adipose tissue: Replenishment of oxaloacetate, *J. Lipid Res.* **8:**73–79.
68. R. M. Denton, 1975, Hormonal regulation of fatty acid synthesis in adipose tissue through changes in the activities of pyruvate dehydrogenase and acetyl-CoA carboxylase, *Proc. Nutr. Soc.* **34:**217–224.
69. N. Begum, H. M. Tepperman, and J. Tepperman, 1983, Effect of high fat and high carbohydrate diets on adipose tissue pyruvate dehydrogenase and its activation, *Endocrinology* **110:**1914–1921.
70. N. Begum, H. M. Tepperman, and J. Tepperman, 1984, Effect of dexamethasone on adipose tissue and liver pyruvate dehydrogenase, *Endocrinology* **114:**99–107.
71. Y. B. Lombardo and L. A. Menahan, 1978, Pyruvate dehydrogenase activity in several tissues of genetically obese hyperglycemic mice, *Life Sci.* **22:**1033–1040.

72. T. R. Soderling, 1979, Regulatory functions of protein multisites phosphorylation, *Mol. Cell. Endocrinol.* **16:**157–179.
73. M. Rodbell, 1964, Metabolism of isolated fat cells. I. Effects of hormones on glucose metabolism and lipolysis, *J. Biol. Chem.* **239:**375–380.
74. L. Weiss, G. Loffler, and O. H. Wieland, 1974, Regulation by insulin of adipose tissue pyruvate dehydrogenase, *Hoppe-Seyler's Z. Physiol. Chem.* **355:**367–377.
75. W. M. Bortz, S. Numa, and F. Lynan, 1965, Regulation of fatty acid synthesis at the acetyl CoA carboxylase step, *Adv. Enzym. Reg.* **3:**407–416.
76. K. Bloch and D. Vance, 1977, Control mechanisms in the synthesis of saturated fatty acids, *Ann. Rev. Biochem.* **46:**263–298.
77. L. Yeh, K. Lee, and K. Kim, 1980, Regulation of rat liver acetyl CoA carboxylase. *J. Biol. Chem.* **255:**2308–2314.
78. M. J. Toussant, M. D. Wilson, and S. D. Clark, 1981, Coordinated suppression of liver acetyl CoA carboxylase and fatty acid synthetase, *J. Nutr.* **111:**146–153.
79. G. Sturm and G. Siebert, 1982, A study of acetyl CoA carboxylase in adipose tissues, *Z. Ernahrungswiss* **21:**12–18.
80. R. W. Guynn, D. Veloso, and R. I. Veech, 1972, The concentration of malonyl-coenzyme A and control of fatty acid synthesis in vivo, *J. Biol. Chem.* **247:**7325–7331.
81. J. J. Volpe and P. R. Vagelos, 1976, Mechanisms of regulation of biosynthesis of saturated fatty acids, *Physiol. Rev.* **56:**339–451.
82. B. A. Ashcraft, W. S. Fillers, S. L. Augustine, and S. D. Clarke, 1980, Polymer-protomer transitions of acetyl CoA carboxylase occurs in vivo and in vitro, *J. Biol. Chem.* **255:**10033–10035.
83. K. Lee and K. Kim, 1979, Stimulation by epinephrine of in vivo phosphorylation and inactivation of acetyl coenzyme A carboxylase of rat adipose tissue, *J. Biol. Chem.* **254:**1450–1453.
84. P. R. Desjardins and K. Dakshinamurti, 1970, Rat epididymal adipose tissue acetyl-CoA carboxylase, *Can. J. Biochem.* **48:**915–921.
85. S. Smith, 1981, Longchain fatty acyl-S 4'phosphopantetheine fatty acid synthetase thioester hydrolase from rat, in *Methods of Enzymology* vol. 7, Academic Press, New York, p. 181–188.
86. K. Chakrobarty and G. A. Leveille, 1969, Acetyl CoA carboxylase and fatty acid synthetase activities in liver and adipose tissue of meal-fed rats, *Proc. Soc. Exp. Biol. Med.* **131:**1051–1054.
87. R. J. Martin, D. Stolz, and D. C. Buck, 1979, Diurnal changes in adipose and liver tissue metabolism of lean and obese Zucker rats, *J. Nutr.* **109:**412–417.
88. G. H. Weiss, O. M. Rosen, and C. S. Rubin, 1980, Regulation of fatty acid synthetase concentration and activity during adipocyte differentiation, *J. Biol. Chem.* **255:**4751–4757.
89. D. G. Steffen, G. Phinney, L. J. Brown, and H. J. Mersmann, 1979, Ontogeny of glycerolipid biosynthetic enzymes in swine liver and adipose tissue, *J. Lipid Res.* **20:**246–253.
90. G. B. Stokes, L. W. Poteat, and S. B. Tove, 1975, Distribution of fatty acids incorporated into triacylglycerol by microsome/cytosol preparations from adipose tissue, *Biochem. Biophys. Acta* **380:**245–256.
91. H. G. Nimmo and B. Houston, 1978, Rat adipose tissue glycerophosphate acyltransferase can be inactivated by cyclic AMP-dependent protein-kinase, *Biochem. J.* **176:**607–610.
92. E. G. Bligh and W. J. Dyer, 1959, A rapid method of total lipid extraction and purification, *Can. J. Biochem. Physiol.* **37:**911–917.

93. D. M. Schlossman and R. M. Bell, 1976, Triacylglycerol synthesis in isolated fat cells. (Evidence that the sn-glycerol-3-phosphate and dihydroxyacetone phosphate acetyltransferase activities are due catalytic functions of a single microsomal enzyme), *J. Biol. Chem.* **251**:5738–5744.
94. R. Coleman and R. Bell, 1976, Triacylglycerol synthesis in isolated fat cells. (Studies on the microsomal diacylglycerol acyltransferase activity using ethanol-dispersed diacylglycerols), *J. Biol. Chem.* **251**:4537–4543.
95. R. M. Bell and Y. Miller, 1976, A highly sensitive assay for microsomal diacylglycerol acyltransferase activity from isolated fat cells, *Anal. Biochem.* **71**:436–443.
96. D. Saggerson and A. L. Greenbaum, 1970, The regulation of triglyceride synthesis and fatty acid synthesis in rat adipose tissue, *Biochem. J.* **119**:221–242.
97. C. J. Fielding and R. J. Havel, 1977, Lipoprotein lipase, *Arch. Pathol. Lab. Med.* **101**:225–229.
98. G. Bengtsson and T. Olivecrona, 1980a, Lipoprotein lipase: Some effects of activator proteins, *Eur. J. Biochem.* **106**:549–555.
99. T. A. Musliner, P. N. Hebert, and E. C. Church, 1979, Activation of lipoprotein lipase by native and acylated peptides of apolipoprotein C-II, *Biochem. Biophys. Acta* **573**:501.
100. R. O. Scow and T. Olivecrona, 1977, *Biochem. Biophys. Acta* **487**:472–480.
101. G. Bengtsson and T. Olivecrona, 1980, Lipoprotein lipase: Mechanism of product inhibition, *Eur. J. Biochem.* **106**:557–562.
102. A. Cryer and H. M. Jones, 1979, The early development of white adipose tissue, *Biochem. J.* **178**:711–724.
103. L. Steingrimsdotter, J. A. Brasel, and M. R. C. Greenwood, 1980, Diet, pregnancy and lactation: Effects on adipose tissue lipoprotein lipase and fat cell size, *Metabolism* **29**:837–842.
104. H. S. Paik and E. S. Yearick, 1978, The influence of dietary fat and meal frequency on lipoprotein lipase and hormone sensitive lipase in rat adipose tissue, *J. Nutr.* **108**:1798–1803.
105. Y. S. Katayama, R. Bou, and M. Katayama, 1981, Effect of fructose on lipoprotein lipase activity of adipose tissue in rats, *Agric. Biol. Chem.* **45**:1583–1589.
106. A. Cryer, S. E. Riley, E. R. Williams, and D. S. Robinson, 1976, Effect of nutritional status on rat adipose tissue muscle and postheparin plasma clearing lipase activities, *Clin. Sci. Mol. Med.* **50**:213–221.
107. D. Reichl, 1972, Lipoprotein lipase activity in the adipose tissue of rats adapted to controlled feeding schedules, *Biochem. J.* **128**:79–86.
108. P. M. Spooner, S. S. Chernick, M. M. Garrison, and R. O. Scow, 1979, Insulin regulation of lipoprotein lipase activity and release in 3T3-L1 adipocytes, *J. Biol. Chem.* **254**:10021–10029.
109. G. J. Hausman, D. R. Campion, J. McNamara, R. L. Richardson, and R. J. Martin, 1981, Adipose tissue development in the fetal pig after decapitation, *J. Anim. Sci.* **53**:1634–1644.
110. J. McNamara, M. Azain, T. R. Kasser, and R. J. Martin, 1982, Lipoprotein lipase and lipid metabolism in muscle and adipose tissue of Zucker rats, *Am. J. Physiol.* **243**:R258–R264.
111. R. Gruen, E. Hietanen, and M. R. C. Greenwood, 1978, Increased adipose tissue lipoprotein lipase activity during the development of the genetically obese rat (fa/fa), *Metabolism* **27**:1955–1966.
112. A. Boulange, E. Planche, and P. de Gasquet, 1979, Onset of genetic obesity in the absence of hyperphagia during the first week of life in the Zucker rat (fa/fa), *J. Lipid Res.* **20**:857–862.

113. A. D. Hartman, 1981, Lipoprotein lipase activities in adipose tissues and muscle in the obese Zucker rat, *Am.J. Physiol.* **241**:E108–E112.
114. J. P. McNamara and R. J. Martin, 1982, Muscle and adipose tissue lipoprotein lipase in fetal and neonatal swine as affected by genetic selection for high or low backfat, *J. Anim. Sci.* **55**:1057–1061.
115. G. Fredrikson, P. Nilsson, and P. Belfrage, 1981, Hormone sensitive lipase of rat adipose tissue, *J. Biol. Chem.* **256**:6311–6315.
116. D. Steinberg, 1982, The role of cyclic AMP in the activation of hormone sensifine lipase of adipose tissue, in *Advances in Cyclic Nucleotide Research,* vol. 7, P. Greengard and G. A. Robinson, eds., Raven Press, New York p. 47–62.
117. L. S. Wise and R. L. Jungas, 1978, Evidence for a dual mechanism of lipolysis activation by epinephrine in rat adipose tissue, *J. Biol. Chem.* **253**:2624–2629.
118. P. Belfrage, G. Fredrikson, H. Olsson, and P. Stralfors, 1983, Control of adipose tissue lipolysis by phosphorylation/dephosphorylation of hormone sensitive lipase, in *The Adipocyte and Obesity,* A. Angel, C. Hollenberg, and D. Roncari, eds., Raven Press, New York, p. 217–224.
119. J. C. Khoo, M. Yamamoto, M. B. Kawomura, and D. Steinberg, 1983, Hormone sensitive lipase system and insulin stimulation of protein phosphatase activities in 3T3-L1 adipocytes, in *The Adipocyte and Obesity,* A. Angel, C. Hollenberg, and D. Roncari, eds., Raven Press, New York, p. 225–244.
120. T. W. Burns, B. E. Terry, P. E. Langley, and G. A. Robison, 1979, Insulin inhibition of lipolysis of human adipocytes, *Diabetes* **28**:957–961.
121. B. Lambert and C. Jacquemin, 1979, Inhibition by insulin of the adrenaline-stimulated adenylate cyclase in rat adipose tissue, *FEBS Lett.* **105**:19–24.
122. J. C. Khoo and D. Steinberg, 1975, Hormone-sensitive triglyceride lipase from rat adipose tissue, in *Methods in Enzymology,* vol. 35, J. M. Lowenstein, ed., Academic Press, New York, pp. 181–189.
123. R. K. Ockner, J. A. Manning, R. B. Poppenhausen, and W. K. Lo, 1972, A binding protein for fatty acids in cytosol of intestinal mucosa, liver, myocardium and other tissues, *Science* **177**:56–58.
124. N. C. Fournier, M. Zuker, R. E. Williams, and Jan C. P. Smith, 1983, Self association of the cardiac fatty acid binding protein. Influence on membrane-bound fatty acid dependent enzymes, *Biochemistry* **22**:1863–1872.
125. M. A. Lunzer, J. A. Manning, and R. K. Ockner, 1977, Inhibition of rat liver acetyl-CoA carboxylase by long chain acyl-CoA and fatty acid, *J. Biol. Chem.* **252**:5483–5487.
126. P. J. A. O'Doherty and A. Kuksis, 1975, Stimulation of triacylglycerol synthesis by Z protein in rat liver and intestinal mucosa, *FEBS Lett.* **60**:256–258.
127. N. Iritani, E. Fukuda, and K. Inoguchi, 1980, A possible dietary control of hepatic triacylglycerol synthesis, *J. Nutr. Sci. Vitaminol.* **26**:271–277.
128. S. C. Jamdar, 1979, Hepatic lipid metabolism: Effect of spermine, albumin, Z protein on microsomal lipid formation, *Arch. Biochem. Biophys.* **195**:81–94.
129. R. Hag and E. Shrago, 1983, Dietary induction of rat liver fatty acid binding protein, *Nutr. Res.* **3**:329–333.
130. F. Morrow, 1983, Effect of altered hepatic lipid metabolism on cytoplasmic levels of FABP, Ph.D. Dissertation, University of Georgia, Athens.
131. R. K. Ockner, N. Lysenko, J. A. Manning, S. E. Monroe, and D. A. Burnett, 1980, Sex steroid modulation of fatty acid utilization and FABP concentration in rat liver, *J. Clin. Invest.* **65**:1013–1023.
132. F. D. Morrow and R. J. Martin, 1983, Quantitation of hepatic fatty acid binding proteins by post-chromatographic ligand binding assay, *J. Lipid Res.* **24**:324–331.

133. M. Jomain and R. W. Hanson, 1969, Dietary protein and the control of fatty acid synthesis in rat adiposc tissuc, *J. Lipid Rcs.* **10**:674 680.
134. S. Rous, S. Bas, and S. Sengupta, 1980, Contribution of leucine in the fatty acid synthesis and ketogenesis in mice adipose tissue, *Int. J. Biochem.* **11**:337–340.
135. M. E. Tischler and A. L. Goldberg, 1980, Leucine degradation and release of glutamine and alamine by adipose tissue, *J. Biol. Chem.* **255**:8074–8081.
136. G. P. Frick, L. Tai, L. Blinder, and H. M. Goodman, 1981, Leucine activates branch chain ketoacid dehydrogenase in rat adipose tissue, *J. Biol. Chem.* **256**:2618–2620.
137. G. A. Leveille and R. W. Hanson, 1966, Quantitative aspects of glutamate utilization by rat adipose tissue and liver in vitro: Effect of periodicity of eating, *Can. J. Physiol. Pharmacol.* **44**:275–285.
138. J. Bellido and E. Herrera, 1978, Utilization of pyruvate, alanine and glutamate by isolated fat cells and their effects on glycerol metabolism, *Rev. Esp. Fisiol.* **34**:429–436.
139. R. E. Miller and D. A. Carrino, 1981, An association between glutamine synthetase activity and adipocyte differentiation in cultured 3T3-L1 cells, *Arch. Biochem. Biophys.* **209**:486–503.
140. S. W. Cushman, 1970, Structure-function relationships in the adipose cell. I. Ultrastructure of the isolated adipose cell, *J. Cell Biol.* **46**:326–341.
141. L. Jarett, 1974, Subcellular fractionation of adipocytes, in *Methods in Enzymology,* vol. 31A, S. Fleischer and L. Packer, eds., Academic Press, New York, pp. 60–71.

5

In Vitro Metabolism of Adipocytes

Mario DiGirolamo and Susan K. Fried
Emory University

The adipose organ provides essential body functions that are critical for survival, particularly during prolonged food deprivation. Since in most species adipose tissue is distributed throughout the organism, and considerable morphologic and metabolic regional differences exist, it is difficult and impractical to assess precisely the in vivo function of the adipose organ the way it is usually done for organs such as the liver, heart, or kidney. For these reasons, most of the advances in the understanding of adipose tissue functions have been derived from a variety of in vitro studies.

The purpose of this chapter is to review the historical development of knowledge on adipose tissue and the advantages or disadvantages of the various in vitro preparations of adipose tissue. It is also meant to provide the reader with a practical view of the important variables involved in the in vitro assessment of adipocyte metabolism. This brief review will deal only with white adipocytes and will not be concerned with brown adipose tissue cells.

Until the 1930s, the concept was generally held that adipose tissue was for inert storage of lipid derived from excess calories ingested. The dynamic nature of adipose tissue was suspected after the demonstration in 1936 by Schoenheimer and Rittenberg of a rapid turnover of adipose lipid. Wertheimer and Shapiro in 1948 (1) and Hausberger and co-workers in 1954 (2) used adipose tissue slices from rodents to demonstrate a rapid uptake of glucose and conversion to tissue triglycerides. The discovery of adipose tissue as a source of metabolic fuels that could be rapidly mobilized during fasting and other states of increased metabolic demands had to await the demonstration by Dole in 1956 that free fatty acids (FFA) are rapidly mobilized from adipose tissue and circulate in the blood bound to albumin (3). The stage was then set for a series of comprehensive investigations spanning from 1958 to the late 1960s that clarified the major metabolic pathways of adipocyte metabolism and the exquisite hormonal regulation of glucose transport, lipid synthesis, and lipid mobilization: Winegrad and Renold; Ball and Jungas; Vaughan and Steinberg; Cahill and Renold; Rudman and DiGirolamo; and many others (4–8).

It soon became apparent that species differences existed in the responsiveness of adipose tissue to lipolytic and lipogenic hormones (8, 9) and that, within each species, regional differences existed in both cellular morphology and in the rates of metabolic activities (9–12). Furthermore, it became known that the metabolic behavior and the hormonal responsiveness of adipose tissue varied with the age and the nutritional status of the animals (10, 13–15). Older, fatter rats showed larger fat cells, a less rapid de novo synthesis of fatty acids, a greater rate of fatty acid reesterification, and a diminished response to hormones such as insulin and, in some instances, to lipolytic hormones (10, 13, 16–20).

Following the introduction by Rodbell in 1964 (21) of methods to isolate fat cells from the stromal-vascular constituents of adipose tissue, it became possible to compare populations of fat cells from different animal species, different adipose regions, or (in the same region) from different animals at variable ages and stages of development, in terms of morphology, basic metabolic functions, and hormonal responsiveness. The common denominator in the comparison became the adipocyte number, rather than the less precise measures of adipose tissue wet weight, protein content, or deoxyribonucleic acid (DNA) content (see below). The lack of validity of a comparison made on the basis of wet tissue weight derived from the observation that equal weights of tissue may contain cells of different sizes and consequently a different number of cells (22, 23); the comparison on the basis of protein content was determined to be imprecise by the finding that cell protein content changed with cell size and possibly with regional sampling (19, 20); the comparison on the basis of DNA (known to be fixed in a given tissue cell and also in the adipocytes) was weakened by the difficulty of totally removing stromal-vascular DNA even from isolated fat cell preparations washed several times (24).

As more experience accumulated with the use of isolated fat cell preparations (which offered the considerable advantage of repeated sampling of nearly identical numbers of suspended cells), control of other variables became important. For instance, Schwabe and co-workers (25) demonstrated that the cell concentration in the incubation medium could influence the rate of lipolysis, with the amount of lipolysis per cell being reduced by a high cell density. An increased cell concentration has also been found to influence adipocyte glucose metabolism by enhancing the fatty acid synthesis and the insulin responsiveness of the cells (26).

In more recent years, additional differences have been described in the preferred lipogenic substrates of various species. Thus, species such as ruminants, and even the herbivorous rabbit, utilize acetate rather than glucose as the main triglyceride precursor (27, 28). Additional studies with other substrates such as pyruvate and lactate, alone or in combination with glucose, have also shown the importance of a more physiological substrate milieu in the proper quantitativc assessment of adipocyte lipid synthetic capacity (29).

GENERAL METHODOLOGY

Although the techniques most commonly used to study the metabolism of adipose tissue in vitro employ tissue fragments (13, 30, 31), isolated adipocytes (21, 32), adipocyte ghosts (33), or subcellular fractions (34), several investigators have tried to bridge in vitro to in vivo studies with techniques that would physiologically maintain "perfusion" or "perifusion" of the adipose pad or isolated adipocytes.

The main advantage of such preparations, described below, resides in the maintenance of the flow of either blood or perfusing fluid to the tissue components, thus avoiding accumulation in the medium of toxic products or of metabolic products that could influence the metabolic activity of the incubated adipose tissue.

Perfusion of Adipose Tissue

Several preparations have been described. In the rat, though difficult to set up and maintain, preparations from epididymal (35) or parametrial (36) fat pads have been used. In the dog, Rosell (37) and DiGirolamo and coworkers (38) have used an isolated, innervated, and perfused preparation of the subcutaneous epigastric fat pad.

This technique is carried out in the anesthetized dog, following dissection of the subcutaneous adipose tissue in the inguinal region and isolation of artery, vein, and nerve branches. After ligation and/or cauterization of all small blood vessels on the surface of the tissue and a suitable equilibration period, the afferent artery and efferent vein are cannulated and the adipose pad is perfused with blood from the same animal previously collected in a heparinized reservoir and diluted with Ringer solution (20% v:v).

The perfusion system allows control of the perfusion pressure, which can be set at approximately 100 mm Hg or can be varied in a stepwise fashion. Venous effluent flow can be measured by collecting the blood in graduated tubes and/or by recording flow changes on the oscillograph. Pressure-flow curves can be calculated from the recording, and blood flow is expressed as milliliters per 100 g tissue per minute. The venous concentrations of glucose, glycerol, FFA, or other metabolites can be assessed at specified time intervals and related to analogous measurements in the arterial blood. Extraction or release by the tissue of metabolic substrates can be related to the wet weight of the tissue or to adipose cell size and number, estimated by further analysis of the perfused tissue (38) at the end of the experiment. Similar preparations have been used in the rabbit (39) and in the rat (35, 36) for both measurements of blood flow and analysis of metabolic exchanges.

Perifusion of Adipose Tissue or Isolated Fat Cells

The "perifused" preparation consists of one or more (40, 41) chambers (sometimes simply made up of a 1-ml insulin syringe) equipped with tubing to deliver 95% O_2–5% CO_2 and kept at the desired temperature by immersion in water circulating from a temperature-regulated bath. Following introduction of the tissue fragments or cells (usually 1 million/ml) in the perifusion chamber, buffered medium containing albumin (from 1 to 4%) flows through the chamber at a controllable flow rate, usually 1 ml/min. At 1-min intervals, perifusate is collected and analyzed for FFA, glycerol, or other metabolic products (40, 41, 42).

This technique, although somewhat more complex and delicate than simple incubation with tissue slices or isolated cells, has been found to be valid and to offer several distinct advantages: (a) The flowing system provides fresh medium to the perifused cells and prevents accumulation of intermediary metabolites, such as adenosine or free fatty acids, that could alter the cell metabolic activities and the hormonal response (25, 43, 44). (b) The perifusion system allows assessment of the dynamic cellular response to variables under scrutiny. (c) In studies of cellular responsiveness to hormones, dose-response relationships can be easily and precisely obtained with the use of a multichannel pump. Lipolytic rates in the absence or presence of lipolytic hormones can be determined by measurement of released glycerol. (d) The preparation is particularly useful for studying the interaction of two hormones added at the same time or sequentially, such as the addition of epinephrine and insulin (40) to study the antilipolytic effect of insulin in fat cells. (e) This technique has also been used to measure lipoprotein lipase release from isolated fat cells (45) or cultured adipocytes (46). (f) Recently, glucose metabolism has also been measured in a modified perifusion system (44).

Incubation of Tissue Fragments

Early investigations with adipose tissue fragments in the early 1960s utilized slices of adipose tissue (from 30 to 200 mg) removed from the epididymal or other region of the animal's adipose tissue (e.g., 4, 13). After the tissue slices are rinsed in saline, blotted, and weighed, they are placed in incubation vessels containing buffered medium, albumin, and labeled and/or unlabeled substrate. Before the incubation vessels are capped, a gas mixture of 95% O_2–5% CO_2 is usually applied for 10–20 sec to provide an optimal gas phase during the incubation. At the end of the incubation period, varying from 30 to 120 min, the incubation is arrested simply by removal of the tissue slices and collection of the medium. When ^{14}C-glucose or other labeled substrate is present in the incubation medium and $^{14}CO_2$ collection is desirable, incubation is terminated by the addition of 0.25 ml 1 N sulfuric acid to the medium and addition of 0.2 ml Hyamine-10 x to a plastic central well containing a piece of filter paper (21). After 15–20 min of additional incubation, the center well can be transferred to a scintillation vial for counting.

The main advantage of this preparation resides in the simplicity of tissue sample preparation and the limited manipulation of the tissue, which is usually placed in the incubation vessel within 1–2 min after the donor animal is killed. Disadvantages of this kind of preparation reside in the variability of the samples incubated and related means of comparison. This variability results in part from tissue thickness, cellular composition (i.e., relative cell size and number), the region of the pad from which the slice is cut, and relative proximity to the vascular tree. Use of the wet weight of the tissue (which may contain variable cell number or protein content) as a means for comparing the metabolic activity of groups of slices from the same adipose depot, different regions, or different animals, may provide an additional limitation (see means of comparison below). Furthermore, both adipose and nonadipose cells are usually contributing to the determined variables in incubations with tissue fragments.

Recently, Liebel and co-workers (31) and Liebel and Hirsch (47) have developed a sensitive radioisotopic method for measuring lipolysis and reesterification in small (5–50 mg) aliquots of unweighed needle biopsies of human adipose tissue.

Incubation of Isolated Adipocytes

Following introduction by Rodbell in 1964 (21) of the collagenase technique to disperse adipose cells by digestion of the stromal-vascular tissue components, the isolated fat cell preparation has become preferred by the majority of the investigators interested in adipose tissue morphology and metabolism.

The reasons for this are multiple:

(a) The isolated fat cell preparation is largely devoid of stromal-vascular cell contamination. Thus, the metabolic activity recorded reflects the adipocytes' metabolism and does not include the metabolic activity of nonadipose cellular constituents.

(b) Fat cell suspensions make it possible to mix the cells by gently swirling the vessel such that repeated samples of the same fat cell population are almost identical in the number and size distribution of cellular constituents (21).

(c) The adipocytes float above the medium due to the density of the cell lipid. Thus, it is possible to wash out debris or other tissue components by gentle centrifugation (1,000 $\times$ g for 1 min) or even by just allowing the cells to stand and rise to the top of the aqueous medium.

(d) The cells assume a spherical shape when they are free in suspension, and microscopic assessment of transverse cell diameter (which reflects to a large extent the diameter of the cell lipid droplet in cells with diameters of 25 μm or greater) allows calculation of cell surface area (important in studies of glucose transport or membrane binding of hormones) and cell volume (48).

(e) A variety of methods based on cell count, determination of cell mean diameter, or lipid content, coupled to cell fixation by glutaraldehyde or osmium tetroxide, has permitted calculation of cell number in aliquots of the cell suspension with a satisfactory degree of accuracy (23, 48).

Thus, the ability to express metabolic activities on the basis of cell number and in relation to cell size (diameter or volume) has facilitated the comparison among different tissue regions of the same animal species, in animals of different species, or (in the same tissue region) in animals at different stages of development, growth, and nutritional status.

Among the possible limitations of the isolated fat cell preparation, one needs to include the additional time needed for collagenase incubation and for separation of the cells, the possible damage to the cells produced by excessive collagenase exposure, and the fragility of the cells during a prolonged incubation. This fragility, which appears to be somewhat greater for the larger fat cells of obese animals or humans than for cells from leaner subjects, can in part be prevented or reduced by careful and gentle handling of the cells, incubation at low cell concentrations (< 100,000 fat cells/ml), and avoidance of marked and rapid temperature shifts. Recently, Pedersen and co-workers (32) suggested several modifications of the original Rodbell procedure (21), including the use of 5% albumin in the incubation medium, that have minimized cell breakage even in fragile human fat cells from obese patients (49). These techniques have been successful as well with very large fat cells from obese Zucker rats (Fried et al., unpublished observations).

Separation of the Fat Cells from Aqueous Medium

Certain metabolic studies, particularly those related to the transport of substrates inside the cell or to the binding of hormones to cell surface receptors, require rapid or instantaneous separation of the aqueous medium from the cells. For cells with densities heavier than water, this is usually accomplished by rapid centrifugation and removal of the supernatant medium or by filtration of the medium through millipore filters. These techniques may have limited application to the fat cells in view of the cells' fragility and flotation characteristics, and, also, high nonspecific binding to filters.

In 1972, Gliemann and co-workers (50) described a procedure for measuring distribution spaces in isolated fat cells through the use of an oil, dinonyl phthalate, of density intermediate between that of water and cell lipid. When aliquots of the cell suspension are rapidly centrifuged (12,000 × *g*) for 1–15 seconds in plastic 400-ul microfuge tubes containing 0.1–0.2 ml of the oil, three layers are rapidly produced. They are, from the top: a packed cell layer, an oil layer, and an aqueous layer. By cutting the plastic tube through the oil layer and collecting the cell layer above, one can then measure, at various time intervals, the association between labeled metabolic substrate or labeled hormone and the packed cells, which are almost entirely devoid of aqueous medium contamination. In order to correct, however, for the small amount of medium trapped between the cells (usually not exceeding 0.2% of the extracellular water), ^{14}C-sucrose or ^{14}C-insulin can be used (50, 51). This technique has proved to be remarkably versatile and uniquely adept in studies of the transport of glucose (or glucose analogs), or in studies directed to the measurement of hormone binding by the fat cells (52, 53).

Means of Comparison

In marked contrast to most tissues of the body, which maintain a relatively constant weight and cellular composition, adipose tissue is unique in its extraordinary capacity to increase or decrease the tissue mass by a combination of changes in fat cell size and number.

Accurate comparison of metabolic activity between two fragments of tissue (from the same depot, different regions in the same animal, or different animals) demands that the comparison be carried out on the basis of an equal or comparable number of fat cells. Knowledge of cell size will provide additional and valuable information in the comparison.

Wet Weight. Earlier studies with adipose tissue slices were based on a comparison of the tissue wet weight, for lack of a better parameter. In view of the variable tissue content of protein, water, lipid and, most important, adipocyte numbers, such comparisons are to be considered imprecise and unsatisfactory, even when fragments of tissue from the same depot are compared.

Protein or Fat-Free Dry Weight. Comparison of the metabolic activity of fragments of adipose tissue on the basis of fat-free dry weight has been carried out (13). Initially, this method appeared to be an improvement over using the wet weight of tissue. Tissue protein content, or fat-free dry weight, however, are not truly representative of the metabolically active units, the adipocytes, for two main reasons: (a) tissue protein of the stromal-vascular compartment contributes to a variable degree, depending on animal species, adipose region, and nutritional-developmental stage of the animals (10); and (b) adipocytes not only vary the lipid and water content when they enlarge (51), but also increase their protein content (20, 54). Discrepant results have been obtained in this regard, however (19). Expressing the results on the basis of tissue protein content may still be valid for a relative comparison within the same adipose depot.

Deoxyribonucleic Acid. Deoxyribonucleic acid is a fixed and stable component of eukaryotic cells, with a concentration of around 0.7 pg/cell. It would then seem to be the ideal means of comparing two or more fragments of adipose tissue. But this is not the case since in a fragment of adipose tissue, DNA from stromal-vascular components can account for up to 40–60% of total DNA, depending on the nutritional state of the animal and adipocyte size (22, 55, 56).

In isolated fat cell preparations, following collagenase digestion, a large portion of cellular debris and stromal-vascular components is removed by the washing procedure. There are reports, however, that indicate the persistent contamination of isolated adipocyte preparations with DNA derived from vascular strands and nuclei. Since stromal-vascular cells are much smaller than adipocytes in the suspended cell population, even a small amount of DNA resulting from incomplete washing and representing cells other than adipocytes could significantly limit the accuracy of the DNA as an accurate marker of adipocyte number (24).

Fat Cell Number. Assessment of fat cell number with a fair degree of accuracy has been the goal of many investigators of adipose tissue cellularity and metabolism. Rodbell has shown that the metabolic activity of cells other than adipocytes accounts for less than 1% of glucose metabolism (21), although the oxygen consumption of stromal-vascular cells can be considerable (57). Advances in methodology during the last 20 years have allowed researchers to measure fat cell size with a good degree of precision. The degree of precision encountered in the determination of cell number is somewhat less and varies with the technique used. Moreover, attempts to define fat cell number in a given tissue fragment, adipose depot, or even in the whole body, are seriously limited by the inability to identify, and thus count, adipose cells with a diameter smaller than 6–10 μm (48, 58) or, in some techniques, smaller than 25 μm (23).

Adipose cell number can be measured directly after fixation of a tissue fragment in osmium tetroxide and counting of the fixed cells in a Coulter counter (23). Alternatively, cell number can be estimated by measuring fat cell lipid content (calculated from cell diameter and lipid density) and relating it to the total lipid content of a tissue fragment or an isolated fat cell preparation (48).

The introduction of fairly accurate techniques for estimating adipose cell number in a given tissue slice or an isolated adipocyte preparation has allowed a meaningful comparison of metabolic activities and hormonal responses on the basis of equal numbers of cells. This simple methodological improvement has contributed markedly to the understanding of metabolic alteration with adipocyte enlargement resulting from age and ad libitum feeding (16). Regional differences (19, 59–61), sex differences (62), and species differences (9) in adipose tissues have also been elucidated.

Adipocyte Morphology—Water Content

Adipocytes are specialized and differentiated cells whose main function is one of synthesis and storage of fuel for subsequent release at times when energy fuel is necessary to maintain body economy.

Before reaching adult size (i.e., > 30 μm in diameter) and the typical signet ring appearance, adipocytes go through various stages of lipid repletion and differentiation. The fibroblast-preadipocyte initially presents multilocular lipid inclusions that later fuse to give a trilocular, bilocular, or unilocular appearance. Once the adipocyte has reached a mature appearance with a single lipid droplet surrounded by a rim of cytoplasm and the offset nucleus, cell size can continue to increase from 30 μm to 120–150 μm, mostly by increases in triglyceride storage. In parallel with lipid content increase, however, there is also accretion of protein (20) and accumulation of small amounts of water (51). A cell that doubles its diameter from 40 to 80 μm will increase its surface area (and protein content) by a factor of 3 to 4 and its volume by a factor of 8. Water content will also increase by a factor of 3 to 4.

It has to be emphasized that, in many studies, fat cell size has been used as a marker of age, adiposity, or lipid accumulation to indicate possible metabolic alterations that accompany or follow the altered lipid accumulation. Expressing metabolic results in relation to cell number and cell size has helped to better define the experimental conditions underlying many investigations with adipose tissue. From time to time, cell size has been overinterpreted as having its own causative effect on metabolic activities or hormonal responses. But caution in interpreting the importance of cell size is needed when several variables (age, nutrition, physical activity, adipose region, etc.) may have an independent influence on the observed metabolic behavior of the cells (59, 63). Expression of metabolic data as per unit of fat cell surface area (48) provides a basis for comparing large and small fat cells, which corrects for differences in cell size (63). In addition, surface area may be the more appropriate functional unit for comparing dose-response relationships that may depend on the coupling of receptors to effector systems (63).

METABOLIC FUNCTIONS OF ADIPOCYTES AND RELATED METHODOLOGY

Adipose tissue is the main site of fuel storage. In the lean adult male, this storage may represent 80,000–100,000 kcal, and in the lean adult female, 110,000–130,000 kcal. Almost all fat storage is in the form of triglyceride, a neutral fat with glycerol and fatty acids as constituents.

The main metabolic functions of adipocytes are: (a) lipid synthesis and accumulation from a variety of substrates and by a variety of mechanisms; and (b) triglyceride breakdown and lipid mobilization. These important processes are influenced by species, sex, adipose region, developmental stage, cell size, diet, hormones, and other factors. The relative activity of lipid deposition and lipid removal may vary from individual to individual in response to nutritional and endocrine factors, and may ultimately be responsible for individual differences in body composition and degree of adiposity.

Carbohydrate Metabolism and Lipid Synthesis

Glucose uptake by the adipose cell precedes conversion of glucose to glycogen, fatty acid, α-glycerophosphate, lactate, and other intermediary compounds. Since no triglyceride can enter the fat cell directly from the circulation, triglycerides are synthesized in situ from available fatty acids and from available α-glycerophosphate. Fatty acids can be synthesized de novo from glucose or other substrates, or they can be derived through hydrolysis of stored triglycerides. Triglyceride fatty acids are derived primarily from the action of lipoprotein lipase (LPL) on circulating lipoproteins (64). In addition, FFA derived from circulating FFA-albumin complexes can, to a minor extent, be taken up by the fat cells (65). Esterification or reesterification of FFA requires availability of α-glycerophosphate. Until recently, glucose was thought to be the main precursor of α-glycerophosphate, but recent studies have shown that lactate and, to a minor extent, pyruvate may be alternate precursors of considerable potential (29).

At its physiological concentration, 3–8 mM, glucose is rapidly transported from the bathing fluid into fat cells by a stereospecific process of "facilitated diffusion." Recent studies by Suzuki and Kono (66) and by Cushman and Wardzala (67) have shown that glucose transporters are recruited from an intracellular pool to the plasma membrane in response to insulin. In addition, altered responsiveness of glucose transport to insulin can be explained by changes in the cellular distribution of transporters with changing nutritional and hormonal status (68).

Once transported inside the cell, glucose is rapidly phosphorylated and then processed by various metabolic pathways. In the fed young rat's fat cell, the main products of glucose metabolism, in the absence of insulin, are: carbon dioxide (30%), glycogen (3–5%), fatty acids (20%), α-glycerophosphate (30%, recovered as glyceride-glycerol), and lactate (10%)(69). In the fat cell from older, fatter fed rats, there is a shift in the main products of glucose metabolism with a decline in fatty acid synthesis (reduced to 2–4%) and carbon dioxide (15%), and a marked increase in conversion of glucose to α-glycerophosphate (40%) and lactate (40%) (69, 70). Fasting and other nutritional interventions can alter these proportional rates of glucose metabolism to a markedly greater extent in adipocytes from young rats than in those from older, fatter rats (26).

Lipoprotein Lipase Activity

The enzyme LPL catalyzes the hydrolysis of circulating triglycerides so that FFA can be taken up by the cell for esterification, triglyceride synthesis, and storage (64). Lipoprotein lipase is actively synthesized by the fat cells, then secreted to the endothelial cell surface, where it acts upon circulating chylomicra or very low density lipoproteins (VLDL) (64).

Lipoprotein lipase activity in white adipose tissue falls with fasting and food deprivation, and is elevated by feeding and insulin stimulation (64). These changes in enzymatic activity result in shunting circulating lipoprotein-derived triglycerides toward triglyceride storage in adipose tissue in the fed state, and toward muscle in the fasted state (64).

Lipoprotein lipase activity is assayed as serum-stimulated lipolytic activity utilizing labeled artificial triglyceride emulsions. Extracts of acetone-ether powders, adipose tissue homogenates (assessing total tissue activity), or the heparin-releasable LPL fraction from small tissue pieces, are usually used as sources of the enzyme (71). The latter fraction is thought to reflect the "functional," extracellular enzyme activity at the endothelial cell surface. A newly developed method using detergent (0.5% deoxycholate) extraction of whole tissue homogenates provides a much-improved yield of LPL activity over previous methods (71), presumably due to disruption of the microsomes where LPL is stored intracellularly. An important practical point with LPL assays is to use a postheparin plasma or a purified bovine milk LPL standard to correct for daily variability in the preparation of the sonicated substrate.

Lipolysis

The large amount of energy stored in adipose cells in the form of triglyceride provides the substrate fuel for the metabolic needs of the organism during protracted starvation and also during daily periods of fasting.

Since no triglyceride can leave the fat cell intact, triglyceride breakdown (hydrolysis) leads to stepwise enzymatic conversion to diglyceride, monoglyceride, and, finally, glycerol and FFA. The initial hydrolysis of triglyceride is promoted by lipolytic hormones and is mediated by activation of adenylate cyclase, which converts adenosine 5′-triphosphate (ATP) to cyclic 3′, 5′-AMP. This latter compound promotes activation of a lipase enzyme called "hormone-sensitive" lipase, which initiates the series of enzymatic events resulting in the production of glycerol and FFA (72). Glycerol is water soluble and leaves the cell, since only very limited reuse of it is possible by very low intracellular glycerokinase activity. The fate of the liberated FFA is multiple: Some are oxidized in the cells; a large proportion of them are reesterified with available α-glycerophosphate; and, depending on the availability of FFA acceptors (mostly albumin) outside the cells, an additional fraction can be removed and bound to circulating albumin, to reach peripheral tissues and provide energy by oxidation, or be resynthesized to triglycerides in the liver.

Since release of glycerol by the cells is proportional to the hydrolysis of triglyceride, this parameter has been used extensively to measure the rate of lipolysis in adipocytes. Glycerol can be assayed by standard spectrophotometric (73) or fluorometric (75) methods (adequate for rat fat cells), or more sensitive bioluminescent assays (75, 76). A new radioisotope procedure has also been described that offers comparable sensitivity (31, 47) but is based on certain assumptions that need to be tested in the preparation used.

Recent evidence underlines the importance of measuring lipolysis in the presence of adenosine deaminase (76, 77) to prevent feedback inhibition of the rate of lipolysis by adenosine accumulation in the medium. This is particularly important in the comparison of fat cells from subjects in different nutritional conditions, or at different ages, which may alter cellular responsiveness to the antilipolytic effects of adenosine (76, 77). An additional consideration is the effect of collagenase digestion on phosphodiesterase activity (78). Thus, some workers prefer tissue fragments for lipolysis studies (78).

Oxygen Consumption

The oxygen consumption in adipose tissue has been studied in tissue slices and, more recently, in isolated fat cell preparations (5, 57, 79). Microcalorimetry has also been used recently to measure heat production in fat cells to estimate overall rates of metabolism (80). In the early studies, when oxygen consumption in adipose tissue was expressed on the basis of 100 mg wet weight of the tissue, the most striking finding was the influence of the age of the rat that served as a tissue donor. Ball and Jungas reported values of 50–60 μl/100 mg/hr in ad libitum fed young rats, which fell to 10–20 μl when the animals reached 250 g in weight (5).

Some of the "age effect" on oxygen consumption, however, may be due to the fact that equal weights of adipose tissue from older animals include larger and therefore fewer adipose cells. When expressed on a per cell basis, the negative age effect persists, but in groups of rats of the same age, a positive and direct effect of cell size is observed. In adipocytes, oxygen consumption mostly reflects the utilization of ATP for the reesterification of the FFA produced under conditions of basal or of stimulated lipolysis. Large adipocytes present a greater rate of basal lipolysis and reesterification (15, 16). This observation may explain the increased oxygen consumption of large fat cells observed in our laboratory (57) and by others (79).

Among other factors capable of influencing oxygen consumption in adipose tissue are the following:

1. Addition of substrates. The addition of glucose raises oxygen consumption by only 10%, and a variety of other substrates, including malate, lactate, and FFA bound to albumin, have only a 10–20% incremental effect on oxygen consumption by adipose tissue. In contrast, the addition of pyruvate can produce a threefold increase in oxygen consumption by rat adipose tissue, and oxaloacetate produces a similar but slightly lower increase (5).
2. Fatty acid synthesis and medium buffer. As was mentioned above, the addition of glucose to the incubation medium with phosphate buffer produces only a small stimulatory effect (10%) on oxygen consumption. But if a bicarbonate buffer is used together with tissue from young rats, a greater stimulation (70–80%) of oxygen consumption is seen in the presence of glucose and insulin. This stimulation is believed to accompany rapid fatty acid synthesis favored by the bicarbonate buffer in the presence of glucose and insulin (9, 70).
3. Effect of lipolytic stimulation. Initiation and acceleration of lipolysis produces rapid and marked increases in oxygen consumption, particularly in the presence of glucose, which provides the α-glycerophosphate needed for FFA esterification (5).
4. Dietary changes. Fasting followed by refeeding a high carbohydrate diet raises the basal oxygen consumption by three to fourfold. This increase appears to be linked to the increased rate of lipolysis, FFA reesterification, and fatty acid synthesis (5).

Binding of Hormones

The adipocyte is the target cell of many hormones that regulate the main functions of lipogenesis and lipolysis.

The availability of techniques to isolate adipocytes, coupled with rapid flotation through oil to remove or drastically reduce the aqueous medium from the cell layer, has made this preparation a preferred one in studies of the last two decades directed toward understanding the kinetics of hormone binding and relationships of binding to hormonal action. Preparations of radioactively labeled hormones that have unaltered affinity for their receptors (e.g., monoiodinated A^{14}-tyrosyl-insulin (52, 81)) should be used. Bacitracin has been added to prevent extracellular insulin degradation when insulin binding is assessed at 37°C (49). In addition, characterization of the fat cell population in terms of cell size (diameter, surface area, and volume) and cell number has facilitated expression of the hormone binding characteristics in terms of cell number or cell surface area (52, 82, 83).

Adipocytes bind insulin in a specific, temperature- and concentration-dependent manner (52). Binding of insulin is a prerequisite for internalization of the insulin-receptor complex, which leads, among other things, to insulin degradation (52). Insulin binding is proportional to insulin concentration in the medium, even at concentrations that exceed the maximally stimulated biological action by a factor of from 10^3–10^6. The possible meaning of this "spare receptor binding" is unclear, but it may serve a protective role and a role of temporary storage for the insulin molecule in the target cell. Although discrepant results exist in the literature about changes in insulin binding with adipocyte enlargement and obesity (52, 82–84), there is fairly good agreement that the decreased responsiveness to insulin seen in large fat cells from obese animals or human subjects reflects alterations in postreceptor metabolic events to a greater extent than alterations in membrane receptor binding per se.

Adipocytes have been shown to specifically bind other types of hormones, such as growth hormone (85), glucagon (86), catecholamines (87, 88), and many others. The advantages of the isolated fat cell preparation have been recognized by many investigators interested in the relationship between hormone binding and hormone action.

Response to Insulin

Adipocyte preparations from adipose tissue of young lean rats (100–150 g) respond rapidly and markedly to insulin. Gliemann (30) has shown that the rate of glucose metabolism changes within 1–4 min after the addition of insulin. The relative effect of insulin on glucose uptake and metabolism is dependent on the glucose concentration in the medium. This effect is maximal at glucose concentrations of 0.1–1.0 mM (30) and drops sharply when the glucose concentration exceeds the physiological concentration of 5 mM. This result is probably best explained by the effect of insulin on acceleration of glucose transport. At low glucose concentrations, transport is rate-limiting and insulin can exert a more marked effect. At high glucose concentrations, glucose metabolism becomes rate-limiting and the insulin effect is less marked.

The response to insulin in adipocytes can be measured in a variety of ways.

Glucose Disappearance. Reduction of glucose concentration in the incubation medium during an in vitro incubation, in the absence or presence of graded doses of insulin, can be an effective way to determine overall glucose consumption by the cells. In these types of experiments, it is important to have sufficient cells to remove at least 10% of the glucose present. But if an excessive number of cells is present, the rapid decline in glucose concentration could influence the assessment of the insulin effect on glucose disappearance. Glucose concentrations of 1 to 3 mM are usually employed in this type of assay (89).

Tracer Studies of Glucose Transport. The effect of insulin on the transport of glucose analogs (^{14}C-labeled 3-0-methylglucose (53), 2-deoxyglucose (90), arabinose (91), etc.) has been measured at very short time intervals (seconds) by the use of a modified oil flotation technique using a phloretin "stop" (53).

A new technique utilizing the uptake of trace (300 nM) uniformly labeled glucose (U-^{14}C-glucose) during a 1-hour incubation at 37°C has the advantage of being an easy and inexpensive index of glucose transport when kinetic parameters are not required (49). This method relies on the fact that at trace glucose concentrations, no significant glucose oxidation occurs and, presumably, no metabolic products are lost from the cell. This finding has been confirmed in human adipocytes and in rat adipocytes (except in rats younger than 6 weeks of age; S. K. Fried, unpublished observations). Recent evidence also indicates that 2-deoxyglucose uptake is not a valid indicator of glucose transport (90).

Tracer Studies of Glucose Metabolism. Glucose labeled in the carbon moiety (either U-^{14}C; or labeled in specific carbons, 1-^{14}C or 6-^{14}C) has been extensively used to study overall glucose metabolism (with U-^{14}C), or to study preferential uses in the pentose shunt (with 1-^{14}C), or in the Krebs cycle (with 6-^{14}C) (87, 92). However, independent assessment of glucose utilization is required if the contribution of the pentose shunt is to be quantitated (92).

By incubating cells at the desired glucose concentration in the presence of ^{14}C-glucose, and at various insulin concentrations (from 0 to 1,000 microunits/ml (μU/ml)), it is possible to determine the effect of insulin in terms of minimal effective dose (MED), maximal response (V_{max}), and sensitivity to insulin (i.e., the concentration of insulin at which 50% of the maximal effect is observed, ED_{50}).

The insulin effect, expressed as percent increase over basal metabolic activity, varies from pathway to pathway. For example, the ranking for the effect of insulin on glucose conversion to specified products in a small, insulin-sensitive fat cell is: FA synthesis > CO_2 = glycogen > glyceride-glycerol. Since it is generally accepted that insulin exerts its main effect on glucose transport, this ranking may represent a nonspecific variation due to the various degrees of pathway saturation, or it may represent additional specific effects of insulin on intracellular enzymatic activities (7, 93).

Antilipolytic Effect of Insulin. Insulin inhibits the lipolysis accelerated by lipolytic hormones. This action of insulin appears to be independent of the insulin effect on glucose metabolism and thus represents a separate insulin action (94).

The antilipolytic effect of insulin is usually determined by incubating fat cells with a submaximal amount (i.e., 0.1 μg/ml) of epinephrine or norepinephrine in the absence of insulin or in the presence of insulin in progressively increasing concentrations. This effect of insulin may be altered by nutritional intervention (76) or by age, stage of development, and other factors. More studies are needed in this area.

Response to Lipolytic Hormones

The adipocyte is the principal body cell for fuel storage. Lipolysis and FFA mobilization are critical for life in states of partial or total food deprivation. It is, therefore, not surprising to find a "redundancy" of hormonal stimulation of lipolysis. In each animal species, several classes of hormones can, at pharmacological doses, be shown to accelerate lipolysis (8). In some species, lipolysis has been accelerated with physiological concentrations of the lipolytic hormones (8).

Species differences exist in the responsiveness of the cells to lipolytic stimuli. This has been studied extensively by Rudman and DiGirolamo (8) and Rudman and co-workers (95). In the rabbit, peptide hormones of the pituitary, such as adrenocorticotropin (ACTH), lipotropin, and melanocyte stimulating hormone (MSH etc.), can markedly stimulate lipolysis. In the human, catecholamines (and possibly PTH) are the main stimulators of lipolysis. In other species, such as rodents, catecholamines are the predominant stimulators but peptide hormones retain a partial effect (8, 95).

Although rat adipocytes appear to possess only beta-receptors (which accelerate lipolysis), other species (hamster, human) also have alpha 1 and alpha 2 receptors, which are antilipolytic (88, 96, 97). Therefore, careful dose-response studies of the lipolytic effects of physiological catecholamines (epinephrine, norepinephrine), which have predominantly beta or alpha effects depending on the relative concentrations of these receptor subpopulations, have to be carried out. These studies are of particular importance because these receptor concentrations vary with depot, and with age or nutritional state of the tissue donors (76, 77). Several reviews of this complex topic have recently appeared (96–99).

VARIABLES AFFECTING ADIPOCYTE METABOLISM

Buffer - Albumin - Collagenase

A variety of buffers have been used to study in vitro adipocyte metabolism. The Krebs-Ringer medium is one of the most commonly used at pH 7.4, with the addition of either phosphate (KRP) or bicarbonate (KRB). The KRB buffer has been found to be particularly suited for incubations with glucose and insulin in which fatty acid synthesis is very active (9). In a direct comparison of the two types of buffer at two different glucose concentrations (1.39 and 13.9 mM) in four animal species, only a slight advantage of the KRB buffer was found in the conversion of glucose to CO_2 and fatty acids (9), and this was seen best at the lower concentration of glucose. Addition of 10–25 nM HEPES can be useful for increasing the buffering capacity (32, 49).

Albumin has traditionally been added to the incubation medium in concentrations varying from 1 to 5% (10 to 50 mg/ml). The reasons for having albumin in the medium are at least three:

(a) A desire to mimic the physiological milieu of the plasma by adding a simple protein at near normal concentrations.

(b) To add a suitable "acceptor" of FFA so that a portion of the FFA resulting from hydrolysis of triglycerides can leave the cells and reduce the intracellular accumulation of FFA with consequent inhibition of the lipolytic process.

(c) To minimize loss of insulin or other hormones by adherence to glass walls of the incubation vessels or to plasticware. Insulin has the property of "sticking" readily to a variety of surfaces. This sticking may pose a problem in in vitro incubations lasting 60–90 min, in which dose response measurements are carried out with as little as 1 μU/ml (0.04 ng/ml) insulin. Addition of albumin to the incubation medium has been found to stabilize the buffer and to preserve the added hormone.

Experience has shown that it is advisable to use batches of the more purified albumin (Fraction V), in order to minimize the possible metabolic effects of contaminants. Furthermore, "fatty acid-poor" albumin is preferred in order to provide a greater buffering capacity to accept FFA liberated from the fat cells. In an experiment in which albumin with variable addition of FFA was used (100), a significant effect of the added FFA on cellular metabolism was observed, suggesting that increased availability of exogenous FFA (or reduced capacity to remove endogenous FFA) may influence adipocyte glucose conversion to fatty acids and other products.

Collagenase treatment of adipose tissue to liberate the adipocytes from stromal-vascular components can be damaging to the liberated cells unless certain precautions are taken:

1. Choice of the appropriate type of collagenase. Collagenase preparations contain a variety of proteases, and it is important to choose the type that is recommended by the supply company for use with adipose tissue.
2. The concentration of collagenase in the incubation medium should vary between 0.2 and 2 mg/ml, depending on the time of incubation, type of adipose tissue (from lean young, or fatter, older animals), quantity of adipose tissue (usually 1 g/2 mg collagenase), and prior preparation of the tissue (mincing by scissors, etc.). Fine mincing should be avoided with larger fat cells (i.e., >200 pl).
3. Thorough washing of the liberated cells is recommended to eliminate even traces of collagenase from the resuspended cell preparation. A minimum of four washings is recommended, with at least 3 times as much fresh medium as the volume of packed cells.

It is also important to emphasize that all aspects of tissue and fat cell preparation require careful and expeditious handling with particular attention to avoiding of temperature differences, to the use of mild centrifugation (1,000 × *g* for 1 min) for cell washing, and to gentle swirling to produce an even suspension of cells before they are dispensed into the incubation flasks. During the incubation, usually carried out in a Dubnoff metabolic shaker (or a similar apparatus), shaker motion should not exceed 60–80 revolutions per minute, to minimize breakage of the cells. A high albumin concentration (5%) is recommended for the more fragile human fat cells (32, 49), or fat cells from obese rodents. Relatively low fat cell concentrations (<100,000 per pl) are also recommended with these cells.

The selection, by a given laboratory, of the type of albumin and collagenase is probably responsible for much of the variability reported from laboratory to laboratory, when similar experimental conditions are used. It is recommended that small samples of the new albumin or of the new collagenase be tested periodically (when the need arises) by comparing each of them with the ones previously employed in that laboratory. Criteria for satisfactory quality of the new batches of reagents are: (a) They should not alter baseline metabolism significantly; (b) They should not affect, positively or negatively, the expected hormonal response to a significant degree; (c) The supply company should be willing to put aside large amounts of the same lot of albumin (3–5 kg) or collagenase (4–5g) for 1–3 weeks until testing has provided sufficient information to choose the lot that is closest to the response obtained with the previous batch in use. In this fashion, one can maintain internal laboratory consistency and reproducible results. When these precautions have been used, reproducible insulin effects have been consistently observed on glucose metabolism (i.e., overall stimulation of glucose conversion to CO_2 and triglycerides by 200–300% at 6 mM glucose, and by 600–800% at 1 mM glucose).

Substrate Concentration

It has already been mentioned that the concentration of glucose in the medium (from 0.1 to 20 mM) influences not only the rate of glucose metabolism, but also the hormone response and the fate of glucose metabolized. This is particularly true for small (40–50 pl) adipocytes from young rats fed ad lib as compared with larger (500–600 pl) cells from fatter, older rats (16).

In small cells, increasing glucose concentrations in the medium (or the addition of insulin) will raise the absolute and relative conversion of glucose to fatty acids and CO_2, and increase the absolute conversion of glucose to glyceride-glycerol (but reduce its rate of conversion relative to the other two products). This capacity to shift glucose conversion to a specified product with changes in substrate concentration or insulin addition is nearly totally abolished in large fat cells (> 500 pl) of older, fatter rats. Not only are these large adipocytes minimally responsive to insulin, they also seem to have a "fixed" partition of glucose into fatty acids (very low, less than 1–3% of total), CO_2, or glyceride-glycerol (very high, representing 50–60% of glucose conversion to CO_2 and triglyceride).

The utilization of other substrates, that is, acetate, pyruvate, and lactate, has been tested in in vitro preparations of small and large rat adipocytes (29). The metabolism of some of these other substrates is also influenced by animal species, variable substrate concentrations and hormone addition (27–29).

Recently, this laboratory (69), as well as others (70, 101), has recognized that a previously neglected product of glucose metabolism, lactate, can represent as much as 10% (in small cells) or 45–50% (in large cells) of all glucose metabolized by adipocytes removed from fed rats. In human fat cells (102), lactate production has been found to account for 60–65% of all glucose metabolized.

The importance of lactate production from glucose in adipose tissue is now under investigation, particularly in obese states, in which lactate production from glucose in large fat cells is of significant magnitude. This finding, coupled with a possibly greater number of fat cells in an expanded adipose mass, raises the possibility that the adipose organ is a major site of lactate production of previously unrecognized magnitude in the human and other species.

Cell Size

Cell size has already been alluded to here as an important determinant of cellular metabolism and hormonal responsiveness. In studies with adipose tissue from many mammalian species (9, 16, 26, 69), at different stages of growth and development, cell size has been an important marker of metabolic activities, although caution is necessary in attributing to cell size a "causative" role, since many other factors may concomitantly be involved.

In the rat, small cells from young animals (100–150 g) are associated with a greater conversion of glucose to fatty acids and a more marked responsiveness to insulin than larger cells from older, fatter rats (500–800 g). Additional studies by several investigators have also revealed variable metabolic activities in subsets of the same population of cells separated by differential flotation (54).

Cell Concentration

In most in vitro incubations with isolated fat cells, the number of cells in the incubation medium varies considerably, from 2×10^4 to 1×10^6/ml. Cell concentration is a critical variable in the assessment of glucose metabolism, lipolysis, and hormonal responsiveness (25, 26, 99, 103).

Schwabe and co-workers (25) have shown that, at a high cell concentrations ($> 5 \times 10^5$/ml), the rate of lipolysis per cell is markedly reduced, possibly because of the accumulation of adenosine and FFA in the medium and/or the cells. Elevated intracellular FFA have been shown to inhibit lipolysis, and adenosine has been shown to have an insulinlike effect on glucose metabolism and an antilipolytic effect of its own (25, 99). Cell concentration effects could also be in part due to accumulation and release of other metabolic products such as prostaglandins (96).

This laboratory has studied (26) the effect of cell concentration on glucose metabolism and insulin response. At low cell concentrations, when the rate of lipolysis is maximal, glucose conversion to lipid is lowest and insulin response is reduced. Increasing cell concentrations are associated with increases in glucose metabolism, fatty acid synthesis, and responsiveness to the effect of insulin on glucose utilization.

Efforts should be made in future studies with isolated fat cells to monitor and keep constant (in a given series of experiments) the number of cells present in the incubation medium, in order to avoid marked variability of metabolic assessment from day to day and from experiment to experiment.

Regional Differences

Morphologic and metabolic differences have been described in the various regions of adipose tissue from different species. Different regions may differ in the mean cell size within a given subject (9, 104–106), and also in the capacity to increase or decrease mean cell size as the result of nutritional intervention (82, 105).

In animals and in man (9, 59, 60, 106), adipose tissue from different regions has also been found to differ in the capacity to respond to lipogenic and lipolytic hormones. Changes in body composition and in fat distribution occurring at different stages of growth and development (i.e., puberty) probably result from differential effects of hormones on different adipose regions. Similarly, in altered metabolic states, such as obesity or Cushing's syndrome, altered regional metabolic and hormonal responses may underlie the observed variations in body fat distribution. More studies, however, are needed to clarify both the mechanisms and the quantitative contribution of the various adipose districts to the total adipose organ mass and to the overall adipose metabolic activities under normal and abnormal conditions of nutrition, growth, and disease processes.

Species Differences

Major species differences have been recorded in utilization of substrates for adipose tissue metabolism and in the responsiveness to hormones. The reader is referred to more detailed reviews of this topic (8, 9, 95).

CONCLUSIONS

The present review has emphasized the important variables that have to be controlled if meaningful data are to result from in vitro incubation of adipose tissue fragments, or of its cellular constituents.

In spite of the remarkable progress achieved in the understanding of adipose tissue metabolic functions in the last 40 years, several considerations limit the extrapolation from in vitro data and an accurate estimate of the contribution of the adipose organ to body fuel economy.

Among these are:

1. The ubiquity of adipose tissue cells in the body, and the reported regional differences in morphology and metabolic activities. The reliability of overall estimates of body cell number and estimates of overall adipose functions (such as glucose uptake and metabolism) is seriously limited by the fragmentary observations gathered from a few samples from selected adipose regions.
2. In vitro incubation suffers from the inability to remove the accumulation of cellular metabolic products (such as adenosine and FFA) that are constantly removed in vivo by the tissue perfusion with blood. This is partly obviated in the "perfused" and "perifused" preparations of adipose

tissue, which allow bathing of the cells with renewable physiological buffers.

3. Most in vitro incubations of adipose tissue have been carried out with one or two substrates in the presence of subphysiological (0 20 mg/ml) or physiological (20–50 mg/ml) concentrations of albumin. Preliminary information with incubations containing two or more substrates at concentrations mimicking a more physiological metabolic milieu have revealed interesting effects of the various substrates on the metabolic activity of the tissue (29).
4. An additional limitation in extrapolating from in vitro to in vivo data may reside in the incomplete recovery of products of adipocyte metabolism not previously recognized. For example, the recent finding that lactate is a major exportable product of fat cell glucose metabolism may provide additional insight into variables controlling production and release of lactate into the circulation and its potential systemic regulatory role.

Even with these cautionary remarks, however, the last 40 years have seen an extraordinary explosion of information confirming the essential and important role provided by the adipose organ in the fuel storage and supply, and in the adaptation to the metabolic needs of the organism. It is hoped that the in vitro data accumulated by several investigators of adipose tissue physiology can be further integrated by novel in vivo techniques, in the near future, to provide an even better understanding of the physiological and pathological role of the adipose organ.

REFERENCES

1. E. Wertheimer and B. Shapiro, 1948, The physiology of adipose tissue, *Physiol. Rev.* **28**:451–464.
2. F. X. Hausberger, S. W. Milstein, and R. J. Ratman, 1954, The influence of insulin on glucose utilization in adipose and hepatic tissues in vitro, *J. Biochem.* **208**:431–438.
3. V. P. Dole, 1956, A relation between non-esterified fatty acid in plasma and the metabolism of glucose, *J. Clin. Invest.* **35**:150–154.
4. A. I. Winegrad and A. E. Renold, 1958, Studies on rat adipose tissue in vitro. I. Effects of insulin on the metabolism of glucose, pyruvate and acetate, *J. Biol. Chem.* **233**:267–272.
5. E. G. Ball and R. L. Jungas, 1965, Net gas exchange and oxygen consumption, in *Handbook of Physiology,* Section 5, Adipose Tissue, A. E. Renold and G. F. Cahill, eds., American Physiological Society, Washington, D.C., pp. 355–362.
6. M. Vaughan and D. Steinberg, 1965, Glyceride biosynthesis, glyceride breakdown and glycogen breakdown in adipose tissue: Mechanisms and regulations, in *Handbook of Physiology, Section 5,* Adipose tissue. A. E. Renold and G. F. Cahill, Jr., eds., American Physiological Society, Washington, D.C., pp. 239–252.
7. G. F. Cahill, Jr., and A. E. Renold, 1965, Regulation of adipose tissue metabolism with

in the intact organism: a summary, in *Handbook of Physiology, Section 5,* A. E. Renold and G. H. Cahill, Jr., eds., American Physiological Society, Washington, D.C. pp. 681–684.

8. D. Rudman and M. DiGirolamo, 1967, Comparative studies on the physiology of adipose tissue, *Adv. Lipid Res.* **5**:35–117.
9. M. DiGirolamo and D. Rudman, 1966, Species differences in glucose metabolism and insulin responsiveness of adipose tissue, *Am. J. Physiol.* **210**:721–727.
10. W. Benjamin, A. Gellhorn, M. Wagner, and H. Kandel, 1961, Effect of aging on lipid composition and metabolism in the adipose tissues of the rat, *Amer. J. Physiol.* **201**:540–546.
11. J. Hirsch and P. W. Han, 1969, Cellularity of rat adipose tissue: Effects of growth, starvation and obesity, *J. Lipid Res.* **10**:77–82.
12. E. Shafrir and E. Wertheimer, 1965, Comparative physiology of adipose tissue in different sites and in different species, in *Handbook of Physiology,* Section 5, Adipose Tissue, A. E. Renold and G. F. Cahill, Jr., eds., American Physiological Society, Washington, D.C., pp. 417–430.
13. M. DiGirolamo and D. Rudman, 1960, Variations in glucose metabolism and sensitivity to insulin of the rat's adipose tissue, in relation to age and body weight, *Endocrinology* **82**:1133–1141.
14. O. Zinder, R. Arad, and B. Shapiro, 1967, Effect of cell size on the metabolism of isolated fat cells, *Israel J. Med. Sci.* **3**:787–791.
15. M. F. Reardon, R. B. Goldrick, and N. H. Fidge, 1973, Dependence of rats of lipolysis, and free fatty acid release in isolated fat cells on age, cell size, and nutritional state, *J. Lipid Res.* **14**:319–326.
16. M. DiGirolamo, M. D. Howe, J. Esposito, L. Thurman, and J. S. Owens, 1974, Metabolic patterns and insulin responsiveness of enlarging fat cells, *J. Lipid Res.* **15**:332–338.
17. M. DiGirolamo, 1972, Fat cell size, metabolic capacities and hormonal responsiveness of adipose tissue in spontaneous obesity in the rat, *Israel J. Med. Sci.* **8**:807.
18. A. D. Hartman, A. I. Cohen, C. J. Richane, and T. Hsu, 1971, Lipolytic response and adenyl cyclase activity of rat adipocytes as related to cell size, *J. Lipid Res.* **12**:498–505.
19. L. B. Salans and J. W. Dougherty, 1971, The effect of insulin upon glucose metabolism by adipose cells of different size. Influence of cell lipid and protein content, age, and nutritional state, *J. Clin. Invest.* **50**:1399–1410.
20. M. DiGirolamo, J. L. Owens, J. G. Patrick, and J. F. Kuo, 1977, Protein kinase activity of isolated rat adipocytes as related to cell size, *Proc. Soc. Exp. Biol. Med.* **154**:513–516.
21. M. Rodbell, 1964, Metabolism of isolated fat cells. I. Effect of hormones on glucose metabolism and lipolysis, *J. Biol. Chem.* **239**:375–380.
22. R. B. Goldrick, 1967, Morphological changes in the adipocytes during fat deposition and mobilization, *J. Clin. Invest.* **212**:777–782.
23. J. Hirsch and E. Gallian, 1968, Methods for the determination of adipose cell size in man and animals, *J. Lipid Res.* **9**:110–119.
24. B. J. Klyde and J. Hirsch, 1979, Isotopic labeling of DNA in rat adipose tissue: Evidence for proliferating cells associated with mature adipocytes, *J. Lipid Res.* **20**:691–704.
25. U. Schwabe, R. Ebert, and H. C. Erbler, 1973, Adenosine release from isolated fat cells and its significance for the effects of hormones on cyclic 3′5′ AMP levels and lipolysis, *Nanyn-Schmiedeberg's Arch. Pharmacol.* **276**:133–148.
26. M. DiGirolamo, 1981, Effects of variable glucose and fat cell concentration on glucose metabolism and insulin responsiveness by adipocytes of different sizes, *Int. J. Obes.* **5**:671–677.

27. R. G. Vernon, 1980, Lipid metabolism in the adipose tissue of ruminant animals, *Prog. Lipid Res.* **19**:23–106.
28. S. Smith, 1975, Lipogenesis in rabbit adipose tissue. *J. Lipid Res.* **16**:324–331.
29. A. A. Francendese and M. DiGirolamo, 1981, Alternative substrates for triacylglycerol synthesis in isolated adipocytes of different size from the rat, *Biochem. J.* **194**:377–384.
30. J. Gliemann, 1968, Glucose metabolism and response to insulin of isolated fat cells and epididymal fat pads, *Acta Physiol. Scand.* **72**:481–491.
31. R. L. Liebel, J. Hirsch, E. M. Berry, and R. K. Gruen, 1984, Radioisotopic methods for the measurement of lipolysis in small samples of human adipose tissue, *J. Lipid Res.* **25**:49–57.
32. O. Pederson, E. Hjollund, H. Beck-Nielson, H. Linskov, O. Sonne and J. Gliemann, 1981, Insulin receptor binding and receptor-mediated insulin degradation in human adipocytes, *Diabetologia* **20**:636–641.
33. M. Rodbell, 1967, Metabolism of isolated fat cells. V. Preparation of "ghosts" and their properties: Adenyl cyclase and other enzymes, *J. Biol. Chem.* **242**:5744–5750.
34. L. Jarett, 1974, Subcellular fractionation of adipocytes, *Methods Enzymol.* **31**:60–71.
35. R. H. Ho and H. C. Meng, 1964, A technique for cannulation and perfusion of isolated rat epididymal fat pad, *J. Lipid Res.* **5**:203–209.
36. R. O. Scow, 1977, Metabolism of chylomicrons in perfused adipose and mammary tissue in the rat, *Fed. Proc.* **36**:182–185.
37. S. Rosell, 1966, Release of free fatty acids from subcutaneous adipose tissue in dogs following sympathetic nerve stimulation, *Acta Physiol. Scand.* **67**:343–351.
38. M. DiGirolamo, W. S. Skinner, Jr., H. G. Hanley, and R. G. Sachs, 1971, Relation of adipose tissue blood flow to fat cell size and number, *Am. J. Physiol.* **220**:932–937.
39. M. DiGirolamo and J. Esposito, 1975, Adipose tissue blood flow and cellularity in the growing rabbit, *American J. Physiol.* **229**:107–112.
40. S. S. Solomon and W. C. Duckworth, Effect of antecedent hormone administration on lipolysis in the perifused isolated fat cell, *J. Lab. Clin. Med.* **88**:984–994.
41. D. O. Allen, K. J. Long, and J. T. Majors, 1979, Characterization of the multiple chamber perifusion fat cell system, *J. Lipid Res.* **20**:1036–1040.
42. L. Sjostrom, U. Smith, P. Bjorntorp, B. Jacobson, and P. Hallgren, 1977, Human adipose tissue maintained in a continuous flow system, *J. Biol. Chem.* **252**:8833–8839.
43. S. S. Solomon and W. C. Duckworth, 1980, Comparative studies of the antilipolytic effect of insulin and adenosine in the perifused isolated fat cells, *Horm. Metab. Res.* **12**:601–604.
44. R. D. Harper, 1985, A modified perifused incubation system for isolated fat cells, *Biochem J.* **224**:235–239.
45. S. K. Fried, 1985, A modified perifusion system to study lipoprotein lipase secretion from isolated rat fat cells, *Int. J. Obes.* **9**(4), Abstract A36.
46. C. Vannier, E. T. Amri, J. Etienne, R. Negrel, and G. Ailhaud, 1985, Maturation and secretion of lipoprotein lipase in cultured adipocytes, *J. Biol. Chem.* **260**:4424–4431.
47. R. Liebel and J. Hirsch, 1985, A radioisotopic technique for analysis of fatty acid reesterification in human adipose tissue, *Am. J. Physiol.* **248**:E140–E147.
48. M. DiGirolamo, S. Mendlinger, and J. W. Fertig, 1971, A simple method to determine fat cell size and number in four mammalian species, *Am. J. Physiol.* **221**:850–858.
49. A. Kashiwagi, M. A. Verso, J. Andrews, B. Vasquez, G. Reaven, and J. Foley, 1983, In vitro insulin resistance of human adipocytes isolated from subjects with non-insulin dependent diabetes mellitus, *J. Clin. Invest.* **76**:1246–1254.
50. J. Gliemann, K. Osterlind, J. Vinten, and S. Gammeltoft, 1972, A procedure for the measurement of distribution spaces in isolated adipocytes, *Biochem. Biophys. Acta* **286**: 1-9.

51. M. DiGirolamo and J. Owens, 1976, Water content of rat adipose tissue and isolated adipocytes in relation to cell size, *Am. J. Physiol.* **231**:1568–1572.
52. J. Gliemann and O. Sonne, 1978, Binding and receptor-mediated degradation of insulin in adipocytes, *J. Biol. Chem.* **253**:7857–7863.
53. R. R. Whitesell and J. Gliemann, 1979, Kinetic parameters of transport of 3-0-methyl glucose and glucose in adipocytes, *J. Biol. Chem.* **254**:5276–5283.
54. P. Bjorntorp and L. Sjostrom, 1972, The composition and metabolism *in vitro* of adipose tissue fat cells of different sizes, *Eur. J. Clin. Invest.* **2**:78–84.
55. M. Rodbell, 1964, Localization of lipoprotein lipase in fat cells of rat adipose tissue, *J. Biol. Chem.* **239**:753–755.
56. C. H. Hollenberg and A. Vost, 1968, Regulation of DNA synthesis in fat cells and stromal elements from rat adipose tissue, *J. Clin. Invest.* **47**:2485–2498.
57. M. DiGirolamo, 1975, Obesity and adipose tissue blood flow, oxygen consumption and metabolic functions: A reevaluation based on morphological characteristics of the tissue, in *Recent Advances in Obesity Research: 1. Proceedings of the First International Congress on Obesity,* A. Howard, ed., Newman Publishing Ltd., pp. 93–97.
58. F. D. Demartinis and A. Francendese, 1982, Very small fat cell populations: Mammalian occurrence and effect of age, *J. Lipid Res.* **23**:1107–1120.
59. D. Lemonnier, 1972, Effect of age, sex, and site on the cellularity of the adipose tissue in mice and rats rendered obese by a high-fat diet, *J. Clin. Invest.* **51**:2907–2915.
60. S. K. Fried, M. Lavau, and F. X. Pi-Sunyer, 1982, Variations in glucose metabolism from three adipose depots of the rat, *Metabolism* **31**:876–883.
61. A. D. Hartman and D. W. Christ, 1978, Effect of cell size, age, and anatomical location on lipolytic response of adipocytes, *Life Sci.* **22**:1087–1096.
62. M. Guerre-Millo, A. Leturque, J. Girard, and M. Lavau, 1985, Increased insulin sensitivity and responsiveness of glucose metabolism in adipocytes from female versus male rats, *J. Clin. Invest.* **76**:109–116.
63. J. E. Foley, A. L. Laursen, O. Sonne, and J. Gliemann, 1980, Insulin binding and hexose transport in rat adipocytes. Relation to cell size, *Diabetologia* **19**:234–241.
64. A. Cryer, 1981, Tissue lipoprotein lipase activity and its action in lipoprotein metabolism, *Int. J. Biochem.* **13**:525–541.
65. N. M. Rodbell and R. O. Scow, 1965, Chylomicron metabolism: Uptake by perfused adipose tissue, in *Handbook of Physiology.* Adipose Tissue, A. E. Renold and G. E. Cahill, Jr., eds., American Physiological Society, Washington, D.C., pp. 491–498.
66. K. Suzuki and T. Kono, 1980, Evidence that insulin causes translocation of glucose transport activity to the plasma membrane from an intracellular storage site, *Proc. Natl. Acad. Sci. (USA)* **77**:2542–2545.
67. S. W. Cushman and L. J. Wardzala, 1980, Potential mechanism of insulin action on glucose transport in the isolated adipose cell. Apparent translocation of intracellular transport systems to the plasma membrane, *J. Biol. Chem.* **255**:4758–4762.
68. P. J. Hissin, E. Karnieli, I. A. Simpson et al., 1982, A possible mechanism of insulin resistance in the rat adipose cell with high fat/low-carbohydrate feeding, *Diabetes* **31**:589–592.
69. C. Crandall, S. K. Fried, A. A. Francendese, M. Nickel, and M. DiGirolamo, 1983, Lactate release from isolated rat adipocytes: Influence of cell size, glucose concentration, insulin and epinephrine, *Horm. Metab. Res.* **15**:326–329.
70. M. P. Czech, 1976, Cellular basis of insulin insensitivity in large rat adipocytes, *J. Clin. Invest.* **57**:1523–1532.
71. P. Iverius and F. D. Brunzell, 1985, Human adipose tissue lipoprotein lipase: Changes with feeding and relation to postheparin plasma enzyme, *Amer. J. Physiol.* **249**:E107–E114.

72. P. Belfrage, G. Fredirikson, L. Olsson, and P. Stralfors, 1983, Control of adipose tissue lipolysis by phosphorylation/dephosphorylation of hormone-sensitive lipase, in *The Adipocyte and Obesity: Cellular and Molecular Mechanisms,* A. Angel, C. Hollenberg, and D. Roncari, eds., Raven Press, New York, pp. 217–224.
73. O. Wieland, 1983, Glycerol, in *Methods of Enzymatic Analysis,* 3rd ed., vol. 6, Hans Ulrich Bergmeyer, Jurgen Bergmeyer, and Marianne Grassl, eds., Verlag Chemie, Weinheim, Germany, and Academic Press, Deerfield Beach, Florida, pp. 504–510.
74. S. S. Chernick, 1969, Determination of Glycerol in Acyl Glycerols, *Methods Enzymol.* **14**:627–630.
75. H. Kather and Wieland, Luminometric Method, in *Methods of Enzymatic Analysis,* 3rd ed., vol 6, Hans Ulrich Bergmeyer, Jurgen Bergmeyer, and Marianne Grassl, eds., Verlag Chemie Publishers, Weinheim, Germany, and Academic Press, Deerfield Beach, Florida, pp. 510–518.
76. H. Kather, E. Wieland, B. Fischer, A. Wirth, and B. Schlierf, 1985, Adrenergic regulation of lipolysis in abdominal adipocytes of obese subjects during caloric restriction and reversal of catecholamine action by adenosine deaminase, *Eur. J. Clin. Invest.* **15**:30–37.
77. B. B. Hoffmann, H. Chang, Z. Frahbakrish, and G. Reaven, 1984, Inhibition of lipolysis by adenosine is potentiated by age, *J. Clin. Invest.* **74**:1750–1755.
78. P. Engfeldt, P. Arner, and J. Ostman, 1980, Influence of adipocyte isolation by collagenase on phosphodiesterase activity and lipolysis in man, *J. Lipid Res.* **21**:443–448.
79. P. Hallgren, E. Raddatz, C-H. Bergh, Pl. Kucera, and L. Sjöstrom, 1984, Oxygen consumption in collagenase-liberated adipocytes in relation to cell size and age, *Metabolism* **33**:897–900.
80. M. Monti, P. Nilsson-Ehle, R. Sorbris, et al, 1980, Microcalorimetric measurement of heat production in isolated human adipocytes, *Scand. J. Clin. Invest.* **40**:581–587.
81. J. Gliemann, O. Sonne, S. Linde, and B. Hansen, 1979, Biological potency and binding affinity of monoiodoinsulin with iodine in A14 or A19, *J. Biol Chem.* **253**:7853–7857.
82. M. DiGirolamo, 1983, Effects of age and nutrition on adipose tissue growth, metabolism and responsiveness to hormones, in *Controversies in Obesity,* B. Hansen, ed., Endocrinology and Metabolism Series, Praeger Publishers, New York, **5**:91–121.
83. S. W. Cushman, D. Noda and L. B. Salans, 1981, Adipose cell size-function relationships: Insulin binding and degradation, *Am. J. Physiol.* **240**:E166–E174.
84. J. M. Olefsky, 1976, The effects of spontaneous obesity on insulin binding, glucose transport,and glucose oxidation of isolated rat adipocytes, *J. Clin. Invest.* **57**:842–851.
85. K. Fagin, S. L. Lackey, C. R. Reagan, and M. DiGirolamo, 1980, Specific binding of growth hormone by rat adipocytes, *Endocrinology* **107**:608–615.
86. J. N. Livingston, P. Cuatrecoses, and D. H. Lockwood, 1974, Studies of glucagon resistance in large rat adipocytes: ^{125}I-labeled glucagon binding and lipolytic capacity, *J. Lipid Res.* **15**:26–32.
87. R. J. Cabelli and C. C. Malbon, 1979, Characterization of (-)-H-dihydroprenolol binding sites on isolated rat fat cells, *J. Biol. Chem.* **254**:8903–8908.
88. M. Lafontan, M. Berlan, and A. Villeneuve, 1983, Preponderance of α-2 over B1-adrenergic receptor sites in human fat cells is not predictive of the lipolytic effect of physiological catecholamines, *J. Lipid Res.* **24**:429–440.
89. J. Olefsky, 1977, Mechanisms of decreased insulin responsiveness of large adipocytes, *Endocrinology* **100**:1169–1177.
90. J. E. Foley, R. Foley, and J. Gliemann, 1980, Glucose-induced acceleration of deoxyglucose transport in rat adipocytes, *J. Biol. Chem.* **255**:9674–9677.
91. J. E. Foley, S. W. Cushman, and L. B. Salans, 1978, Glucose transport in isolated rat adipocytes with measurements of L-arabinose uptake, *Am. J. Physiol.* **234**:E112–E119.

92. J. Katz, B. R. Landau, and B. E. Bartsch, 1966, The pentose cycle, triose phosphate isomerization and lipogenesis in rat adipose tissue, *J. Biol. Chem.* **241**:727–740.
93. S. R. Sooranna and E. D. Saggarson, 1975, Studies on the role of insulin in the regulation of glyceride synthesis in rat epididymal adipose tissue, *Biochem. J.* **150**:441–451.
94. J. Fain, and L. Rosenberg, 1972, Antilipolytic action of insulin on fat cells, *Diabetes* **21**:414–425.
95. D. Rudman, M. DiGirolamo, M. F. Malkin, and L. A. Garcia, 1965, The adipokinetic property of hypophyseal peptides and catecholamines: A problem in comparative endocrinology, *Handbook of Physiology,* Section 5, Adipose Tissue, A. E. Renold, and G. F. Cahill, Jr., eds., American Physiological Society, Washington, D.C., pp. 533–540.
96. H. Kather, 1981, Hormonal regulation of adipose tissue lipolysis in man: Implications for the pathogenesis of obesity, *Triangle* **20**:131–143.
97. J. N. Fain, and J. A. Garcia-Sainz, 1983, Adrenergic regulation of adipocyte metabolism, *J. Lipid Res.* **24**:945–966.
98. E. D. Saggerson, 1972, The regulation of glyceride synthesis in the isolated white fat cells. The effects of palmitate and lipolytic agents, *Biochem. J.* **128**:1057–1067.
99. J. N. Fain and Weisner, 1975, Effects of adenosine deaminase on cyclic adenosine monophosphate accumulation, lipolysis and glucose metabolism of fat cells, *J. Biol. Chem.* **250**:1027–1034.
100. S. W. Cushman, and M. Rizak, 1970, Structure-function relationships in the adipose cell. III. Effects of bovine serum albumin on the metabolism of glucose and the release of nonesterified fatty acids and glycerol by the isolated adipose cell, *J. Cell. Biol.* **46**:354–361.
101. H. Kather, M. Rivera, and K. Brand, 1972, Interrelationship and control of glucose metabolism and lipogenesis in isolated fat cells, *Biochem. J.* **128**:1097–1102.
102. A. Kashiwagi, M. A. Verso, J. Andrews, et al., 1983, *In vitro* insulin resistance of human adipocytes isolated from subjects with noninsulin-dependent diabetes mellitus, *J. Clin. Invest.* **72**:1246–1254.
103. R. B. Goldrick, Ashley, B. C. E., and M. L. Lloyd, 1969, Effects of prolonged incubation and cell concentration on lipogenesis from glucose in isolated omental fat cells, *J. Lipid Res.* **10**:253–259.
104. P. R. Johnson and J. Hirsch, 1972, Cellularity of adipose depots in six strains of genetically obese mice, *J. Lipid Res.* **13**:2–11.
105. I. Faust, P. R. Johnson, J. S. Stern, and J. Hirsch, 1978, Diet-induced adipocyte number increase in adult rats: A new model of obesity, *Am. J. Physiol.* **235**:E279–E285.
106. U. Smith, 1983, Regional differences and effect of cell size on lipolysis in human adipocytes, in *The Adipocyte and Obesity: Cellular and Molecular Mechanisms,* A. Angel, C. H. Hollenberg, and D. A. K. Roncari, eds., Raven Press, New York, pp. 245–250.

6

Techniques for Measuring Hormone Binding to Adipocytes

Terry D. Etherton
The Pennsylvania State University

QUANTIFYING ADIPOCYTE HORMONE RECEPTOR AFFINITY AND NUMBER

It is now accepted that the first step in protein hormone action at the target cell is recognition of the hormone by specific receptors in the plasma membrane (1, 2). This finding has led to numerous studies that have assessed the relationship between hormone binding (recognition) and biological action in a wide variety of cell types. Interest in this area has increased dramatically in the last 15 years in an attempt to determine the mechanisms by which protein hormones produce chemical signals that alter intracellular metabolism. The recognition that hormone binding is the first step of hormone action has ushered in a new scientific era in endocrinology. This era has led to the concept that hormone-resistant states are often associated with marked changes in hormone binding (3, 4). In animal agriculture, the concentration of many plasma hormones is not highly correlated with different growth rates and carcass composition (5, 6). This observation has resulted in studies addressing the issue that differences in cellular growth and metabolism may be associated with alterations in tissue sensitivity to circulating hormones. The ability to quantify hormone binding to adipocyte receptors has resulted in the discovery that binding of a hormone to its receptor can be regulated by the hormone that is homologous for the receptor and, also, other hormones that are not bound to that receptor (7–9).

Based on the above information, it is evident that tissue responsiveness to a hormone is a function of not only plasma concentration but also of binding to the target cell and subsequent generation of chemical signals that affect cellular metabolism. Because of these considerations, the ability to quantify hormone binding accurately and reliably represents an important research tool in many experiments examining the relationship between hormone binding and the regulation of intracellular metabolism.

Other important applications of receptor biology have been the development of radioreceptor assays (RRA), which have been used for measuring hormone concentrations in biological fluids and assessing the quality of iodinated hormones. With respect to this latter point, this is accomplished by determining the extent to which various fractions eluted from the column, used for purification of the tracer, bind to receptor preparations.

TECHNIQUES FOR MEASURING PROTEIN HORMONE BINDING

In any study assessing binding of a protein hormone to adipocytes, there are three major considerations: (a) the purity (and potency) of the hormone and the effects iodination has on the structure of the hormone, (b) the receptor preparation to use (i.e., isolated cells, crude membrane preparations, purified membrane preparations, or solubilized receptor preparations), and (c) the techniques employed to conduct the assay.

With respect to the hormone it is important that the hormone used for iodination be purified to homogeneity. This purification is easier with some hormones than it is with others. The problem of working with a protein containing other impurities is that one may be measuring binding of a mixture of proteins to a variety of different hormone receptors. After the hormone of interest is obtained, various schemes (depending on the hormone) are employed for solubilization of the hormone. The mode of solubilization is fundamentally important since it is imperative that solubilization of the hormone not affect its biological activity. After solubilization, the hormone may be aliquoted to containers for storage. The temperature, buffer, and containers used for storage need to be ascertained for the respective hormone. However, it is not difficult to determine this since numerous reports exist that enumerate the conditions used for storing hormones.

After the protein hormone is solubilized, some quantity of hormone needs to be radiolabeled for the binding assay. Typically, ^{125}I is used for labeling protein hormones. A wide variety of techniques exist for iodinating protein hormones by chloramine T (10), lactoperoxidase (11), and iodogen (12). The best way to establish the technique of choice is to determine what has been used previously by other investigators. If the protein has not been studied, then an evaluation of iodination techniques is mandatory. With respect to iodination, particular care should be used to not overiodinate the protein. With chloramine T this laboratory typically uses 5 μg of hormone and 2–3 μg of Chloramine T. The reaction is allowed to proceed for 20–30 sec, at which point it can be stopped by dilution and transfer to a column used for separation of bound from free ^{125}I. Alternatively, sodium metabisulfite can be added to the reaction mixture prior to chromatography. This reducing agent effectively stops the oxidization of iodide (^{125}I) to iodine by Chloramine T (13). Determining the extent to which iodination affects the biological activity of the hormone (discussed later) is particularly important.

Once the protein is iodinated, the iodinated hormone needs to be quickly separated from the free ^{125}I, aggregated hormone, and any fragments that may have been produced during iodination. Depending upon the hormone, this separation may be accomplished in a variety of ways (e.g., High Performance Liquid Chromotagraphy [HPLC], ion exchange chromatography, gel permeation chromatography, or affinity chromatography). The reader needs to evaluate which method is satisfactory for the hormone of interest. The importance of purifying the tracer can not be overemphasized. Many protein hormones aggregate to some extent during iodination. The aggregated labeled hormone (at least insulin and growth hormone) has not been shown to bind well to the receptor or antibodies. Following purification of the tracer, it is important to determine the biological and immunological "health" of the iodinated hormone. This determination is important simply because iodination of the hormone may alter its biological potency and ability to bind to the receptor. For example, monoiodinated insulin is composed of a mixture of isomers, since iodination can occur at either of the four tyrosyl amino acids located at positions A14, A19, B16, and B26 (14, 15). The A19 isomer has diminished binding and the B26 isomer has enhanced binding to target cells (14, 16).

To assess the effects of iodination on the biological potency of the hormone, several different evaluations can be made. The first evaluation is simply to determine the elution profile of the labeled hormone after gel chromatography. From this profile, it is possible to determine the extent to which the hormone has aggregated and/or fragmented. Furthermore, the elution position of the monomeric hormone can be estimated and, as a result, molecular weight. Selecting the fractions corresponding to the protein peak for subsequent evaluation reduces the problem of free ^{125}I or aggregated hormone.

Samples from the tubes corresponding to the monomeric iodinated hormone can be subjected to trichloroacetic acid (TCA) precipitation to estimate the integrity of the hormone. Routinely, in this laboratory a small (10 μl) aliquot of such a fraction is taken and added to 1 ml of 4°C Krebs-Ringer bicarbonate buffer that contains 3% bovine serum albumin. This step is followed by the addition of 1 ml of ice-cold 10% TCA. The tube is swirled and subjected to low-speed centrifugation. The supernatant is decanted and both the supernatant and the pellet are counted in a gamma counter. The proportion of radioactivity that is precipitated by TCA is considered to represent intact hormone. In actuality, this method is not accurate. After exposure of ^{125}I-insulin to pig adipocytes, the amount of intact insulin remaining has been found to differ considerably depending on whether the estimate of degradation was made with TCA or immunoprecipitation (17). This laboratory's estimates of hormone integrity by TCA precipitation always exceed the estimates obtained by the ability of the hormone to bind to an antibody.

Another way to estimate the immunological integrity of the labeled hormone is to determine the proportion of tracer that binds to excess first antibody. This determination is made simply by incubating trace quantities (0.1–0.2 ng) of labeled hormone with excess first antibody (sufficient antibody to bind all of the tracer). The first antibody-antigen complex can be precipitated by the second antibody, and the amount of tracer bound can be estimated by using standard approaches employed for radioimmunoassay. Theoretically, all of the tracer should bind to the antibody if the iodination has not affected the hormone. In practice, this seldom occurs. For example, iodination of insulin with chloramine T results in a tracer that is 93–95% precipitable by excess first antibody. With porcine and bovine growth hormone our experience is that after purification they are 92–98% precipitable in TCA, whereas about 80–85% binds to excess first antibody.

Another approach for evaluating iodinated hormone is to assess the ability of the tracer to bind to the receptor. Typically, crude membrane (microsome) preparations have been used to determine the relative capacity for binding. In this context, the use of a radioreceptor assay allows one to determine the fraction of iodinated hormone after purification that possesses the greatest ability to bind to the receptor.

The most direct way to establish whether iodination has affected the biological activity of the hormone is to compare the potency of the native and labeled hormone in a bioassay. For insulin, this can be done by assessing the ability of insulin to stimulate glucose metabolism in isolated adipocytes. In adipose tissue cultured for 48 hr growth hormone has been found to antagonize the ability of insulin to maintain lipogenic capacity in a dose-dependent manner. This laboratory has used this finding to compare the biological potency of ^{125}I-growth hormone and native hormone. No difference has been found in the biopotency of native versus iodinated porcine growth hormone when iodogen has been used to iodinate porcine growth hormone.

After the tracer has been established as suitable for receptor binding studies, it is imperative to accurately determine the mass of labeled hormone. This determination is particularly important, because a small mass of labeled hormone (0.1–0.6 ng/ml) is used in binding assays and when small concentrations of unlabeled competing hormone are present in the tube a considerable proportion of the total hormone is labeled. If the hormone mass is not accurately established, then the estimates of binding affinity and receptor number of the high-affinity receptor population are not going to be accurate. A convenient way to determine hormone mass is to use a self-displacement assay (18). In this assay, a Radioimmunoassay (RIA) is run conventionally. In addition, increasing amounts of radioactivity are added to several tubes. Initially, a mass is used that corresponds to the amount of radioactivity in the conventional RIA and then 2×, 4× and 6× the radioactivity is added to respective tubes. In this assay the total counts per minute (cpm) added to each tube increase and, as a result, the cpm bound to the antibody increase; however, the ratio of bound (B) to total (T) cpm in the tube decreases. This B/T ratio can then be read on the standard curve used for the conventional RIA where increasing amounts of cold rather than hot hormone were added.

ISOLATION OF ADIPOCYTES

After the quality and quantity of radiolabeled hormone have been established, it is necessary to validate the procedure for isolating adipocytes. Validation in this sense means that the adipocytes after isolation are at least as metabolically active as the cells in tissue slices that have not been exposed to the isolation procedure and that the isolated cells retain their sensitivity to the hormone of interest. Techniques for isolating rat (19), pig (20), and bovine (21) adipocytes that meet these criteria have been published. For adipocyte isolation this laboratory prefers that all beakers, flasks, tubes, and mesh be made of polypropylene. In addition, the source of collagenase must be carefully evaluated. Many studies have typically used collagenase from Worthington; however, it is important to screen the various lots to determine what works best. In addition, this laboratory has found Armour CRG-7 (bovine serum albumin (BSA)) to be superior to other brands. The importance of this choice is illustrated by the findings that pig adipose tissue is quite sensitive to insulin in vitro if Armour BSA is used, whereas little stimulation of glucose metabolism is seen with other preparations of BSA (22).

As has been noted above, the technique for adipocyte isolation has been discussed in detail. Recently, however, a considerable amount of insulin degrading activity has been found to be released into the buffer during the last step in the isolation of pig adipocytes (transfer of adipocytes from test tubes to a flask used for pooling adipocytes prior to in vitro assay). This finding needs to be evaluated in each laboratory prior to binding or metabolic studies. The above mentioned reference (17) describes how to eliminate the insulin-degrading activity from the media.

ADIPOCYTE BINDING PROTOCOL

In initial studies of hormone binding to adipocytes, several points need to be determined in order to validate the binding assay. These factors include the time of incubation, temperature, pH, cell concentration, buffer composition, specificity of binding, extent of hormone degradation, and reversibility of binding. In this laboratory, hormone binding studies with adipocytes are conducted in polypropylene test tubes with a total incubation volume of 1 ml. Before binding studies are conducted, it is important to understand nonspecific binding (NSB) of the hormone. This understanding is important because a hormone can bind specifically to receptors and nonspecifically to the cells (adsorbed to membrane or trapped in cell pellet after centrifugation). To determine NSB, a small amount of tracer (0.1–0.3 ng) is incubated in the presence of a vast excess of unlabeled hormone (about 5,000 to 20,000 times more unlabeled than labeled tracer). The amount of tracer bound to the cell pellet in the presence of the vast excess of cold hormone is hormone nonspecifically bound. This determination is based on the law of mass action, which indicates that the probability that the labeled hormone can bind to the receptor is negligible given that the tracer concentration is fleetingly small versus that of the cold hormone (i.e., all receptors are occupied by cold hormone). To calculate hormone specifically bound, one determines the cpm of hormone bound to cells for each respective hormone concentration (that is, tracer bound in the presence of zero to the quantity of hormone used for NSB) and subtracts from that value the cpm bound in the NSB tube. The resulting value is expressed as a proportion of the total cpm in the tube (which enables one to calculate the percent of hormone specifically bound).

To terminate the binding study 200 μl of cell suspension is removed (in triplicate) from the incubation tube and added to a microfuge tube that contains 100 μl of silicone oil. The tube is quickly spun in a microfuge for

30 sec and removed. Three layers are observed in the tube: the adipocyte layer on top, the oil layer in the middle, and the buffer with free hormone in the bottom. The tube is cut through the oil layer, the resulting top of the tube is placed in another tube, and the tube is capped and counted in a gamma counter.

The first point to address is whether binding of tracer amounts of hormone reaches equilibrium (or steady state). This is important in order to do accurate Scatchard analyses of binding data. Binding must be measured at equilibrium. Figure 6-1 illustrates the time necessary for insulin binding to reach equilibrium for pig adipocytes at 30°C. Data presented in Figure 6-2 are necessary to document that binding changes as a linear function of cell number. This is critical since it is impossible to know the number of cells being added to the incubation before the binding study is performed (see Ref. 20 and 21 for details of how to fix and count isolated adipocytes). Thus, data obtained must be standardized to some reference number of cells per milliliter. It is important to not presume that the number of cells in a

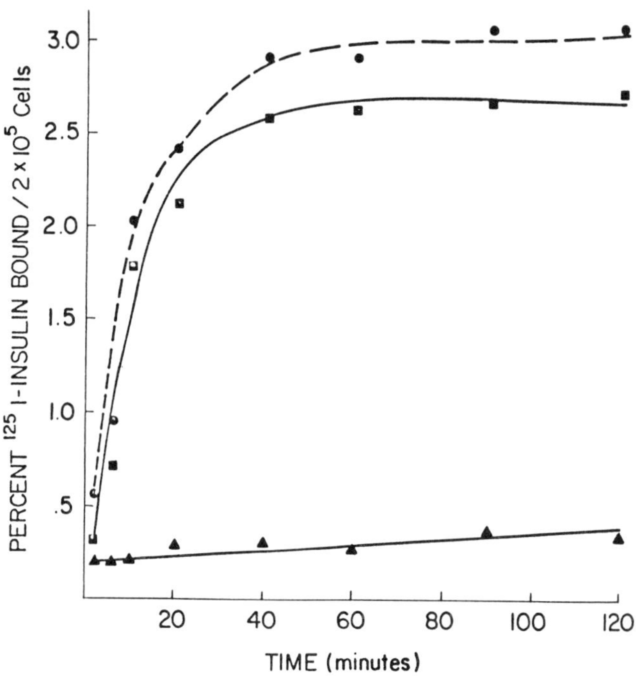

Figure 6-1. A typical curve demonstrating the time necessary to achieve equilibrium. ^{125}I-insulin was incubated with porcine adipocytes at 30°C. Total binding (●—●) and specific binding (■—■) increase and then plateau. Nonspecific binding (▲—▲) is 10% of specific binding. (*Adapted from Etherton and Walker, 1982, pp. 1720–1724.*)

250 μl aliquot of adipocytes is one-half that in a 500 μl aliquot. For studies determining the relationship between binding and cell number, 100 to 500 μl of cells are added and an aliquot of cells is always counted for each respective volume. The number of cells in an aliquot has been found to increase as the volume of the aliquot increases; however, the change in cell number is not proportional to the change in volume.

The pH to use and the buffer composition have been described for a variety of hormones and will not be discussed here. Many different temperatures have been used to study hormone binding to adipocytes. For the most part (at least for insulin), most studies have used temperatures below 37°C. As temperature is decreased the rates of hormone internalization and degradation are decreased. At 37°C it is essential to determine what proportion of radioactivity "bound" to the cell is intact hormone versus larger molecular weight species or smaller fragments.

When the above parameters have been established the specificity of the binding sites must be determined. Specificity is determined by incubating the radiolabeled hormone with increasing concentrations of unlabeled hormones. It can be seen in Figure 6-3 that increasing the concentration of insulin results in a decrease in ^{125}I-insulin binding. In contrast, incubation of pig adipocytes with proinsulin results in a displacement that is shifted markedly to the right. This shift indicates that proinsulin can bind to the

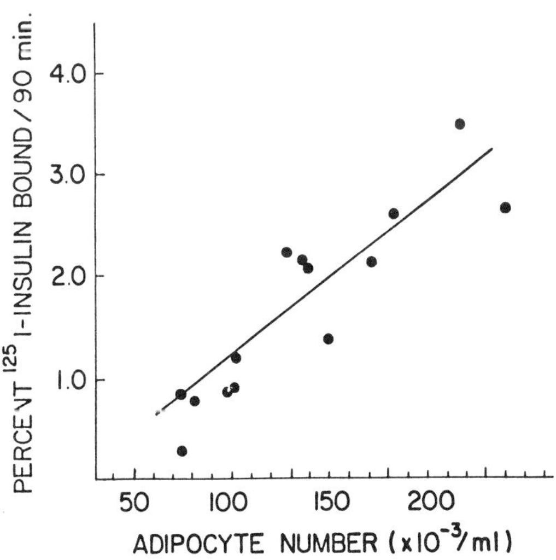

Figure 6-2. The relationship between increasing adipocyte number and insulin binding. *(Adapted from Etherton and Walker, 1982, pp. 1720–1724.)*

insulin receptor; however, the concentration required to decrease tracer binding by 50% (ED_{50}) is much greater. Thus, the cross-reactivity (measured as ED_{50} for insulin/ED_{50} for proinsulin) of the insulin receptor for proinsulin is negligible. Glucagon does not affect insulin binding, which indicates that glucagon does not bind to the insulin receptor. To determine specificity the hormone of interest is iodinated and displacement curves are established for the hormone and other structurally similar hormones.

In order to determine whether binding is reversible (which is necessary in order to do a Scatchard plot accurately), dissociation studies should be done. These studies can be accomplished by allowing insulin to associate with adipocytes for various time periods and then removing the buffer by centrifugation and resuspending the cells in the same volume of insulin-free medium. At various times after the initiation of the dissociation study, cells are collected and radioactivity is determined to ascertain the rate at which the cpm of bound insulin decreases.

In order to determine whether differences in binding are due to a change in the number of receptors or in receptor affinity, the binding data are typically subjected to Scatchard analysis. A typical Scatchard plot for insulin binding to pig adipocytes is illustrated in Figure 6-4. In order to conduct Scatchard analysis of binding, all one needs to determine are the mass of bound hormone (B), mass of free hormone (F), and the ratio of B/F. The mass of bound hormone is simply calculated by multiplying the percentage

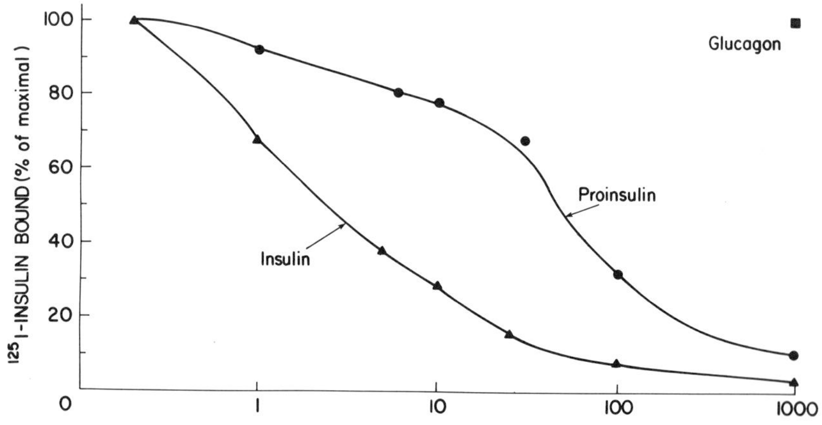

Figure 6-3. Specificity of insulin binding to porcine adipocytes. ^{125}I-insulin was incubated in the presence of various competing proteins. (*Adapted from Etherton and Walker, 1982, pp. 1720–1724.*)

of radioactive hormone bound (after correction to some standardized number of adipocytes; e.g., 2 x 10^5, 1 x 10^6, etc.) by the total mass of hormone in the tube (both hot and cold). Free hormone is simply calculated by the difference between the total and bound. A recent review for conducting Scatchard plots has been published (23). The analysis makes use of a computer program (LIGAND) that is now available for use on various microcomputers.

The previous discussion has given an overview of approaches for measuring protein hormone binding to adipocytes. It is particularly important to verify that the isolated adipocytes are viable and that iodination of the hormone does not affect its biological potency. With these considerations in mind and careful attention to the other points discussed in this chapter, one can conduct hormone binding studies accurately.

REFERENCES

1. J. R. Roth, 1973, Peptide hormone binding to receptors. A review of direct studies in vitro, *Metabolism* **22:**1059–1073.
2. P. Cuatrecasas, 1974, Membrane receptors, *Annu. Rev. Biochem.* **43:**169–214.

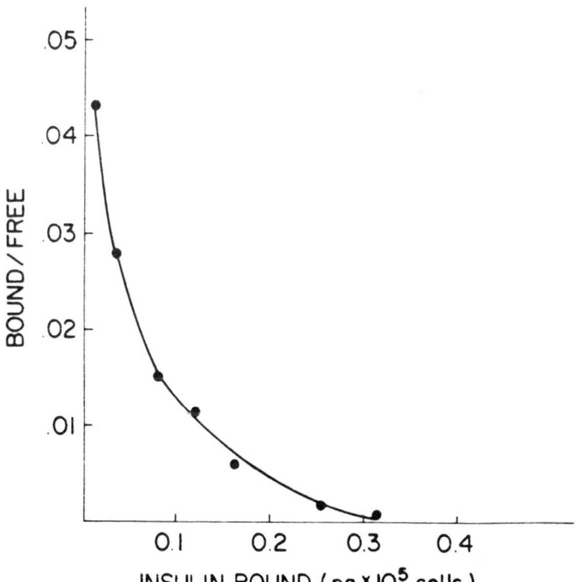

Figure 6-4. A Scatchard plot of insulin binding to porcine adipocytes. (*Adapted from Etherton and Walker, 1982, pp. 1720–1724.*)

3. J. R. Roth, M. Lesniak, R. Bar, M. Muggeo, K. Megyesi, L. Harrison, J. Flier, H. Wachslicht-Rodboard, and P. Gorden, 1979, An introduction to receptors and receptor disorders, *Proc. Soc. Exp. Biol. Med.* **162**:3–12-.
4. C. R. Kahn, 1979, The role of insulin receptor and receptor antibodies in states of altered insulin action, *Proc. Soc. Exp. Biol. Med.* **162**:13–21.
5. A. Trenkle and D. Topel, 1978, Relationships of some endocrine measurements to growth and carcass composition of cattle, *J. Anim. Sci.* **46**:1604–1609.
6. T. D. Etherton and R. S. Kensinger, 1984, Endocrine regulation of fetal and postnatal meat animal growth, *J. Anim. Sci.* **59**:511–528.
7. J. Roth, C. R. Kahn, M. Lesniak, P. Gorden, P. DeMeyts, K. Megyesi, D. Neville, Jr., J. Gavin III, A. H. Soll, P. Freychet, I. Goldfine, R. Bar, and J. Archer, 1975, Receptors for insulin, NSILA-S and growth hormone: Application to disease states in man, *Recent Prog. Horm. Res.* **31**:95–139.
8. R. S. Bar, L. Harrison, M. Muggeo, P. Gorden, C. R. Kahn, and J. Roth, Regulation of insulin receptors in normal and abnormal physiology in humans, *Adv. Intern. Med.* **24**:23–52.
9. J. E. Pessin, W. Gitomer, Y. Oka, C. Oppenheimer, and M. Czech, 1983, β-Adrenergic regulation of insulin and epidermal growth factor receptors in rat adipocytes, *J. Biol. Chem.* **258**:7386–7394.
10. F. C. Greenwood, W. M. Hunter, and J. S. Glover, 1963, The preparation of ^{131}I-labelled human growth hormone on high specific radioactivity, *Biochem. J.* **89**:114–123.
11. J. Thorell and B. Johansson, 1971, Enzymatic iodination of polypeptides with ^{125}I to high specific activity, *Biochim. Biophys. Acta.* **251**:363–369.
12. C. S. Chung, T. D. Etherton, and J. P. Wiggins, 1985, Stimulation of swine growth by porcine growth hormone, *J. Anim. Sci.* **60**:118–130.
13. S. Linde, B. Hansen, and A. Lernmark, 1983, Preparation of stable radioiodinated polypeptide hormones using polyacrylamide gel electrophoresis, *Methods Enzymol.* **92**:309–335.
14. S. Linde, B. Hansen, O. Sonne, J. Holst, and J. Gliemann, 1981, Tyrosine A14[^{125}I] monoiodoinsulin. Preparation, biologic properties, and long-term stability, *Diabetes* **30**: 1–8.
15. B. Frank, D. Peavy, C. Hooker, and W. C. Duckworth, 1983, Receptor binding properties of monoiodotyrosol insulin isomers purified by high performance lipid chromatography, *Diabetes* **32**:705–711.
16. D. Podlecki, B. Frank, M. Koa, H. Horikoshi, G. Freidenberg, S. Marshall, T. Ciaraldi, and J. M. Olefsky, 1983, Characterization of the receptor binding properties of monoiodoinsulin isomers and the identification of different insulin receptor specificities in hepatic and extrahepatic tissues, *Diabetes* **32**:697–704.
17. T. D. Etherton, C. S. Chung, and J. P. Wiggins, 1984, Receptor-dependent and independent degradation of insulin by isolated swine adipocytes, *J. Anim. Sci.* **59**:366–375.
18. V. R. Thurlow and H. J. Puxley, 1976, Preparation of ^{125}I-thyroxine and ^{125}I-triiodothyronine of high specific activity for the radioimmunoassay of serum total T3 and T4, *Ann. Clin. Biochem.* **13**:364–368.
19. M. Rodbell, 1964, Metabolism of isolated fat cells. I. Effects of hormones on glucose metabolism and lipolysis, *J. Biol. Chem.* **239**:375–380.
20. T. D. Etherton and C. S. Chung, 1981, Preparation, characterization, and insulin sensitivity of isolated swine adipocytes: Comparison with adipose tissue slices, *J. Lipid Res.* **22**:1053–1059.
21. R. Vasilatos, T. D. Etherton, and P. J. Wangsness, Preparation of isolated bovine adipocytes: Validation of use for studies characterizing insulin sensitivity and binding, *Endocrinology* **112**:1667–1673.

22. P. E. Walton, T. D. Etherton, and J. F. Rebhun, 1984, Stimulatory effects of insulin on glucose metabolism in swine adipose tissue, *J. Anim. Sci.* **59**(Suppl. 1):211.
23. P. J. Munson, 1983, LIGAND: A computerized analysis of ligand binding data, *Methods Enzymol.* **92**:543–576.
24. T. D. Etherton and O. A. Walker, 1982, Characterization of insulin binding to isolated swine adipocytes, *Endocrinology* **110**:1720–1724.

7

Primary Cell Culture of Adipose Tissue

Jan E. Novakofski
University of Illinois

A major question in the study of adipose tissue concerns the regulation of adipocyte hyperplasia and hypertrophy. Postnatal growth of somatic tissues, including adipose tissue, is primarily regulated by the endocrine system. However, it is apparent that in various tissues local interactions between cell types, or paracrine mechanisms, play a role in regulating and coordinating cell proliferation. Available evidence also suggests that local mechanisms regulate adipose tissue development. A primary piece of circumstantial in vivo evidence comes from the study of developing hyperplastic obesity. If adult animals are induced to deposit excessive fat, typically through palatable or high-fat diets, existing adipocytes enlarge. If excessive caloric intake continues, preadipocytes proliferate and differentiate into new fat cells. The newly formed adipocytes presumably provide increased lipid storage potential, reducing the need for enlargement of existing adipocytes. This process occurs in the absence of general somatic growth and is therefore unlikely to be regulated by endocrine-directed mechanisms, such as the somatomedins, that affect general somatic growth.

Questions about endocrine regulation may be approached with a variety of experimental techniques. Examining local regulation, on the other hand, has been difficult with in vivo techniques. In vitro culture has been a major tool for investigating questions about cell proliferation in many tissues, because culture techniques allow separation of different cell types as well as providing the control of cellular microenvironments necessary to study interactive mechanisms of regulation. However, culture of the cells present in adipose tissue presents several problems that until recently have prevented the wide use of tissue culture to investigate adipose growth and differentiation. Adipose tissue is composed of several cell types (see Chapter 1). To examine local interactions, culture systems for the different cellular components of adipose tissue must be available. These culture systems must maintain the cells in a physiological state similar to the in vivo. Mature adipocytes are differentiated cells; therefore, a culture system for studying adipocytes must maintain this differentiated state. Such a culture system must not only provide an adequate hormonal and nutritional environment, but must also keep the bouyant lipid-filled adipocytes from floating away. Culture systems for preadipocytes present additional problems. There is as yet no satisfactory technique for identifying preadipocytes before the appearance of adipocyte-specific function or for separating preadipocytes from other adipose stromal cells. Therefore, a culture system for preadipocytes must support eventual differentiation of the preadipocytes to demonstrate the actual identity of the cells.

This chapter describes techniques for the primary cell culture of three fractions of adipose tissue: (a) mature adipocytes; (b) the stromal cell fraction, which contains preadipocytes; and (c) adipose capillary endothelium. The use of stromal cultures to measure preadipocyte growth rate in mixed cell preparations is also described. Several important considerations affecting interpretation of culture data are discussed, and, finally, culture data suggesting interactions between adipocytes and other cells in adipose tissue are presented.

Although the purpose of this chapter is to describe primary culture techniques, it is important to recall that a number of adipose-like cell lines have been developed, several of which are widely used as models. Cell lines have the distinct advantage of being simpler to use than primary cultures. On the other hand, continuously cultured lines may not have exactly the same characteristics as the tissue they were derived from or the tissue they are modeling. Endless debate is possible about the relative merits of each type of culture as related to a specific experiment; however, few would debate that both types of culture have some uses. For the sake of comparison, a brief list of adiposelike cell lines is presented in Table 7-1.

MATERIALS AND METHODS

Mature Adipocyte Culture

Two strategies are commonly used for maintaining adipose cells in culture. The most widely used technique (24–43) is referred to as explant culture.

Table 7-1. Adiposelike Cell Lines[a]

Cell Line	*Source*	*References*
1243	Mouse teratocarcinoma	1, 2
3T3-L1	Swiss mouse	3–6
3T3-F442A		See also chap. 9
Balb/3T3-T	Balb/C mouse	7–9
MC3T3-G2/PA6	Mouse calvaria	10–12
MS2-2A	Mouse bone marrow	13
MS3-2A		
Ob17	Ob/ob mouse epididymal fat	14–17
REC-A16	Rat tibial rudiments (SV-40 transformed)	18
ST13	Mouse mammary carcinoma	19, 20

[a]Listed references are intended for comparison only. There are also cell lines which develop adipocyte-like characteristics when induced by serum or azacytidine (21–23).

Explants are small pieces or slices of adipose tissue placed in a culture flask along with medium. Because the adipose tissue pieces are bouyant, they are usually mechanically retained in the culture flask with coverslips or nylon screens. This approach is not a recent development, being first described by Foot in 1912 (24). In these early experiments, explants of subcutaneous and pericardial fat from chickens were cultured in clotted plasma. Plasma was prepared by quickly chilling and centrifuging fresh whole blood. Explants were placed in this plasma, which coagulated when warmed. The clot served as both medium and mechanical support. Because the clotted plasma medium could not be renewed, experiments were necessarily short-term, and Foot concluded that he was unsuccessful in culturing the adipose explants since he was unable to observe ''considerable growth.'' He did, however, observe flattened, elongated cells spreading from the explants and many of these cells contained numerous lipid droplets. In retrospect, it is apparent from Foot's drawings and photographs that the cells growing from the explants were stromal cells and preadipocytes; therefore, Foot had successfully, if only briefly, maintained adipose explants in vitro.

The advent of renewable liquid culture media—first serum (25), then synthetic media (26, 44), and finally, the contemporary mixture of serum and synthetic media (30)—saw increasingly successful cultures. Adipose explants have maintained a differentiated lipid-filled morphology for as long as 30 weeks (31). Various appropriate hormone effects have also been reported (Table 7-2), demonstrating that explants are viable and physiologically responsive.

Explant techniques have several advantages. Explants have been extensively characterized; furthermore, preparation is simple and may be done

Table 7-2. Response of Adipose Explants to Hormones

Compound	*Effects*	*References*
Insulin	↑ Cell diameter	71
	↑ Lipogenesis	35, 42, 46, 47
	↑ Lipolysis; chronic exposure	34, 35
	↓ Lipolysis; acute exposure	36
	Maintains glucose oxidation	47
Norepinephrine	↑ Lipolysis; acute exposure	37, 39
	↓ Acute lipolysis with chronic exposure	37
	↓ Cyclic AMP (cAMP) with long-term exposure	37
	↓ Cell size	28
Dexamethasone	↑ Deoxyribonucleic acid (DNA) content	29,30
Hydrocortisone	↓ Lipogenesis	39
	↓ Effects of insulin	39
	↑ DNA content	29, 30
Growth hormone	↓ Lipogenesis	38
	↓ Cell diameter	28
ACTH	↓ Cell diameter	28
Thyroid stimulating hormone	↓ Lipogenesis	46
β-Lipotropin	↓ Cell diameter	45
Dibutyryl cyclic-AMP	↑ Lipolysis	37

quickly. As long as the explants are small enough to allow adequate diffusion on media, explants can be maintained for long periods while appropriate morphology and responsiveness to hormones in vitro is maintained. (The distinction between culture and in vitro incubation is arbitrary at best.)

Unfortunately, for the purposes of examining cell interactions, explant cultures have several of the drawbacks of in vivo preparations. Adipose explants contain several cell types besides adipocytes, including preadipocytes, endothelial cells, fibroblasts, and smooth muscle cells. These stromal cells may have functions such as the aromatization of androstenedione by adipose fibroblasts (48), which might interfere with assessment of interactions. The second strategy for adipocyte culture, the culture of isolated cells, is intended to overcome these problems. These cultures are prepared by isolating adipocytes through collagenase digestion. The isolated cells, which are also buoyant, must be mechanically retained in a culture flask as for explants. This retention is typically achieved by incubating the cells under or between coverslips (49–53). Smaller adipocytes will float and may attach to the bottom of the coverslip, while larger cells are presumably held between the coverslip and the lower surface. Considering the fragility of mature adipocytes, it seems likely that larger cells are disrupted by the weight of the coverslip and the remaining lipid is merely a physical spacer. Also, given the minimal contact area of a spherical adipocyte as well as the physical obstacle of a large lipid droplet that would restrict the cell spreading necessary for attachment, it is surprising that any adipocytes attach to the culture plate or coverslip. Mature adipocytes may also be cultured with standard cell culture technique (54–60). To use this technique, isolated cells are plated with a minimal amount of medium so that even floating cells would presumably be in contact with the culture substrate. Although this technique may appear less satisfactory than the coverslip approach since very few floating adipocytes actually attach to the substrate (56), it is unlikely there is actually much difference.

In contrast to the situation with explants, isolated adipocytes cultured with these techniques generally lose the morphological characteristics of mature adipocytes. Dividing fibroblastic cells rapidly appear in the cultures; this process is referred to as dedifferentiation (49–58, 60).

Adipocyte Dedifferentiation

The apparent dedifferentiation of mature adipocytes raises a major question. If adipocytes may resume proliferation, then we must reconsider the view that adipocytes are in a state of terminal differentiation (61). Alternatively, there is the possibility that the dedifferentiation phenomenon is an artifact of the culture technique rather than a result of cell culture.

Evidence suggesting that adipocytes may retain a mitotic capability despite having differentiated morphology comes from experiments with the 3T3 preadipocyte-like cell line. Scott and his co-workers (8, 9) propose a model in which differentiation is a multistep process. In an early step, preadipocytes become nonterminally differentiated. In other words, these cells begin to acquire morphological and biochemical characteristics of mature adipocytes, but may lose these characteristics and resume proliferation. Terminal differentiation reflects a loss of the capacity to resume replication, not an acquisition of specific morphology. Adipocyte dedifferentiation in primary culture could then result from isolation of adipocytes that were nonterminally differentiated in vivo and reverted to a proliferative state in response to the culture conditions.

Adipose dedifferentiation occurs in media similar to ones adequate for maintaining the differentiated state. Furthermore, these media are similar to those supporting differentiation of preadipocytes, creating more of a paradox.

Preadipocyte Culture

Historically, the development of culture techniques for adipocyte precursors or preadipocytes has paralleled that for adipose explants since the preadipocytes were a component of the explants. However, specific attempts to culture precursor cells first occurred much later than explant culture. Sidman (44) used explant techniques to culture intrascapular mesenchyme from 15-day-old rat embryos. At this stage, intrascapular tissue is morphologically undifferentiated, but in culture it developed into recognizable adipose tissue. Although other workers had previously observed the development of lipid droplet-containing cells in explant cultures of adipose tissue, this was the first instance in which these cells developed from as yet undifferentiated tissue, providing support for the theory that adipocytes were derived from unique precursor cells and were not merely modified connective tissue cells.

Culture of isolated preadipocytes was first described by Ng and co-workers (49). Stromal cells were isolated by collagenase digestion and cultured by using standard technique for isolated cells. The stromal cells proliferated and some of them eventually accumulated lipid, assuming a morphology resembling that of mature adipocytes. No special treatments or conditions were necessary for this morphological differentiation, and preadipocytes have subsequently been shown to accumulate lipid sufficient for morphological differentiation in various media. Several workers have used lipid enrichment (62, 63) of the media to enhance morphological development, but this may be inadvisable because cell types present in preadipocyte-stromal cultures may fill with lipid under certain culture conditions. Fibroblasts (64) and myoblasts (65, 66) may be induced to accumulate sufficient lipid to morphologically resemble adipocytes. Therefore, demonstrating that cultures of adipose tissue stromal cells contain preadipocytes with the potential to differentiate into adipocytes requires additional evidence besides morphology.

Several enzymes that collectively are fairly indicative of adipose tissue, or at least of lipid storage function (Table 7-3), are elevated in preadipocyte/stromal cell cultures. These enzymes provide evidence that the lipid-filled cells are probably preadipocytes.

Although the enzymatic characteristics of preadipocyte/stromal cell cultures indicate the presence of preadipocytes, several problems still remain with interpreting experiments with these preparations. The problems center on what parameters of preadipocyte growth (proliferation, differentiation, or hypertrophy) are of interest.

First of all, despite a consensus that preadipocytes are developing in preadipocyte/stromal cell cultures, it is clear that other cell types are present as well.

Table 7-3. Characteristics of Preadipocyte/Stromal Cultures Following Differentiation or Hormonal Stimulation

Stimulus	*Effects*	*Reference*
Differentiation (or comparison with fibroblast cultures	↑ Lipid content	50, 62, 68
	↑ Glucose incorporation into lipid	51, 56
	↑ Lipoprotein lipase	53, 59
	↑ Fatty acid synthetase	53
	↑ Triglyceride synthetase	53
	↑ Glycerole-3-phosphate dehydrogenase	67,69
Insulin	↑ Lipid content	68, 69
	↑ Glycerol-3-phosphate dehydrogenase	69
	↑ Phosphotidate phospho-hydrolase	70
	↑ Diacylglycerol acyltransferase	70
	↑ Fatty acid CoA lipase	67
17β Estradiol	↑ Cell proliferation	71
Dexamethasone	↑ Lipid content	70
	↑ Glycerol-3-phosphate dehydro-genase	70
Prostaglandin E_1	↑ Number of lipid-filled cells	63
	↑ Triglyceride/cell	69
	↑ Glycerol-3-phosphate dehydrogenase	69
	↑ AMP	56
Very low density lipoproteins (VLDL)	↑ Lipid content	67, 68, 72, 73
	↑ VLDL hydrolysis	74
	↑ Number of lipid-filled cells	68
Fatty acid (linoleic) (oleic)	↑ Lipid content	62
	↑ Glycerol-3-phosphate dehydrogenase	75

Because of the mixed cell types, certain choices must be made when growth parameters are measured. The entire culture may be examined if the measurements being made are indicative of adipocytes or preadipocytes. If a growth parameter is being measured, one must assume that all of the cells are responding in an identical manner. This second assumption is particularly tenuous when one is interested in adipocyte-specific agents. Data refuting this assumption are presented later in Table 7-5. One can attempt to isolate preadipocytes from other cells, either prior to or following treatments. Attempts to purify stromal cells by taking advantage of differences in density have been reasonably successful (63, 67). However, since the variations in density observed are probably the result of lipid filling, separation is based on degree of differentiation, which may or may not alter growth characteristics or physiological responsiveness. As long as the objective of an experiment is not affected by this possibility, purification of preadipocytes based on density is an excellent tool. An alternate approach is to perform experiments with the intent of making measurements on preadipocytes separated after differentiation.

Culture of Adipose Capillary Endothelium

Several reports have described primary culture of capillary endothelium from adipose tissue (76–80). Previous strategies have relied on either separation of isolated endothelial cells from other stromal cells based on density differences (77) or on a digestion process that leaves intact capillary fragments that may be collected on screening (79).

The density technique yields relatively pure endothelial preparations although the cell yield is very low. The technique of collecting capillary fragments on a nylon screen after digestion has a much higher yield of cells. Unfortunately, with this technique, variations in digestion time or collagenase batch may result in substantial contamination of the cultures with vascular smooth muscle cells and fibroblasts. Techniques for culturing endothelial cells from various tissues have recently been reviewed in detail (81).

MATERIALS AND METHODS

The three types of primary adipose-derived cell cultures described may all be prepared from a single digestion as indicated by the scheme in Figure 7-1. In the text, a digestion procedure is described followed by special techniques used for each preparation.

Cell and Culture Techniques

Cells for culturing are obtained from inguinal adipose tissue of 21- to 28-day-old weanling Holtzman or Sprague-Dawley rats. Animals are acclimated in the laboratory for at least 48 hours before use and are given free access to food and water. Rats are anesthetized with sodium pentobarbital (50 mg/kg intraperitoneally (IP)) or ketamine plus acepromazine (100 mg/kg plus 1 mg/kg) and disinfected with iodine and alcohol. Inguinal or epididymal adipose tissue is removed from two to four rats by sterile dissection and pooled to obtain about 2 g of tissue. This tissue is then finely minced with scissors and digested in 20 ml of collagenase (1 mg/ml, Sigma type II), hyaluronidase (0.5 mg/ml, Sigma), and elastase (0.3 mg/ml, Sigma type I) in a digestion buffer containing 0.1 M N-2-Hydroxyethylpiperazine-N[1]-2-hydroxypropanesulfonic acid (HEPES), 0.12 M sodium chloride (NaCl), 0.05 M potassium chloride (KCl), 5 mM glucose, 1 mM calcium chloride ($CaCl_2$), and 1.5% bovine serum albumin described by Björntörp and colleagues (60).

It is important to be aware that obtaining a "suitable" batch of collagenase is not an objective procedure. Lot to lot variations and differences among manufacturers are extensive and difficult to quantify. Selection of

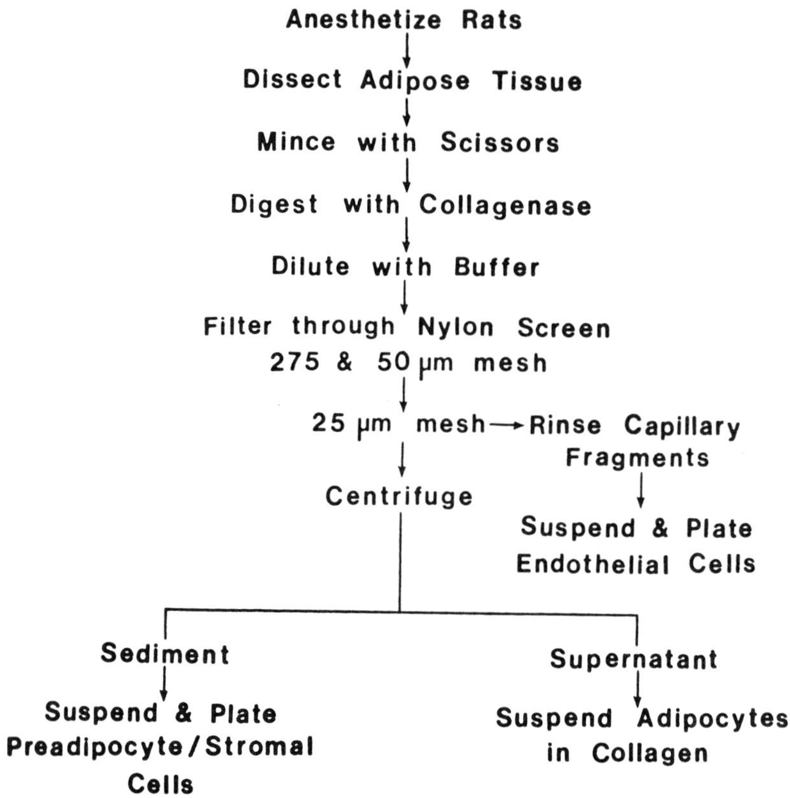

Figure 7-1. Protocol for culture of adipose tissue cells.

a particular batch that yields the most viable and cleanest preparation must be based on trial use. Different batches may produce "good" adipocytes but not endothelial cell preparations and so forth. Part of the problem arises from the fact that collagenase preparations are crude and contain a number of enzymes (82). Purified collagenase alone does not work particularly well.

Digestion is most satisfactory in 50-ml polypropylene Erlenmeyer (Nalgene) flasks, although siliconized trypsinizing flasks are also suitable. After a 45-min incubation in a shaking water bath at 37°C, 20 ml of buffer at room temperature without enzymes is added and the flask contents are filtered through nylon screens (Tetko) with nominal pore sizes of 275, 50, and 25 μm to removed undigested debris, cell aggregates, and capillary fragments.

The small final screen was used to obtain a cell suspension that contained a minimum of adipocytes with attached stromal cells. Although the 25-μm mesh screen undoubtedly disrupted larger adipocytes, cells in excess of 60 μm in diameter were frequently observed in the filtrate (75). The passage of adipocytes larger than the mesh size through the screen was probably facilitated by the hypertonicity (≈ 450 mOsm/liter) of the buffer.

The screening procedure is most conveniently done by using die-cut screens in Millipore filter holders (SXOO 025 00) and a 50-ml syringe as a funnel. The filtered cell suspension is then centrifuged for 10 min at 100 *g* to separate the cells into a floating layer of adipocytes and a pellet of stromal cells. The fractions are then separated and washed twice with complete culture media. In these washes, stromal cells may be centrifuged at 500 *g* while the adipocytes are allowed to float at unit gravity. Aliquots of the adipocyte fraction are then removed and stained with Rappaport's stain (0.1% crystal violet in 0.1 M citric acid) and counted on a hemocytometer. This stain is used in preference to trypan blue because its intense nuclear staining facilitates counting of viable adipocytes.

The question of which is the most advantageous culture medium to use for adipose cell culture has not been well investigated. In general, primary cultures do best in chemically complex media and usually require a serum supplement. Medium 199 is the most commonly used basal medium for adipose tissue cells and a variety of sera have been used. Volumes have been written concerning the choice of antibiotics to use in culture media. The best choice is to use none, although this is impractical in primary cultures. Gentamicin sulfate was used in the experiments described here since it is fairly stable and has antimycoplasmic activity, a consideration since mycoplasmic infections are common in rat colonies. Second-generation, β-lactamase-resistant antibiotics such as the Cephapirin used here are more stable, are less toxic, and have a broader spectrum of activity than the traditional penicillin. Medium 199 with Earles salts (Gibco, Grand Island,

N.Y.) supplemented with serum (usually 10% horse serum [HS]), 50 mg/l Cephapirin, and/or 40 mg/l gentamicin sulfate and 2 mg/l amphotericin B (Fungizone) was used in the experiments described in this chapter.

Mature Adipocyte Culture

Adipocytes are plated in the cold soluble collagen solution described below at a density of 5 x 10^5 cells/ml. Upon warming, this collagen solution sets up and forms a rigid gel. Culture flasks are actually coated with two layers of collagen with cells only in the top layer. Newly reconstituted collagen gels are initially quite fragile, so the first layer (1.5 ml for a 25-cm^2 flask) is gelled overnight or for a minimum of 30 min at 37° C to develop sufficient strength to be handled without damage. Cells were plated by mixing cell suspension (slowly cooled to ≈ 4° C) with the collagen solution and rapidly spreading them on the precoated flasks. The first collagen layer provides sufficient support for the cell-containing collagen layer so that media can be added after gelling for 10 min at room temperature.

The collagen volume of 2.5 ml per 25-cm^2 flask results in a gel about 1 mm thick with the cells in the upper third of the gel layer. Gels with a collagen concentration of less than 2 mg/ml tend to be disrupted easily, while gels with collagen concentrations of greater than 5 mg/ml are difficult to prepare because of the viscosity of the solution.

Preadipocyte Culture

Stromal cells are plated at a density of 1×10^3 to 1×10^4 cells per square centimeter surface area of the culture vessel. Plastic culture flasks or plates from various manufacturers are satisfactory. For long-term observation (more than 2 wk), a hydrated collagen substrate may be used because cells adhere more tenaciously to the collagen than to the plastic substrate, preventing lipid-filled cells from floating off the substrate (see ref. 83 for description of collagen cell interaction).

Tissue preparation and digestion procedures are identical to those for adipose cells. Following the screening procedure and centrifugation step that separates adipocytes and stromal cells, the stromal fraction is resuspended and washed twice with complete media. An aliquot of cells is stained with trypan blue (equal volumes of cell suspension and 0.5% trypan blue in a balanced salt solution) or Rappaport's stain. When trypan blue is used, a two-dye method (incubate equal volumes of cell suspension and 0.04% neutral red, 10 min, 37° C) (84) facilitates identifying live cells. Viable cells exclude the particulate trypan blue dye and are unstained in a single dye solution but are stained by the neutral red in the two-dye method. Stromal cells are plated at 10^3–10^4 viable cells/cm^2 in plastic culture flasks. Collagen-coated plates enhance attachment, proliferation, and longevity, but are not usually necessary for short-duration cultures (<2–3 weeks). Culture media used are generally the same as for adipocyte cultures.

Endothelial Culture Procedure

After the adipose tissue digest is filtered through the nylon screens, endothelial cells are prepared from capillary fragments on the 25 μm screen. The screen is removed from the filter holder and vortexed with 15 ml of medium in a 50-ml centrifuge tube. The resulting cell and fragment suspension is plated directly onto culture plates ($\approx 1 \times 10^4$ cells/cm^2, although accurate counts are hard to obtain). The culture medium used for plating and subsequent growth of the endothelial cells contains 20% serum. Either Fetal bovine serum (FBS), HS, or a 1:1 mixture works satisfactorily. Plasma-derived horse serum (Hyclone Labs) may be a useful supplement since it does not support rapid fibroblast growth. The use of Laminin as a substrate coating also reportedly enhances endothelial cell attachment and growth.

To prevent overgrowth of nonendothelial cells, a proline analog, thiolproline, may be added to the medium (200 μg/ml) for the first 48 hr of culture. This analog apparently blocks collagen synthesis and thereby prevents fibroblasts and smooth muscle cells from growing in the culture. If nonendothelial contamination is minimal, *cis*-hydroxyproline may be used rather than thiolproline since it may be used at higher concentrations for longer periods of time. The disadvantage of using these analogs is that the endothelial cells may also be affected in some manner that is not readily apparent but nonetheless changes experimental results.

Collagen Preparation

Purified collagen was prepared by modifying the method of Kleinman and co-workers (85) as follows. Tail tendons were dissected from mature rats and carefully separated from blood vessels and tendon sheaths. Tendons were then rinsed in 70% ethanol to remove lipid and debris followed by distilled water (2 times). Tendons were finely minced, and then digested in 0.5% acetic acid for 24 to 48 hr at 4° C with stirring (0.5 g tendon/100 ml acid). Insoluble material was then removed by centrifugation (10,000 $g \times$ 20 min). Soluble collagen was then precipitated from the supernatant by adding 10% NaCl (w/v), collected by centrifugation, then dialyzed against 0.05% acetic acid for 72 hr at 4° C (three changes). Insoluble material was again removed by centrifugation and the collagen-containing supernatant was lyophilized. The yield was about 0.20 g/tail. Collagen was sterilized with 95% ethanol for 2 hr, dried under a vacuum, and then dissolved in sufficient sterile 0.1 N HCl to obtain a final concentration of 3 mg/ml. This solution was stored at 4° C. For use, a working solution was made by mixing 8 parts collagen, 1 part 10× medium (without $NaHCO_3$), and 1 part 10× $NaHCO_3$ in 0.08 N NaOH. This solution was kept on ice and made as needed since it gelled rapidly upon warming.

Other Techniques

Glucose oxidation was measured by adding complete culture medium containing 0.4 μCi/ml U-^{14}C-glucose (New England Nuclear; medium-specific activity = 0.4 mCi/nmol). T-25 flask caps were replaced with perforated serum stoppers fitted with plastic absorption wells (A. H. Thomas). After a 4 hr incubation at 37° C, flasks were cooled to 4° C, the medium was acidified, and the absorption wells were filled with hyamine hydroxide. Carbon dioxide was absorbed overnight and the wells were counted. Stimulation of glucose oxidation was measured by adding insulin (200 microunits/ml), Sigma I 5500) to the culture medium at the start of the incubation. The effect of long-term exposure to insulin on adipocyte size was determined by measuring cell diameters on photomicrographs of the cultures. Insulin was added to complete medium at 200 μU/ml and the medium was changed on alternate days. Tetrazolium reductase was demonstrated by incubating washed cultures in Earles' salt solution containing nitro blue tetrazolium (1 mg/ml) for 2 hr at 37° C followed by fixation in Baker's formalin solution.

CHARACTERISTICS OF CULTURED ADIPOCYTES

Mature Adipocytes

Collagen substrates are widely used in the culture of various cell types, including myoblasts, hepatocytes, epithelial cells, chondrocytes, and mammary cells, either as a two-dimensional collagen layer or as a three-dimensional matrix such as has been described here. The use of collagen substrates in cell culture to promote cell growth or enhance differentiated function has been extensively reviewed (86, 87, 88). For the culture of isolated adipocytes, the use of a collagen matrix provides a simple means of retaining the cells in a culture flask. Cells prepared in this manner are handled with no more care than other cell cultures, nor did they require special media or apparatus.

Adipocytes cultured in the collagen gel matrix maintained a rounded, lipid-filled morphology for extended periods (Fig. 7-2). Multiple, small lipid droplets were occasionally observed in the perinuclear region, while extended cell processes were rare. Cultures could be routinely maintained for 2 weeks, and with care they could be kept for as long as 6 weeks. During the first 24–48 hours of culture, the adipocyte number in the matrix declined and free lipid was observed on the surface of the medium. This was presumably the result of autolysis of the nonviable adipocytes present in the cell suspension at plating. The number of adipocytes remaining after 24–48 hours reflects the viable fraction of the initial cell suspension, typically 70 to 80%. Further changes in adipocyte number were not observed.

The collagen volume of 2.5 ml per 25-cm^2 flask results in a gel about 1 mm thick with the cells in the upper third of the gel layer. Gels with a

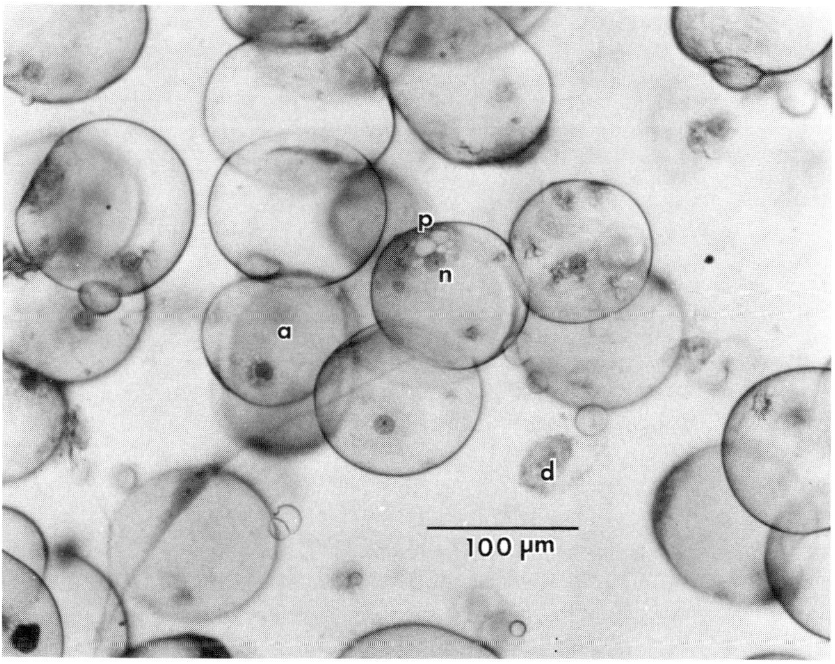

Figure 7-2. Micrograph of mature adipocytes cultured in a collagen gel for 28 days. The upper collagen layer containing ($\approx$ ½ mm thick) adipocytes was separated and fixed with Bouin's and stained with oil red 0 and hematoxylin. Culture was maintained in Medium 199 plus 10% horse serum. Adipocytes (a); residual debris from dead cells (d); adipocyte nucleus (n); basophilic perinuclear cytoplasm (p). Note that usually only a very few cells (or cell features) are in focus at a given time because of the thickness of the adipocyte preparation.

collagen concentration of less than 2 mg/ml tended to be disrupted easily, while gels with collagen concentrations of up to 5 mg/ml could be used with no apparent changes in adipocyte morphology. As the time in culture increased, the collagen layer tended to become thinner, or if the gel was purposely detached, it would shrink in size. Cultures are also prepared in petri dishes, in multiwell plates, and on Thermanox (Lux) coverslips. Collagen gel cultures could not be prepared directly on glass coverslips unless they were first coated with a gelatin solution and then dried.

Adipocytes cultured in media containing porcine serum, horse serum, or a mixture of porcine or horse serum and fetal bovine serum retained a relatively smooth surface morphology. Adipocytes in media with only fetal bovine serum occasionally became roughened and eventually appeared crenulate. This appeared to be caused by specific lots of fetal bovine serum, since it did not always occur. There was no change in morphology of the adipocytes when the concentrations of serum in the media were varied from 5 to 20% although the initial decline in cell number after plating was less in media with higher serum levels. Attempts to culture adipocytes directly in serum-free media were generally unsuccessful, as uninhibited residual collagenase degraded the collagen matrix within the first 24 hr. Cultures could be maintained serum free for several days if serum-containing medium was initially used.

Viability and Function

Cultured adipocyte viability was determined histochemically by demonstrating the presence of tetrazolium reductase (89). Cells reducing nitro blue tetrazolium to insoluble formazan in the absence of any added cofactors or substrate except glucose are assumed to be generating reduced adenine nucleotides and to have functioning electron transport. The viability of cultured adipocytes by this test is usually greater than 97%.

Physiological responsiveness of the cultured adipocytes was assessed by measuring insulin stimulation of glucose oxidation (Table 7-4) and the effect of chronic insulin exposure on cell size. After 14 days of culture in media with no supplements, adipocytes exhibited a small but significant increase in glucose oxidation in response to acute insulin stimulation. Continuously supplementing the culture media with insulin (200 μU/ml) throughout the 14-day culture period caused a greater increase in cell size than occurred in unsupplemented media. Adipocytes in media without insulin also increased in size, however, as was indicated by a higher frequency of larger adipocytes at 14 days compared with that at 2 days of culture.

Table 7-4. Oxidation of U^{-14}C-Glucose to $^{14}CO_2$ by Cultured Adipocytes

Media[a]	nMol oxidized/10^5 cells/4[b]
− insulin	0.82 ± 0.07
+ insulin	1.30[c] ± 0.10

[a]Medium for the entire culture period was Medium 199 plus 10% porcine serum. Insulin (μU/ml) was added along with labeled glucose after 14 days. Insulin was not present during the initial culture period.
[b]Values are the means of three trials using duplicate or triplicate determinations ± (S.D.)
[c]$p(t) < 0.05$ based on paired students t-test.

Contaminating Cells

One notable feature of adipocytes cultured in medium with high serum contents was the occasional appearance of proliferating fibroblastic cells (Fig. 7-3). These cells were generally present as scattered colonies and could be first observed after 5 to 7 days in media with 10 to 15% serum. The location of colonies appeared random and the number was quite variable between different preparations. The cells in all colonies were initially fibroblastlike, although in some of the colonies a portion of these cells accumulated lipid to a degree comparable with preadipocytes cultured in the collagen matrix. Fibroblastic cells were always present simultaneously and in close proximity with mature, lipid-filled adipocytes and were not observed in media with less than 5% serum. Both lipid-filled and nonfilled fibroblastic cells were observed in cultures with media containing fetal bovine, porcine, or horse serum. After 10 to 14 days, the colonies had frequently eroded the collagen matrix in their vicinity, resulting in the formation of a visible hole or depression in the collagen.

The suitability of the collagen matrix for maintaining the viability and physiological function of adipocytes was assessed by several means. Cultured adipocytes remained viable, as was indicated by nitro blue tetrazolium reduction, and maintained a morphology resembling in vivo adipocytes. Moreover, cells tended to increase in size during culture in the basal media and exhibited a greater size change in insulin-containing media. This additional increase in size with insulin exposure should reflect primarily lipogenesis from glucose. The physiological behavior of adipocytes in the collagen was further demonstrated by an acute increase in glucose oxidation upon exposure to insulin. Adipocyte size in cultured explants has previously been reported to decrease in culture (31, 34), although chronic insulin exposure eliminated shrinkage in fat cells of less than 95 μm in diameter. Three factors may account for the discrepancy between cell size changes seen in explant culture and collagen culture adipocytes. First, cell shrinkage in explants occurred in Medium 199 without serum, and hence no endogeneous lipoprotein, while the changes reported here are from cultures in 10% serum. Furthermore, average cell diameters described here are relatively small, while in the explants shrinkage occurred primarily in fat cells larger than 95 μm. Finally, cells for these experiments were obtained from adult humans. The degree of increased glucose oxidation in response to acute insulin exposure of adipocytes in the collagen matrix was similar to that observed in explant cultures (43).

In contrast to explant techniques, the collagen matrix allows culture of a relatively homogeneous adipocyte preparation since fibroblasts and endothelial cells are removed during adipocyte isolation. In contrast to other

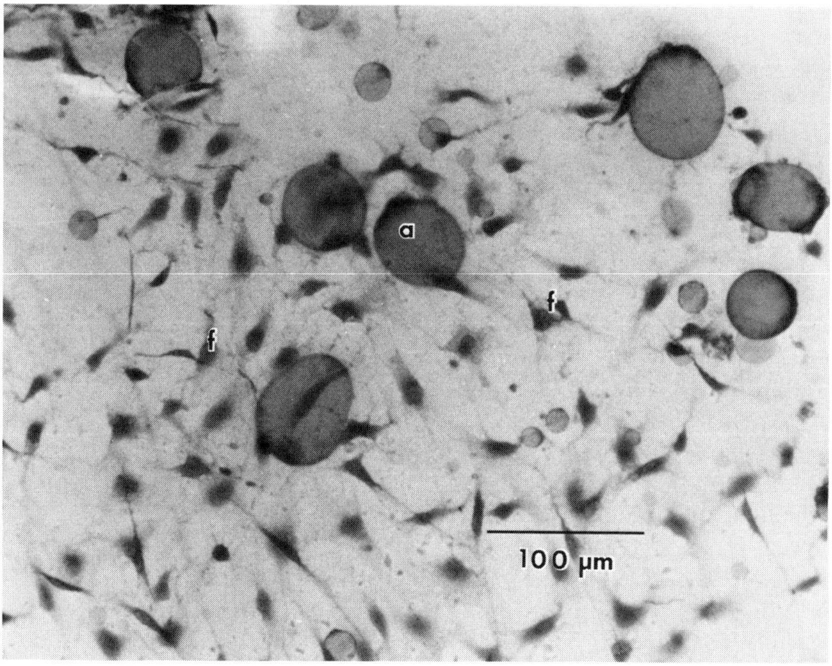

Figure 7-3. Contaminating fibroblastic cells in a mature adipocyte culture, whole mount of the cell containing collagen layer. Cells were cultured for 5 days and then fixed in Bouin's and stained with oil red O and hematoxylin. Culture was maintained in Medium 199 plus 15% horse serum. For the purpose of obtaining this photograph, an adipocyte preparation was filtered only through 275 and 50 μm nylon screens to increase the likelihood of contamination. Cultures were also plated at 1/10th normal density to make seeing fibroblasts easier. Adipocytes (a); fibroblastic cells (f).

techniques for culturing isolated adipocytes, an appropriate morphology and differentiated function are maintained in the collagen matrix. Mature adipocytes retained in culture with glass coverslips in media similar to that used in this study rapidly lose a differentiated appearance and begin to divide (46). Therefore, it seems likely that the collagen matrix functions not only to retain and support adipocytes in a culture flask but also to maintain a differentiated function (87).

Despite extensive digestion and screening procedures, cells other than adipocytes are occasionally observed in the collagen matrix. The nonadipocyte cells present are morphologically similar to fibroblasts and to cells previously described as dedifferentiated adipocytes (49–60). However, several observations suggest that these cells are the result of contamination of the isolated adipocyte preparation with stromal cells. Fibroblastic cells, when observed, are always present with a stable population of mature adipocytes, while the numerous thin cell processes observed on these cells are typical of fibroblasts cultured on hydrated collagen matrices (83). Fibroblastic cells are not present in cultures with low concentrations of serum in the media, nor are they present in every culture preparation. Furthermore, in a given culture not all of the fibroblastic cells filled with lipid, suggesting that they are of heterogeneous origin, an unlikely situation if they are dedifferentiated adipocytes. When the stromal cell fraction obtained during adipocyte isolation was prepared in a manner identical to that for adipocytes, cells grew rapidly. The proliferation of bone marrow-derived preadipocytes in a collagen matrix has been described (90). It is therefore reasonable to assume that any preadipocytes or stromal cells adhering to adipocytes would have grown in the cultures with media containing adequate serum.

The holes or depressions in the collagen matrix observed around many of the nonadipocyte colonies would also be consistent with a fibroblastic origin, since fibroblasts reportedly have a high capacity to degrade reconstituted collagen (91) or to cause thinning of collagen gels by reorganizing the collagen fibrils (92, 93).

It is possible that these cells are related to the rapidly multiplying cells associated with adipocytes described by Klyde and Hirsch (94, 95). They described a population of cells that are difficult to remove from the adipocytes during tissue digestion and are capable of incorporating radioactive thymidine. The portion of such cells decreased as the screen size used in isolation was reduced. Since the screen used in the culture preparation is considerably smaller than that used by Klyde and Hirsch, it would account for the low number of these cells observed. However, the heterogeneous morphology of the nonadipocytes in the collagen matrix would argue against any specific characteristic or function.

If stromal cells are cultured in the collagen matrix to evaluate diffusion of substrate into the gel, the growth rate as measured by DNA content is slightly slower than that in monolayer cultures, although cells eventually reach a higher density than in monolayers.

Characteristics of Cultured Stromal Cells

On the first day after culture, a few cells containing small lipid droplets (2–5 μm diameter) are occasionally observed. These cells, which are presumably small and, therefore, high-density adipocytes, disappear after 2–3 days. It is unlikely that these cells contribute significantly to this culture, even if they do proliferate, because of the low initial numbers. In cultures with fairly high serum levels (and hence rapid growth), cells with one or more very small lipid droplets can be observed after 3–5 days. These droplets expand and lipid-filled cells are clearly apparent after 5–7 days in culture (Figure 7-4*A, B*). As cultures continue to grow (7–21 days), the number of lipid-containing cells increases, roughly paralleling overall growth. The size of lipid droplets also continues to increase (Fig. 4*C, D*), although this process eventually levels off as cells with very large lipid droplets (>20–25 μm diameter) are observed infrequently. Lipid filling of preadipocytes (and adipocytes) occurs primarily by reesterification of fatty acids derived from medium lipids with glucose-derived glycerol (68, 73, 74). Little glucose is synthesized to fatty acid (42, 68).

Lipid-filling cells contain a varied number of lipid droplets. Expansion of the lipid vacuoles appears independent of droplet number. Cells with lipid droplets similar in size to monolocular droplets in other cells occasionally contain numerous smaller droplets as well. The number and distribution of lipid-filled cells developing in a particular preparation is variable (Fig. 7-5). When small numbers of these cells are present, they tend to be more concentrated in certain areas. As the number of lipid-filled cells increases, they appear more randomly dispersed. There are specie differences in this regard. Lipid-accumulating cells in porcine cultures tended to develop in discrete clusters of 10–100 cells (Fig. 7-5), whereas those from rats are more uniformly intermingled with nonfilled cells.

Some light may be shed on the variation seen in the characteristics of cultured preadipocytes by an experiment in which cells were plated at very low density (clonal density). All of the colonies containing lipid-accumulating cells had a fibroblastlike appearance. However, the number and size of cells accumulating lipid was quite variable between colonies and appeared unrelated to colony size (Fig. 7-6). This variability may reflect a difference in animals or a difference between cells from the individual. In either case, it would suggest that preadipocytes may have an inherent vari-

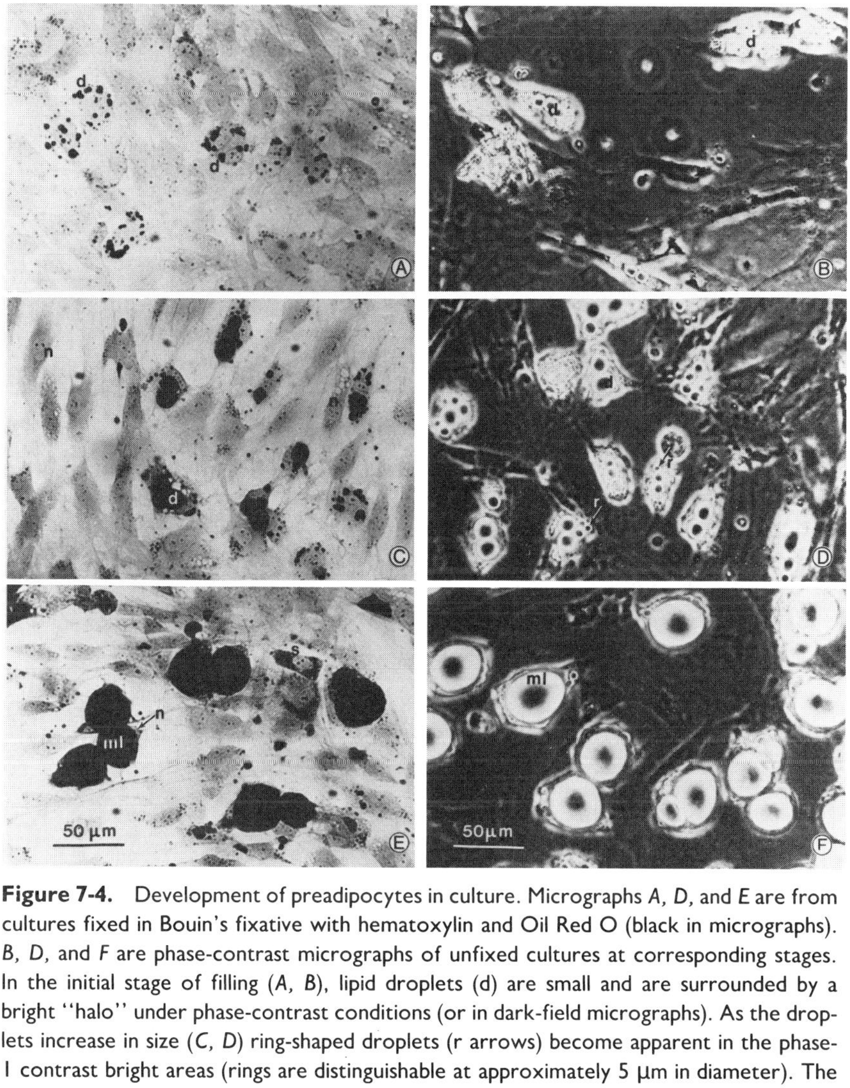

Figure 7-4. Development of preadipocytes in culture. Micrographs *A, D,* and *E* are from cultures fixed in Bouin's fixative with hematoxylin and Oil Red O (black in micrographs). *B, D,* and *F* are phase-contrast micrographs of unfixed cultures at corresponding stages. In the initial stage of filling (*A, B*), lipid droplets (d) are small and are surrounded by a bright "halo" under phase-contrast conditions (or in dark-field micrographs). As the droplets increase in size (*C, D*) ring-shaped droplets (r arrows) become apparent in the phase-contrast bright areas (rings are distinguishable at approximately 5 µm in diameter). The lipid droplets continue to enlarge (*E, F*) and form monolocular lipid-filled cells (ml), with peripherally displaced nuclei (n arrow). Note the cells in *E* (s) with a number of small lipid granules that are present simultaneously with monolocular cells. Micrographs were taken in sparsely populated regions of a culture plate so that individual cells could be distinguished.

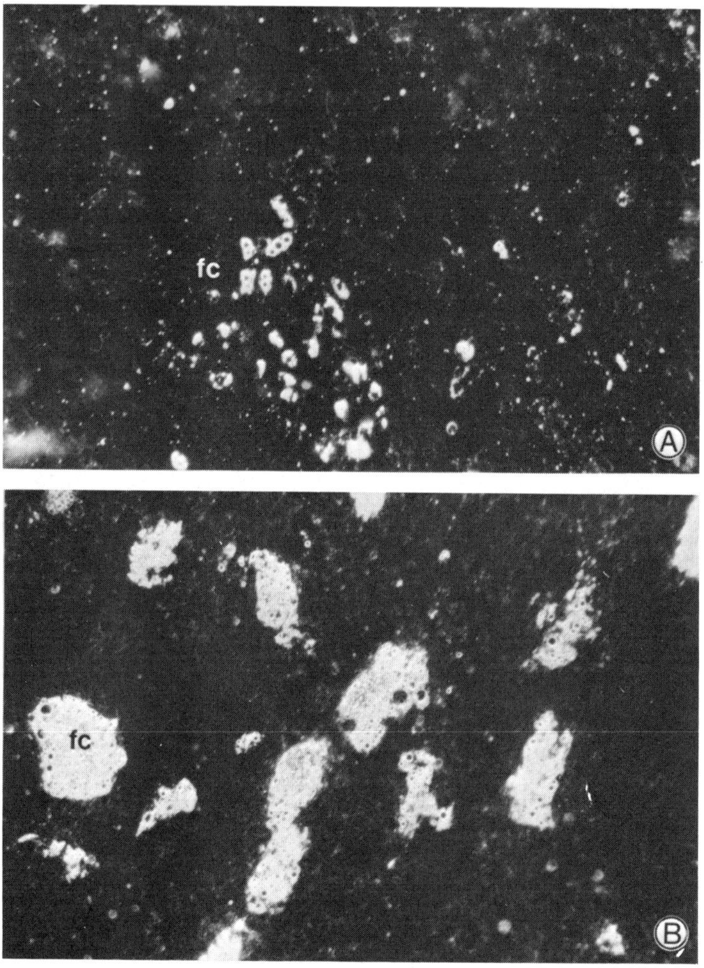

Figure 7-5. Variation in distribution of lipid-filled preadipocytes in a primary preadipocyte/stromal cell culture. Dark-field micrographs of unfixed cultures after 14 days in culture. While individual stromal cells are not apparent, it should be noted that the dark areas contain confluent cell layers. Lipid-filled preadipocytes (fc) tend to develop as isolated cells in rat preadipocyte/stromal cell cultures (*A* and *C*), whereas they develop as large clusters of cells in porcine cultures (*B* and *D*). Even with the more extensive development of lipid-filled preadipocytes in *C* compared to the culture in *A*, the cells are still distributed in a fairly random pattern. *D* is from a culture of porcine cells and illustrates the peeling phenomenon caused by buoyancy of the lipid-filled cell clusters in these cultures. In the areas indicated by p, the preadipocyte cluster sheet has peeled sufficiently to form a large clump (cl, out of focus) of cells.

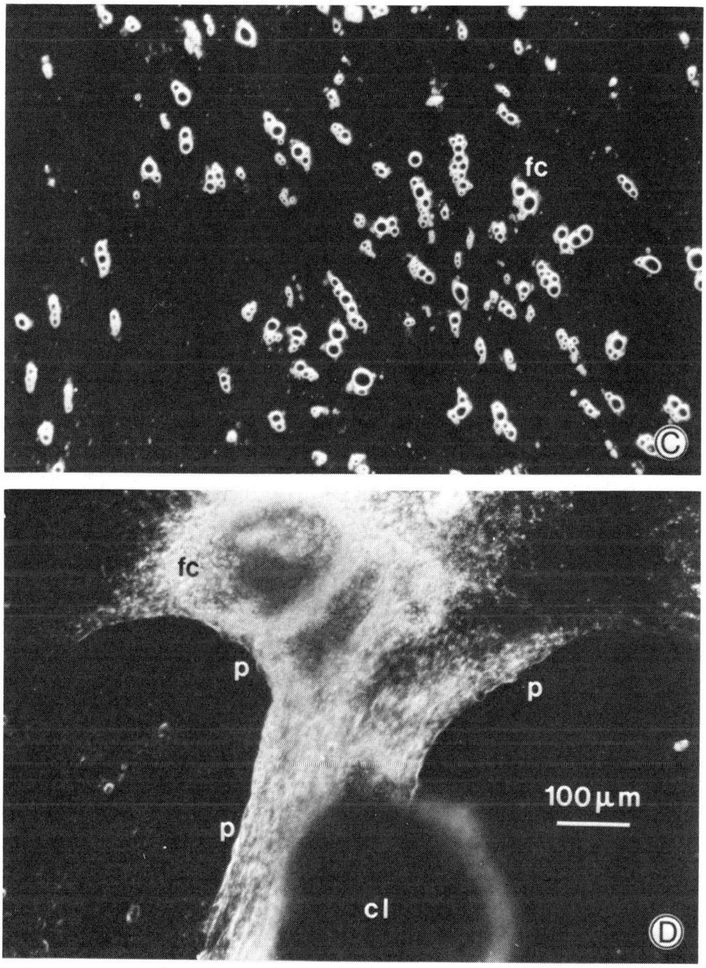

ation in their propensity to fill with lipid since all were in the same medium. This is noteworthy in light of reports indicating that the extracellular environment has an overriding influence on cell size in vivo (96, 97).

Measures of Preadipocyte Growth

As was stated previously, various parameters of preadipocyte growth (proliferation, differentiation, or hypertrophy) may be of interest. The diffi-

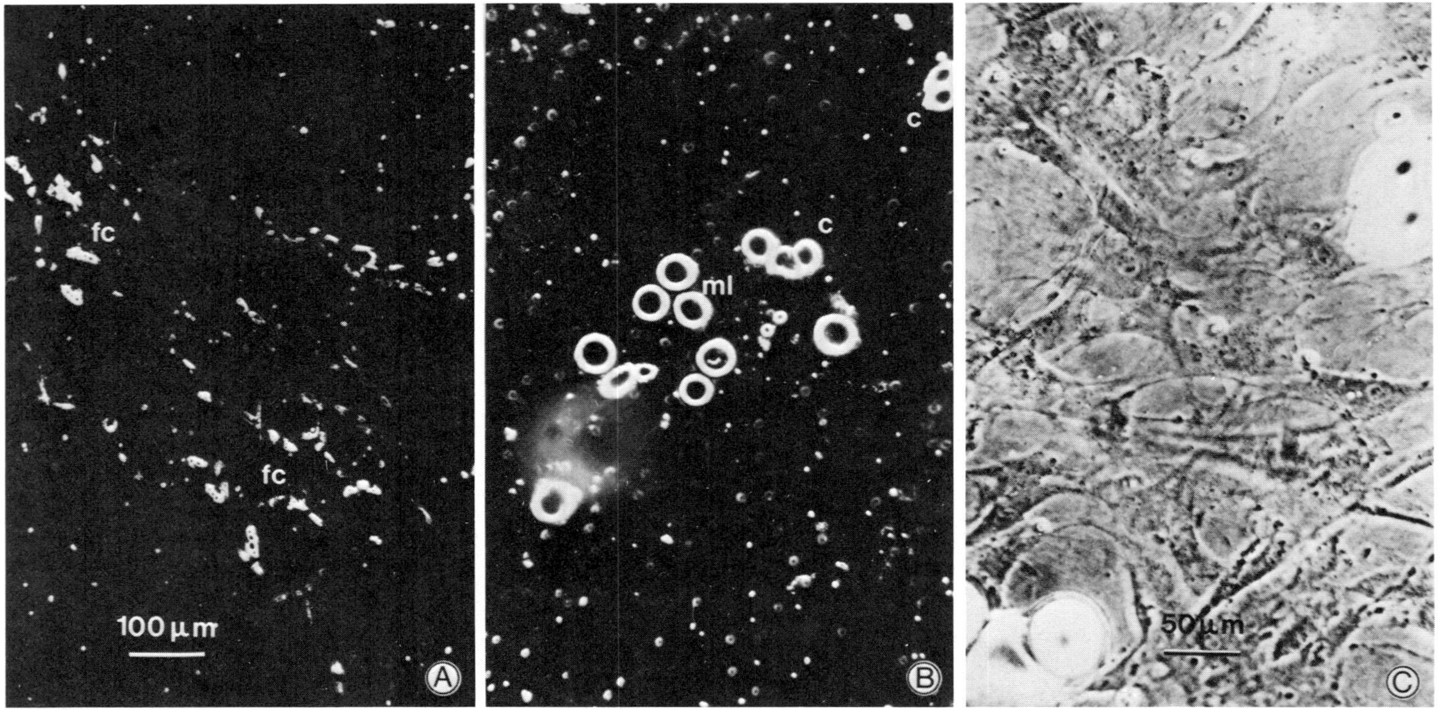

Figure 7-6. Variation in size and lipid filling of clonally derived preadipocytes. *A* and *B* are dark-field micrographs of individual preadipocyte clones taken after 21 days in culture. Clonal cells were made by initially seeding at a density of 1 cell/cm^2 on collagen-coated plates. In *A* the lipid-filled cells (fc) are quite small and multilocular, whereas in the clone in *B*, the cells are very large and monolocular (ml). The very small white spots in *B* are air bubbles in the underlying collagen. The area between the two c's in B is shown enlarged in phase-contrast micrograph C. Note that despite the presence of monolocular lipid-filled cells, the adjacent cells, sharing a common origin, show no indication of lipid filling.

culty in evaluating growth is in selecting a measurement (if possible) that reflects only a single parameter of preadipocyte growth, or at least is minimally confounded by alterations. A number of measurements are typically used for preadipocyte cell cultures. These measurements may be cell counts, of either lipid-filled cells or total cell number. Alternatively, some indication of cell number such as DNA content or thymidine incorporation into DNA may be determined. Some measure of adipocyte-specific character, such as lipid or glycerol-3-phosphate dehydrogenase (GPDH) may also be used to estimate the extent of preadipocyte differentiation. The interpretation of each approach has limitations because of the characteristics of preadipocyte/stromal cell cultures.

Following plating, isolated stromal cells attach and begin to proliferate. Although various cell morphologies may be found, it is not possible to reliably determine which cells are which at this point. The mixed stromal cells grow rapidly and at some point will become confluent or, in other words, the cells will be sufficiently dense that they touch other cells, a condition referred to as a monolayer. At this point, all growth slows down because of contact growth inhibition. It is important to note that primary cultures do not cease growth in the monolayer state but continue proliferating, although more slowly, until they are several cell layers deep as long as adequate nutrients and mitogens are present in the medium. This is in distinct contrast to the growth characteristics of the 3T3 adipose cell line, which by the nature of the selection process used to establish the line is very sensitive to contact inhibition (3, 4). Growth of 3T3 cells is markedly reduced following confluency, and only in high-serum media do these cells form multilayers.

The distinction of slowed growth in primary cultures compared to essentially halted growth in cell lines is important when the differentiation of cultured preadipocytes is considered. In cell lines, differentiation typically occurs following confluency. Changes in triglyceride content and the adipocyte marker enzymes GPDH, which indicate differentiation, are abrupt and rapid (98). This is not the case in primary culture, as both triglyceride content and GPDH activity begins to gradually rise after several days in culture while proliferation is still rapid. Glycerol-3-phosphate dehydrogenase activity in primary culture is highly correlated to cell density and triglyceride content (69); therefore, measurements indicative of adipocytelike characteristics may be affected by a number of growth parameters.

Two of the possible explanations for this difference between primary and cell line cultures have implications for the validity of certain types of experiments with primary preadipocyte/stromal cultures. The rise in GPDH activity, reflecting differentiation while the culture as a whole is still in a proliferative phase, may indicate a heterogeneous mixture of preadipocytes in vivo. Some preadipocytes may be relatively committed when isolated and then differentiate after one or two cell cycles. At the same time, uncommitted preadipocytes may proliferate more extensively before differentiation. This explanation is supported by the different proportions of preadipocytes obtained in fractions of cell preparations separated by density and hence, lipid content (63, 67) as well as by the variation in clones described earlier. Variation in the frequency of adipose conversion of 3T3 clonal variants (5) and the potential to maintain proliferating 3T3-T cells at different stages of differentiation (99, 100) also supports the concept of preadipocyte heterogeneity. This potential heterogeneity must be considered when preadipocytes are evaluated in culture.

Another possibility is that the correlation between lipid filling and cell density means that preadipocytes in a rapidly proliferating culture are less likely to differentiate than those in a slow-growing one. If this is the case, then the density of other stromal cells will influence the proliferation and/or differentiation of preadipocytes after confluency. A treatment that stimulated proliferation of a particular cell type would appear to cause an earlier differentiation of preadipocytes. The actual effect, however, was to cause the culture to reach confluency sooner, which in turn facilitated differentiation by slowing proliferation. Note also that if differentiated, lipid-filled preadipocytes were then counted to estimate growth, the actual number might be reduced because differentiation occurred after fewer cell divisions. Lipid-filled cells would also appear to be less numerous if preadipocytes were stimulated to proliferate but stimulated to a lesser degree than other cells. Treatments that stimulate all cells to proliferate (such as increased serum concentration) would proportionately increase the fraction of apparent lipid-filled cells because the earlier differentiation would allow more time for lipid synthesis before evaluation. These are likely to be extreme cases, but they illustrate the problems of interpreting preadipocyte growth data based on characteristics indicating differentiated cell function.

One additional problem with estimating preadipocyte characteristics by evaluating confluent or differentiated cultures is the variations in experimental results caused by different plating efficiencies, different digestion efficiencies leading to different cell mixtures, and so on. A useful way to avoid these problems and to specifically measure proliferation rates is to conduct experiments before confluency is reached. During exponential growth, rates of proliferation are not subject to variations caused by physical interactions with other cells or by variable differentiation. This solution still does not address the problem of mixed cell types, which means that exponential growth of preadipocytes cannot be measured directly. The following section describes an approach to avoiding this problem as well.

Procedures for Determining Proliferation Rate of Preadipocytes

This section details a procedure in which preadipocytes and stromal cells are labeled by thymidine incorporation during the exponential growth phase, then maintained in culture until preadipocytes have differentiated and finally are separated by density. Measurement of the radioactive thymidine incorporated earlier is used to obtain separate measures of preadipocyte and other stromal cell replication rates. If one assumes that the specific activity of DNA synthesized on any particular day is equal to that for any other day, the curve of changes in thymidine incorporation is the same as the curve for changes in cell number with time.

Stromal/preadipocyte cultures are prepared as described above. Plates or flasks of 25 cm^2 surface area are suitable. Cells are plated at a density of $1 \times 10^3/cm^2$ to allow a longer period of exponential growth. Twenty-four hours following plating, cultures are changed to experimental media. If any of the test media contain lower levels of serum than the plating medium, all cultures should be adapted to the lowest serum level for 24 hr before changing to test media. Each day after changing to test media, replicate or triplicate plates are labeled with 50–100 nCi/ml ^3H or ^{14}C-thymidine (specific activity = 100 Ci/mole) for 24 hours. Following the 24-hour labeling period, plates are rinsed with balanced salt solution and washed for 30 min with media containing 50 μM cold thymidine (a 1,000-fold excess). This effectively eliminates further incorporation of label. Plates are then returned to experimental media and are subsequently changed on alternate days. This schedule results in a series of plates—some labeled on day 1, some labeled on day 2, and so on. Seven days after plating, all culture media are changed to a medium promoting lipid filling of differentiated cells. Empirically, this medium contains 8–10% porcine serum, 2–5% horse serum or fetal bovine serum (or a mixture of horse serum and fetal bovine), plus 1 mM glucose and 200 μU/ml of insulin. This medium is also changed on alternate days. Fourteen to eighteen days after plating, cells are removed from culture and lipid-filled cells are separated from other stromal cells. Cultures are digested with a mixture of trypsin and collagenase (0.5 mg/ml of each) in Hanks' salt solution containing 0.5% bovine serum albumin and adjusted to 310 mOsmolar.

Low-temperature digestion is used to minimize cell breakage and subsequent aggregation. The isolated cells (preadipocyte-derived adipocytes and other stromal cells) are then centrifuged through a solution of Percol in Hanks' salts plus bovine serum albumin having a density of 1.02 (1,000 × *g* for 10 min). Stromal cells sediment in Percol media of this density, while the lipid-filled cells do not. Floating adipocytes and the layer of Hanks' solution are removed and sedimented cells are resuspended in the Percol solution and recentrifuged. The two supernatant fractions are combined and thymidine incorporation into each fraction is measured.

Characteristics of Preadipocyte Replication Rate Assay

Cultures labeled sequentially display the type of incorporation pattern seen in Figure 7-7. The initial lag following plating seen in DNA content is not observed; the cultures are proliferating in an exponential manner at the earliest point measurable. The rate of proliferation slows and then declines at 5-7 days after culturing, indicating that cells are becoming density or contact inhibited. Thymidine incorporation reflects DNA synthesis, so a negative slope in the incorporation curve indicates only that the rate of proliferation is decreasing, not that cells are being lost. At this point replication of the preadipose fraction is decreasing more rapidly than the stromal fraction. This more rapid decrease is probably because some of the preadipocytes are differentiating at this time. These cells are no longer proliferating at all, effectively reducing the rate of thymidine incorporation more quickly for preadipocytes than for stromal cells. Although other stromal cells are not proliferating exponentially, the population of dividing cells is not being reduced by differentiation.

Thymidine incorporation data may be used for several types of comparisons. Thymidine incorporation on a particular day (day 1 is most appropriate) may be compared between treatments. Alternatively, incorporation from several days may be used to determine the actual growth rate expressed as doublings per day or as the frequency of doubling (F). The advantages of determining a value for F are that numbers from different experiments are independent in initial incorporation and are, therefore, directly comparable and not related to cell density. Doubling frequency is determined by regression using the equation for exponential growth: $N_t = N_0 \cdot 2_F \cdot_t$ (101), where N_0 = initial incorporation, N_t = incorporation at time t, t = time (in days), and F = frequency of doubling (expressed as doubling/day). Note that once experimental parameters are established for experimental growth, incorporation need only be measured at two time points. Frequency of doubling is then determined from the equation $F = \log (N_t/N_0)/0.301t$ (101).

Determining an F value for test media or compounds has an additional advantage in that data may be analyzed with an approach analogous to enzyme kinetics (102, 103, 104). If F is determined for several concentrations of test medium or compound, it is possible to calculate a theoretical maximum frequency of doubling, F_{max}, as well as k_m value indicating the concentration of growth factor giving half-maximal growth. This is done with a modification of the Lineweaver-Burke equation, $1/F = k_m/F_{max} \cdot 1/[G] + 0.1/F_{max}$. These values give an indication of the relative potency and specificity of growth factors for certain cell types. This indicator is particularly useful for cells with similar characteristics such as fibroblasts and preadipocytes, which respond to many mitogens in a similar manner (105).

It should be pointed out that cultured cells may not grow in an expo-

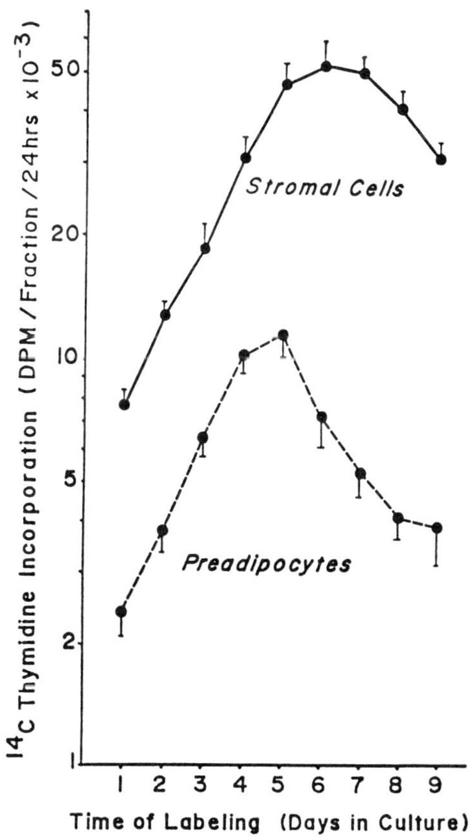

Figure 7-7. Thymidine incorporation characteristics of growing preadipocyte/stromal cell cultures. Preadipocyte/stromal cell cultures were derived from rat inguinal adipose tissue. Thymidine labeling and separation were as described in the text.

nential manner under some conditions (106). Although the exponential model generally fits the types of data obtained with primary preadipocyte/stromal cells, it is important to be aware that deviations from any mathematical model may occur. This is particularly true when culture conditions are changed markedly.

Endothelial Cells

Immediately after plating, the endothelial fragments appear as either elongated cell aggregates or as clearly tubular structures. After 24 hr, cells have attached to the substrate and have begun to spread out from the fragments. These endothelial cell clusters have a clearly defined circular pattern at the edges of cell growth. Fibroblastic cells spreading from tissue fragments have a fairly ragged appearance. The endothelial cells also have a distinctly mono-

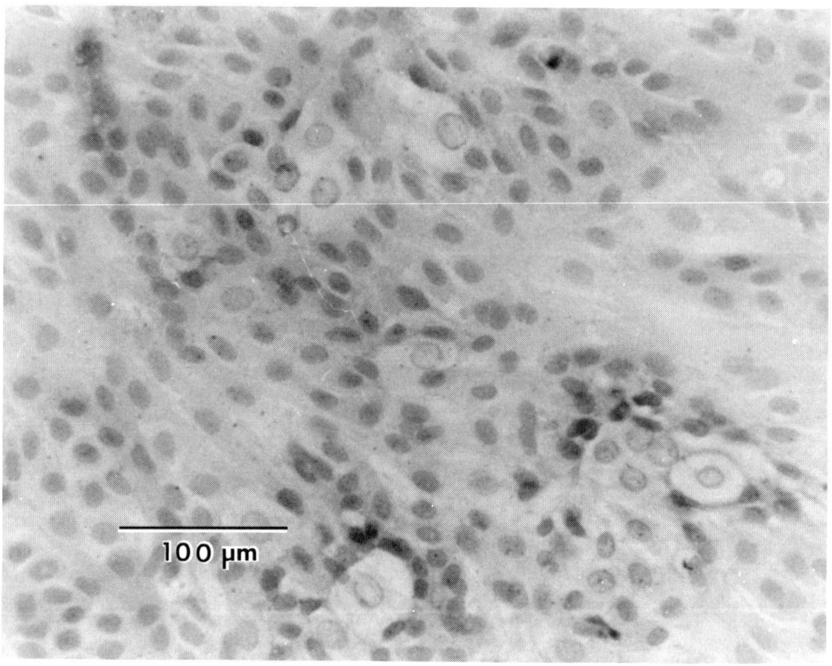

Figure 7-8. Micrograph of adipose tissue-derived endothelial cells. Capillary fragments cultured with 200 µg/ml thiolproline for 48 hr in Medium 199 plus 10% HS and 10% FBS. Cells were fixed and stained with hematoxylin. Cultures are characterized by a monolayer of relative uniformity and a fairly indistinct area of cytoplasm around each nucleus.

layered appearance and do not form indistinct, multilayered foci as do the fibroblastic cells. After 2–3 days in culture, any contaminating cells will be apparent, with the fibroblastic cells forming an apparent boundary or junction where they contact the endothelial cells. If a hydroxyproline analog has been used, most of the fibroblastic cells will have disappeared. If contamination is still present, further use of the analog is not helpful, as endothelial cells also begin to die. After 6–7 days, growth in cultures with minimal contamination slows and cultures form a reasonable uniform monolayer (Fig. 7-8). Cultures remain relatively stable after this time, although subculturing is only moderately successful. Subcultures have a low plating efficiency and frequently become overgrown by any fibroblastic cells that may have been present in small numbers in a quiescent state in the original culture.

As with adipocyte cultures, a major concern with endothelial cell culture is the amount of contaminating cells. Histochemical staining for endothelial antigens, such as factor IV antigen, may be helpful in making this assessment. However, there is no definitive assay to establish purity, so visual evaluation is still the most appropriate criterion.

Evidence for Local Adipose Interactions

As was suggested earlier, one of the potential uses of cell culture is to look at the role of particular cell types in the regulation of physiological functions or characteristics of other cells or tissues. Because of the central role of adipose tissue in energy storage, it seems likely that adipose tissue may exert some sort of regulatory influence on other tissues.

A major advantage of primary cell cultures over cell line cultures is the potential to model in vivo cell-to-cell interactions. Cell lines are derived from or represent a single cell type. To study interactions between different cell types, multiple cell lines must be used. Because the adaption problem of each cell line is then multiplied, interpretation becomes tenuous. In primary culture, different cell types may be derived from the same tissue as described previously in the chapter. Therefore, interpretation of observed interactions as representing in vivo phenomena is more reasonable. In this section, evidence for local adipose interactions is briefly presented.

A classic approach to investigating cell type interactions with tissue culture utilizes conditioned media. Cells in successful culture systems are metabolically quite active, as is clearly indicated by the rapid changes in the pH of media over dense cultures. Furthermore, if the culture system is appropriate, cells have similar specific functions to the same cells in vivo; that is, secretory cells also secrete specific products in culture. Thus cultured cells alter the culture medium. Nutrients may be removed or added, as may growth factors, hormones, or other compounds. Media changed by cultured cells in this manner are referred to as conditioned media.

To evaluate whether mature adipocytes are producing locally active factors affecting the proliferation or development of other cell types, a simple strategy is used. Since mature adipocytes in culture as described are viable, one assumes that if adipocytes are releasing factors in vivo, they are also doing so in culture. If these factors are stable, as might be anticipated from the relatively slow time course of adipose cellularity changes, they will be present in the culture medium.

For the experiments described below, adipocyte-conditioned medium was collected every 48 hours beginning when the cells had been cultured for 6 days. The purpose of the initial delay was to allow soluble debris from cells killed during plating to be completely washed off the culture plate. Conditioned medium was stored frozen at $-70°$ C until used. Nonadipocyte-conditioned medium was collected from preadipocyte-stromal cells cultured in collagen in a manner identical to that for mature adipocytes. Referred to a SV-conditioned medium, this medium was collected only on day 4 through day 10 of culture as the stromal preparations become excessively dense after 10 days. The SV-conditioned medium is assumed to have similar changes in nutrients and metabolites compared with the adipocyte-conditioned medium but not any adipocyte-produced factors.

The micrographs in Figure 7-9 demonstrate that medium conditioned by

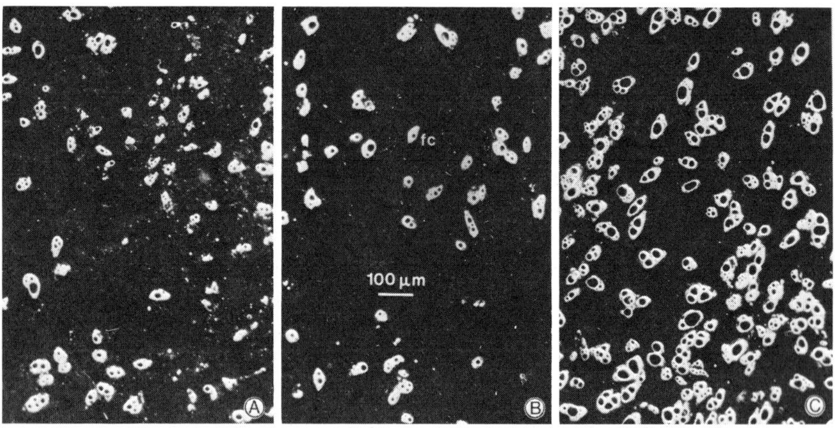

Figure 7-9. Dark-field micrograph of preadipocyte/stromal cell cultures. *A* is from control plate. *B* is from a plate grown in medium conditioned by stromal cells, and *C* is from a plate of cells grown in adipocyte-conditioned medium. The increased number and size of lipid-filled (fc) preadipocytes in *C* is evident. Recall that the dark areas contain confluent stromal cells.

Table 7-5. Growth Coefficients[a] for Preadipocyte/Stromal Cell Cultures for Horse Serum and Adipocyte-Conditioned Media

Test Media	Cell Type	
	Stromal Cells	Preadipocytes
Horse serum[b]	$F_{max} = 1.25$	$F_{max} = 1.23$
10% horse serum medium	$F_{actual} = 0.73$	$F_{actual} = 0.70$
Potential increase by adding serum to 10% horse serum medium[c]	$F_{increase} = 0.52$	$F_{increase} = 0.53$
Adipocyte conditioning of medium[d]	$F_{max} = 0.18$	$F_{max} = 0.72$

[a]The frequency of doubling, F, expressed as doublings/day.
[b]Coefficient from varied levels of horse serum added to Medium 199.
[c]The difference between the maximum growth rate obtainable with horse serum, F_{max}, and the rate actually observed in the 10% horse serum medium.
[d]Coefficient for conditioning above basal growth. 10% horse serum medium conditioned for 48 hr by mature adipocytes. Value for F_{max} determined by assaying dilutions made with basal medium.

adipocytes has an apparent effect on preadipocyte development. These micrographs were taken after 21 days, and the increase in the number of preadipocytes that have differentiated into lipid-filled adipocytes is apparent. Since the total DNA content in the cultures exposed to conditioned medium is lower than that in control cultures (69), it is unlikely that the effect is the result of general growth stimulation. The level of GPDH activity, which corresponds to the apparent increase in cell number, also demonstrates that lipid-filled cells have adipocyte characteristics and are not merely fibroblasts or other cells pathologically lipid filled (69, 107).

When the sequential labeling technique is applied using adipocyte-conditioned medium, the specific effect is on preadipocyte proliferation as is indicated by the potential F_{max} values for conditioned medium. The F_{max} for preadipocytes represents a large increase over control values, while the F_{max} for other stromal cells is small (Table 7-5).

It would appear that adipocytes produce an extracellular factor that stimulates proliferation of preadipocytes and not stromal or vascular cells. The nature of the factor is as yet unknown, but it illustrates the potential for using cell culture systems to evaluate local regulatory mechanisms.

Proliferation is not the only developmental process that can be examined with the "conditioned medium" approach. In newly developing (or enlarging) adipose tissue, vascular development must keep pace with new adipocytes. However, in vascular development cell migration of endothelial cells is a primary event followed later by cell proliferation. Endothelial cell migration can be measured by using cultured cells in a chemotaxis assay (108). Conditioned medium from both differentiated 3T3-L1 cells (109, 110) and cultured mature adipocytes (111) has been shown to have a chemotactic (attraction) and chemokinetic (increased motility) effect on endothelial cells. This would suggest that adipose-derived factors affect at least two steps (preadipocyte proliferation and vascular development) necessary for coordinated development of adipose tissue. The 3T3-L1 factor is reported to be a prostaglandinlike compound (112).

Other conditioning experiments with adipocytelike cell lines support the idea of adipocyte-derived regulatory factors. Both the extracellular matrix and conditioned media from 3T3-L1 cells promote proliferation of primary mammary epithelial cells. Conditioned media from 3T3-C2 cells (a control line that does not differentiate) has a smaller effect (113). On the other hand, differentiated MC3T3-G2/PA6 adipocytes are less able to support the growth of hemopoietic stem cells than are the undifferentiated preadipocytes (12). In a primary co-culture system Björntörp (114) was unable to demonstrate the importance of specific conditions for successfully observing in vivo functions in culture. Alternatively, this result may demonstrate just the ease of generating artifacts with cell culture. In either case it emphasizes the importance of carefully interpreting results from culture experiments.

It is hoped that this final section has demonstrated the potential utility, as well as some of the potential problems, of adipose cell culture in obtaining new information about the growth and development of adipose tissue.

REFERENCES

1. M. Darmon, G. Serrero, A. Rizzino, and G. Sato, 1981, Isolation of myoblastic, fibro-adipogenic, and fibroblastic clonal cell lines from a common precursor and study of their requirements for growth and differentiation, *Exp. Cell Res.* **132**:313.
2. G. Serrero and J. Khoo, 1982, An in vitro model to study adipose differentiation in serum-free medium, *Anal. Biochem.* **120**:351–359.
3. H. Green and O. Kehinde, 1974, Sublines of mouse 3T3 cells that accumulate lipid, Cell. 1:113–116.
4. H. Green and O. Kehinde, 1975, An established preadipose cell line and its differentiation in culture. II. Factors affecting the adipose conversion, *Cell* **5**:19–27.
5. H. Green and O. Kehinde, 1976, Spontaneous heritable changes leading to increased adipose conversion in 3T3 cells, *Cell* **7**:105–113.
6. H. Green and O. Kehinde, 1979, Formation of normally differentiated subcutaneous fat pads by an established preadipose cell line, *J. Cell Physiol.* **101**:169–172.
7. L. Diamond, T. G. O'Brien, and G. Rovera, 1977, Inhibition of adipose conversion of 3T3 fibroblasts by tumorer promoters, *Nature* **269**:247–249.
8. B. Krawisz and R. Scott, 1982, Coupling of proadipocyte growth arrest and differentiation. I. Induction by heparinized medium containing human plasma, *J. Cell Biol.* **94**:394–399.
9. R. Scott, B. Hoerl, J. Wille, Jr., D. Florine, B. Krawisz, and K. Yun, 1982, Coupling of proadipocyte growth arrest and differentiation. II. A cell cycle model for the physiological control of cell proliferation, *J. Cell Biol.* **94**:400–405.

10. H. Kodama, Y. Amagai, H. Koyama, and S. Kassai, 1982, Hormonal responsiveness of a preadipose cell line derived from newborn mouse calvaria, *J. Cell Physiol.* **112**:83–88.
11. H. Kodama, Y. Amagai, H. Koyama, and S. Kassai, 1982, A new preadipose cell line derived from newborn mouse calvaria can promote the proliferation of pluripotent hemopoietic stem cells in vitro, *J. Cell Physiol.* **112**:89–95.
12. H. Kodama, H. Sudo, H. Koyama, S. Kasai, and S. Yamamoto, In virto hemopoiesis within a microenvironment created by MC3T3-G2/PA6 preadipocytes, *J. Cell Physiol.* **118**:233–240.
13. M. Lanotte, D. Scott, T. Dexter, and T. Allen, 1982, Clonal preadipocyte cell lines with different phenotypes derived from murine marrow stroma: Factors influencing growth and adipogenesis in vitro, *J. Cell Physiol.* **111**:177–186.
14. R. Negrel, P. Grimaldi, and G. Ailhaud, 1978, Establishment of preadipocyte clonal line from epididymal fat pad of ob/ob mouse that responds to insulin and to lipolytic hormones, *Proc. Natl. Acad. Sci. (USA), Cell Biol.* **75**:6054–6058.
15. G. Ailhaud, 1982, Adipose cell differentiation in culture, 1982, *Mol. Cell. Biochem.* **49**:17–31.
16. D. Gaillard, R. Negrel, G. Serrero-Dave, C. Cermoclacce, and G. Ailhaud, 1984, Growth of preadipocyte cell lines and cell strains from rodents in serum-free hormone-supplemented medium, *In Vitro* **20**(2):79–88.
17. P. Grimaldi, R. Negrel, J. P. Vincent, and G. Ailhaud, 1979, Differentiation of ob 17 preadipocytes to adipocytes, *J. Biol. Chem.* **254**:6849–6852.
18. S. Yasumoto, 1984, Hormonal regulation of the transformation phenotype in simian virus 40-transformed rat embryonic preadipose cell lines, *Mol. Cell Biol.* **4**(4):712–721.
19. K. Kohno, A. Hiragun, A. Takasuki, G. Tamura, and H. Mitsui, 1980, Tunacamycin inhibits the differentiation of ST 13 fibroblasts to adipocytes with suppression of the insulin binding activity, *Biochem. Biophys. Res. Commun.* **93**:842–899.
20. A. Hiragun, M. Sato, and H. Mitsui, 1980, Establishment of a clonal line that differentiates into adipose cells in vitro. *In Vitro* **16**:685–693.
21. B. S. Hoon, S. Balk, R. S. Mitchell, and D. Le Stourgeon, 1979, Rapid quantitative adipose conversion of chicken fibroblasts by high concentrations of chicken serum or plasma, *Experientia* **35**:1038–1039.
22. Y. Y. Maeda, M. Rokutanda, and N. Yamoto, 1980, Induction of adipose conversion by mouse serum in cultured cells and its characterization, *Exp. Cell Res.* **126**:99–107.
23. B. Nixon and H. Green, 1984, Growth hormone promotes the differentiation of myoblasts and preadipocytes generated by azacytidine treatment of 1OT1/2 cells, *Cell Biol.* **81**:3429–3432.
24. N. C. Foot, 1912, Über das wachstum von knochenmark in vitro. Experimenteller beitragzur entstehung des fettgewebes, *Beitr. Path. Anat. Allg. Path.* **53**:446–476.
25. L. Burkhardt, 1934, Boebachtungen an explantiertem fettgewebe, *Arch. Exp. Zellforsch.* **16**:187–202.
26. O. A. Trowell, 1959, The culture of mature organs in a synthetic medium, *Exp. Cell Res.* **16**:118–147.
27. B. G. Slavin, 1967, Hormonal influences on white and brown adipose tissue in organ culture, *Anat. Rec.* **157**:389–390.
28. B. G. Slavin and J. J. Elias, 1969, The influence of pituitary hormones and norepinephrine on the size of adipose cells in organ culture, *Anat. Rec.* **164**:141–152.
29. C. Esanu and G. Bray, 1969, DNA synthesis in human adipose tissue in vitro. II. Effect of Obesity, *J. Clin. Endocrinol.* **29**:1033–1035.
30. C. Esanu, S. Murakawa, G. A. Bray, and M. S. Raben, 1969, DNA synthesis in human adipose tissue in vitro. I. Effect of serum and hormones, *J. Clin. Endocrinol.* **29**:1027–1032.

31. U. Smith, 1971, Morphologic studies of human subcutaneous adipose tissue in vitro, *Anat. Rec.* **169**:97–104.
32. U. Smith, 1972, Studies of human adipose tissue in culture. I. Incorporation of glucose and release of glycerol, *Anat. Rec.* **172**:597–602.
33. J. Frohlich, A. Vost, and C. Hollenberg, 1972, Organ culture of rat white adipose tissue, *Biochem. Biophys. Acta.* **280**:579–587.
34. U. Smith and B. Jacobsson, 1973, Studies of human adipose tissue in culture. II. Effects of insulin and of medium glucose on lipolysis and cell size, *Anat. Rec.* **176**:181–184.
35. U. Smith, 1974, Studies of human adipose tissue in culture. III. Influence of insulin and medium glucose concentration on cellular metabolism, *J. Clin. Invest.* **53**:91–98.
36. U. Smith, S. Boström, R. Johansson, and G. Nyberg, 1976, Human adipose tissue in culture. V. Studies on the metabolic effects of insulin, *Diabetologia* **12**:137–143.
37. U. Smith, O. Isaksson, G. Nyberg, and L. Sjöström, 1976, Human adipose tissue in culture. IV. Evidence for the formation of hormone antagonist by catecholamines, *Eur. J. Clin. Invest.* **6**:35–42.
38. G. Nyberg and U. Smith, 1977, Human adipose tissue in culture. VII. The long-term effect of growth hormone, *Horm. Metab. Res.* **9**:22–27.
39. M. Cigolini and U. Smith, 1979, Human adipose tissue in culture. VIII. Studies on the insulin-antagonistic effects of glucocorticoids, *Metabolism* **28**:502–510.
40. R. G. Vernon, 1978, Lipogenesis in sheep adipose tissue maintained in tissue culture: Effects of insulin and growth hormone, *Biochem. Soc. Trans.* **6**:988–992.
41. R. G. Vernon, 1978, Maintenance of rat adipose tissue in tissue culture, *Int. J. Biochem.* **9**:502–512.
42. R. G. Vernon, 1979, Metabolism of ovine adipose tissue in tissue culture, *Int. J. Biochem.* **10**:57–60.
43. R. S. Bernstein, 1979, Insulin insensitivity and altered glucose utilization in cultured rat adipose tissue, *J. Lipid Res.* **20**:848–856.
44. R. L. Sidman, 1956, Histogenesis of brown adipose tissue in vivo and in organ culture, *Anat. Rec.* **124**:581–601.
45. B. G. Slavin and J. J. Elias, 1970. Morphologic changes in white and brown adipose tissue by insulin, thyroxin cortisol in organ culture, *Anat. Rec.* **167**:213–224.
46. R. G. Vernon, 1979, Lipogenesis in ovine adipose tissue in tissue culture, *Ann. Rec. Vet.* **10**:399–400.
47. P. E. Walton, T. D. Etherton, and J. F. Rebhun, 1984, Stimulatory effects of insulin on glucose metabolism in swine adipose tissue, *J. Anim. Sci.* **59** (Suppl. 1):211.
48. E. R. Simpson, G. E. Ackerman, M. E. Smith, and C. R. Mendelson, 1981, Estrogen formation in stromal cells of adipose tissue of women: Induction by glucocorticosteroids, *J. Cell Biol.* **78**(9):5690–5694.
49. C. W. Ng, W. J. Poznanski, M. Borowiecki, and G. Reimer, 1971, Differences in growth in vitro of adipose cells from normal and obese patients, *Nature* **231**:445.
50. W. J. Poznanski, J. Rushton, and I. Waheed, 1973, The origin and metamorphosis of the human fat cell, in *The Regulation of the Adipose Tissue Mass,* J. Vague and J. Boyer, eds., Excerpta Medica, Amsterdam, p. 145–150.
51. W. J. Poznanski, I. Waheed, and R. L. R. Van, 1973, Human fat cell precursors. Morphologic and metabolic differentiation in culture, *Lab. Invest.* **29**:570–576.
52. I. Dardick, W. J. Poznanski, I. Waheed, and G. Setterfield, 1976, Ultrastructural observations on differentiating human preadipocytes cultured in vitro. *Tissue Cell* **8**:561–571.
53. R. L. R. Van, C. E. Bayliss, and D. A. K. Roncari, 1976, Cytological and enzymological characterization of adult human adipocyte precursors in culture, *J. Clin. Invest.* **58**:699–704.

54. F. O. Adebonojo, 1973, Synthesis and storage of lipids by cultured adipocytes of a human neonate, *Biol. Neonate* **23**:366–370.
55. R. E. Adebonojo, 1975, Enzymatic adaptations by cultured adipocytes of human infants and children: Effect of obese serum on the activities of lactate-, malate-, and glucose-6-phosphate dehydrogenases, *Pediat. Res.* **9**:889–893.
56. D. Dixon-Shanies and J. L. Knittle, 1976, Effect of hormones on cyclical AMP levels in cultured human cells, *Biochem. Biophys. Res. Commun.* **68**:982–993.
57. D. Dixon-Shanies, J. Rudick, and J. L. Knittle, 1975, Observations on the growth and metabolic functions of cultured cells derived from human adipose tissue, *Proc. Soc. Exp. Biol. Med.* **149**:541–545.
58. F. O. Adebonojo, 1976, Triglyceride mobilization in cultured adipocytes of human infants, *Biol. Neonate* **29**:77–81.
59. G. H. Rothblat and F. D. DeMartinis, 1977, Release of lipoprotein lipase from rat adipose tissue cells grown in culture, *Biochem. Biophys. Res. Commun.* **78**:45–50.
60. M. Tavassoli, 1982, In vivo development of adipose tissue following implantation of lipid-depleted cultured adipocyte, *Exp. Cell Res.* **137**:55–62.
61. F. Wasserman, 1965, The development of adipose tissue, in *Handbook of Physiology,* Section 5: Adipose Tissue, A. E. Renold and G. F. Cahill, Jr., eds., American Physiological Society, Washington, D.C. pp. 87–100.
62. R. L. R. Van and D. A. K. Roncari, 1978, Complete differentiation of adipocyte precursors: A culture system for studying the cellular nature of adipose tissue, *Cell Tissue Res.* **195**:317–329.
63. P. Björntörp, M. Karlsson, P. Petroft, P. Pettersson, L. Sjöström, and U. Smith, 1978, Isolation and characterization of cells from rat adipose tissue developing into adipocytes, *J. Lipid Res.* **19**:316–324.
64. M. D. Rosenthal, 1981, Accumulation of neutral lipids by human skin fibroblasts: Differential effects of saturated and unsaturated fatty acids, *Lipids* **16**:173–182.
65. B. H. Lipton, 1977, A fine structural analysis of normal and modulated cells in myogenic cultures, *Dev. Biol.* **60**:26–47.
66. R. P. Bird, P. K. Basrur, and J. C. Alexander, 1981, Cytotoxicity of thermally oxidized fats, *In Vitro* **17**(5):397–404.
67. A. M. Gaben-Cogneville, Y. Aron, G. Idriss, T. Jahchan, J. Y. Pello, and E. Swierczewski, 1983, Differentiation under the control of insulin of rat preadipocytes in primary culture isolation of homogeneous cellular fractions by gradient centrifugation, *Biochim. Biophys. Acta* **762**(3):437–444.
68. P. Björntörp, M. Karlsson, P. Petterson, and G. Sypniewski, 1980, Differentiation and function of rat adipocyte precursor cells in primary culture, *J. Lipid Res.* **21**:714–723.
69. G. W. Onan, 1984, Effects of Medium Composition and Conditioning on Adipose Development In Vitro, Thesis, University of Wisconsin–Madison.
70. D. A. K. Roncari, E. Y. Mack, and D. K. Yip, 1979, Enhancement of microsomal phosphatidate phospholydrolase and diacylglycerol acyltransferase activity by insulin during growth of rat adipocyte precursors in culture, *Can. J. Biochem.* **57**:573–577.
71. D. A. K. Roncari, 1981, Hormonal influences on the replication and maturation of adipocyte precursors, *Int. J. Obes.* **5**:547–577.
72. M. De La Llera, J. Kempe, F. De Martinis, and G. Rothblat, 1978, Very low density lipoprotein stimulation of triglyceride accumulation in rat preadipocyte cultures, *Biochim. Biophys. Acta* **529**:359–364.
73. M. De La Llera, G. Rothblat, and B. Howard, 1979, Cell triacylglycerol accumulation from very low density lipoprotein isolated from normal and hypertriglyceridemic human sera, *Biochim. Biophys. Acta* **574**:414–422.
74. M. De La Llera, J. M. Glick, and G. Rothblat, 1981, Mechanism of triglyceride accu-

mulation in rat preadipocyte cultures exposed to very low density lipoprotein, *J. Lipid Res.* **22**(2):245–253.
75. S. A. Pommier, 1984, In Vitro Adipose Development, Ph.D. Thesis; University of Wisconsin-Madison.
76. R. C. Wagner and M. A. Matthews, 1975, The isolation and culture of capillary endothelium cells from epididymal fat, *Microvasc. Res.* **10**:286–297.
77. P. A. Kern, A. Knedler, and R. H. Eckel, 1983, Isolation and culture of microvascular endothelium from human adipose tissue, *J. Clin. Invest.* **71**:1822–1829.
78. J. A. Madri and S. K. Williams, 1983, Capillary endothelial cell cultures phenotypic modulation by matrix components, *J. Cell Biol.* **97**(1):153–165.
79. P. Björntörp, G. Hansson, L. Jonasson, P. Pettersson, and G. Sypniewska, 1983, Isolation and characterization of endothelial cells from the epididymal fat pad of the rat, *J. Lipid Res.* **24**:105–112.
80. J. E. Novakofski, 1984, Conditioned media from cultured adipocytes has a chemotactic effect on capillary endothelial cells, *Fed. Proc.* **43**(3):787.
81. B. R. Zetter, 1984, Culture of capillary endothelial cells, in *Biology of Endothelial Cells,* E. A. Jaffe, ed., Martinus Nijhoff Publishers, Boston, Massachusetts, pp. 14–26.
82. Millipore Corporation, 1979, Enzymatic Tissue Dissociation and Cell Harvesting, Publication No. AR810, Millipore Corporation, Bedford, Massachusetts.
83. F. Grinnell and M. H. Bennett, 1982, Ultrastructural studies of collagen-cell interactions, in *Methods of Enzymology,* L. W. Cunningham, and D. W. Frederiksen, eds., **82**:535–544.
84. G. Humason, 1979, *Animal Tissue Techniques,* 4th ed., W. H. Freeman and Co., San Francisco, California, pp. 335–350.
85. H. K. Kleinman, E. B. McGoodwin, S. I. Rennad, and G. R. Martin, 1979, Preparation of collagen substrates for cell attachment: Effect of collagen concentration and phosphate buffer, *Anal. Biochem.* **94**:308–312.
86. H. K. Kleinman, R. J. Klebe, and G. R. Martin, 1981, Role of collagenous mixtures in the adhesion and growth of cells, *J. Cell Biol.* **88**:473–485.
87. S. C. Strom and G. Michalopoulos, 1982, Collagen as a substrate for cell growth and differentiation, in *Methods in Enzymology,* L. W. Cunningham, and D. W. Frederiksen, eds., **82**:544–555.
88. J. Yang and S. Nandi, 1983, Growth of cultured cells using collagen as substrate, *Int. Rev. Cytol.* **81**:249–286.
89. M. K. Patterson, 1978, Measurement of growth and viability of cells in culture, in *Methods in Enzymology,* W. B. Jakoby, and I. H. Pastan, eds., **58**:141–152.
90. M. Lanotte, D. Scott, T. M. Dexter, and T. D. Allen, 1982, Clonal preadipocytes with different phenotypes derived from murine marrow stroma: Factors influencing growth and adipogenesis in vitro, *J. Cell Physiol.* **111**:177–186.
91. R. Laub, G. Huybrechts-Godin, C. Peeters-Joris, and G. Vaes, 1982, Degradation of collagen and proteoglycins by macrophages and fibroblasts. Individual potentialities of each cell type and cooperative effects through the activation of fibroblasts by macrophages, *Biochem. Biophys. Acta* **721**:425–433.
92. R. Grinnell and C. R. Lamke, 1984, Reorganization of hydrated collagen lattices by human skin fibroblasts, *J. Cell Sci.* **66**:51–63.
93. J. M. A. Fausi and J. E. Aubin, 1984, Microfilament rearrangements during fibroblast-induced contraction of three-dimensional hydrated collagen gels, *Cell Motil.* **4**:29–40.
94. B. J. Klyde and J. Hirsch, 1979, Isotopic labeling of DNA in rat adipose tissue: Evidence for proliferating cells associated with mature adipocytes, *J. Lipid Res.* **20**:691–704.
95. B. J. Klyde and J. Hirsch, 1979, Increased cellular proliferation in adipose tissue of adult rats fed on a high fat diet, *J. Lipid Res.* **20**:705–715.

96. C. J. Meade, M. Ashwell, and C. Sowters, 1979, Is genetically transmitted obesity due to an adipose tissue defect? *Proc. R. Soc. Lond. B.* **205**:395–410.

97. M. Enser and M. Ashwell, 1983, Fatty acid composition of triglycerides from adipose tissue transplanted between obese and lean mice, *Lipids* **18**:776–780.

98. R. E. Miller and D. A. Carrino, 1981, An association between glutamine synthetase activity and adipocyte differentiation in cultured 3T3 L1 cells, *Arch. Biochem. Biophys.* **209**(2):486–503.

99. R. E. Scott, Y. Kankatsu, and D. L. Florine, 1983, Differential mitogenic effects of methyl isobutyl xanthine and tumor growth factor on G1-arrested 3T3 T proadipocytes at the predifferentiation G_D state and the growth-factor deficiency G_S state, *Exp. Cell Res.* **143**:405–414.

100. B. J. Hoerl, M. L. Wier, and R. E. Scott, 1984, Biological mechanisms for the loss of the differentiated phenotype by non-terminally differentiated adipocytes, *Exp. Cell Res.* **155**:422–434.

101. C. -H. Chen and S. C. Chen, 1981, Cell growth factor activity, *Exp. Cell Res.* **136**:43–51.

102. K. A. O. Ellem and J. F. Gierthy, 1977, Mechanism of regulation of fibroblastic cell replication. IV. An analysis of the serum dependence of cell replication based on Michaelis-Menten kinetics, *J. Cell Physiol.* **92**:381–400.

103. J. F. Lechner and M. E. Kaighn, 1979, Application of the principles of enzyme kinetics to clonal growth rate assays: An approach for delineating interactions among growth promoting agents, *J. Cell Physiol.* **100**:519–530.

104. B. Azzarone and A. Macieira-Chelho, 1982, Heterogeneity of the kinetics of proliferation within human skin fibroblastic cell populations, *J. Cell Sci.* **57**:177–187.

105. J. E. Novakofski, G. J. Hausman, and R. J. Martin, 1983, Replication rates of preadipocytes in culture, *Int. J. Obes.* **7**:377.

106. P. Skehan and S. J. Friedman, 1984, Non-exponential growth by mammalian cells in culture, *Cell Tissue Kinet.* **17**:335–343.

107. G. W. Onan, J. Novakofski, and R. G. Kauffman, 1982, A preadipocyte growth factor from adipose conditioned medium, *J. Anim Sci.* **55**(Suppl. 1):105.

108. P. M. Gullino, 1981, Angiogenesis factor(s), *Handb. Exp. Parmacol.* **57**:427–449.

109. J. J. Castellot, Jr., M. J. Karnovski, and B. M. Spiegelman, 1980, Potent stimulation of vascular endothelial cell growth by differentiated 3T3 adipocytes, *Proc. Natl. Acad. Sci. (USA)* **77**(10):6007–6011.

110. J. J. Castellot, Jr., M. J. Karnovski, and B. M. Spiegelman, 1982, Differentiation-dependent stimulation of neovascularization and endothelial cell chemotaxis by 3T3 adipocytes, *Proc. Natl. Acad. Sci. (USA)* **79**:5597–5601.

111. J. E. Novakofski, G. J. Hausman, and R. J. Martin, 1982, Preadipocyte development in culture. Influence of insulin and adipocyte conditioned media. *Fed. Proc.* **41**(3):736.

112. D. E. Dobson, J. J. Castellot, and B. M. Spiegelman, 1985, 3T3-adipocytes stimulate angiogenesis via prostaglandin-like molecules, *Fed. Proc.* **44**(5):1462.

113. J. F. Levine and F. E. Stockdale, 1984, T3-L1 adipocytes promote the growth of mammary epithelium, *Exp. Cell Res.* **151**:112–122.

114. P. Björntörp, 1983, Interactions of adipocytes and their precursor cells with endothelial cells in culture, *Exp. Cell Res.* **149**:277–287.

8

Hormonal Regulation of Glutamine Synthetase in Cultured 3T3-L1 Adipocytes

Richard E. Miller, Douglas M. Burns, and Basant Bhandari
Veterans Administration Medical Center Cleveland, Ohio

Certain subclones of cultured Swiss 3T3 fibroblasts (including L1 and F442A) differentiate (convert) with a high frequency into cells with morphological and biochemical characteristics of adipocytes (1–6). Although adipocyte conversion (differentiation) has been reported to occur spontaneously over 2 to 4 weeks following confluence (1), morphological and biochemical differentiation can be accelerated by 20 to 30% serum (2), insulin (3), 1-methyl-3-isobutylxanthine (MIX) (7), prostaglandin $F_2\alpha$ (7), indomethacin (8), dexamethasone (DEX) (9), and biotin (5) in various combinations. A recent study indicates that the adipogenic activity of serum is largely, if not completely, attributable to growth hormone (10). Interferon (11), bromodeoxyuridine (2), and retinoic acid (12) inhibit adipocyte conversion. Adipocyte conversion is accompanied by substantial increments in lipogenesis (1–5) and in the activities of several enzymes (5, 13–34), many of which are involved directly or indirectly in lipid metabolism. Lipid biosynthesis from leucine increases at least 30-fold (32). During differentiation, 3T3-L1 cells also acquire sensitivity to the metabolic effects of insulin and adrenocorticotropic hormone; they also display increased sensitivity to beta-adrenergic stimulation of adenylate cyclase activity (9, 19, 24, 35) and an increase in the activity of type I cyclic adenosine 3′5′-monophosphate (cAMP)-dependent protein kinase (31). They develop increased numbers of insulin receptors (9, 36–41), and loss of choleragen receptors (42), and they undergo substantial alterations in the biosynthesis of a large number of proteins (27, 28, 43, 44). Contrary to earlier indications (9), insulin receptors in cultured 3T3-L1 adipocytes are subject to insulin-induced down-regulation (45, 46). In 3T3 adipocytes, insulin has been shown to stimulate the phosphorylation of ribosomal protein S6 (47, 48) and the enzyme adenosine 5′-triphosphate (ATP) citrate lyase (49). In addition, insulin activates a tyrosine-specific cAMP-independent kinase that catalyzes the phosphorylation of the 92,000-dalton component of the insulin receptor and that may itself be a part of that receptor (50). Enzymes that increase manyfold during adipocyte conversion of 3T3 cells include ATP citrate lyase (5) acetyl coenzyme A (acetyl CoA) carboxylase (5), fatty acid synthase (5, 27, 28), lipoprotein lipase (17, 21, 22), glycerol-3-phosphate acyltransferase (16, 20), malic enzyme (16), glycerol-3-phosphate dehydrogenase (G3PD) (18, 43), pyruvate carboxylase (15, 26), catecholamine-stimulated adenylate cyclase (19), hormone-sensitive particulate cAMP phosphodiesterase (33, 34), hormone-sensitive lipase (30), aldolase (43), glyceraldehyde phosphate dehydrogenase (43), lactate dehydrogenase (43), hexokinase (13, 14), glucose-6-phosphate dehydrogenase (14, 25), and glutamine synthetase (G-S) (13, 14). When injected subcutaneously into Balb C (athymic) nude mice, 3T3-F442A preadipocytes differentiate into fat pads with typical adipose tissue morphology (51).

Differentiation-associated increases and dibutyryl cAMP (Bt_2 cAMP)-mediated decreases in several enzyme activities were shown to correlate, in 3T3-F442A adipocytes, with specific enzyme synthesis rates and with the cellular content of specific functional (translatable) messenger ribonucleic acids (mRNAs) (43, 52). In addition, differentiation-associated decreases in tubulin and actin biosynthesis in 3T3-F442A cells correlate both with decreased tubulin and actin mRNA content (determine by specific cDNA hybridization) (53). Recently, Spiegelman and co-workers (54) cloned mRNA from 3T3-F442A adipocytes and selected recombinants complementary to mRNAs for G3PD, two other major differentiation-dependent proteins of molecular weights 28,000 and 13,000, and actin. They found by hybridization analysis that during adipocyte differentiation of 3T3-F442A cells the cellular mRNA content increased manyfold for glycerol-3-P dehydrogenase mRNA, and for the mRNAs encoding the proteins of molecular weights 28,000 and 13,000. By contrast, actin mRNA decreased during adipocyte conversion. These changes in mRNA content were nonsynchronous. It is noteworthy that the cDNA library prepared by Spiegelman and co-workers from insulin-treated 3T3-F442A adipocytes would be expected to contain a low frequency of recombinants complementary to GS mRNA. This low frequency is because insulin-treated 3T3-F442A cells have a relatively low GS biosynthetic rate (unpublished).

Cultured 3T3 adipocytes have many advantages over rat epididymal adipocytes for the study of adipocyte intermediary metabolism: (a) The 3T3 adipocytes have not been subjected to traumatic isolation procedures involving proteolytic enzymes; (b) the duration of hormonal treatment or nutritional perturbation can be greatly expanded since monolayer cultures are stable for days to weeks, while isolated epididymal adipocytes or adipose tissue are relatively unstable; (c) finally, hormonal and nutritional perturbation can be precisely controlled in cultured cells. By contrast, these are very difficult to define or control in whole animals due to the complexity of in vivo regulatory mechanisms and the functioning of multiple endocrine organs.

At least two problems invariably arise during studies of 3T3 adipocytes. First, adipocyte differentiation is often incomplete and asynchronous. The monolayers begin to differentiate in isolated patches and cells may later be at varying stages of morphological differentiation. This is not a major problem if all or nearly all of the cells differentiate since they appear to become relatively uniform 2 to 4 days after incubation with DEX and MIX. Occasionally, however, substantial areas on a dish retain the preadipocyte morphology even when the cells were uniformly confluent at the time of incubation with DEX and MIX. We have no explanation for this nonuniform differentiation. However, the phenomenon was well described by Green and Kehinde (1, 3, 4) and Green and Meuth (2) and should be expected.

A more perplexing problem is failure of 3T3-L1 cells to differentiate. This happens occasionally for no apparent reason. Failure to differentiate seems to occur more frequently as passage number increases, but there is no consistent relationship between this phenomenon and passage number. The 3T3-L1 cells up to passage 15 are often used for our experiments.

Mature mammalian fat cells have been studied in great detail and are known to respond to a variety of hormones (55). Available data strongly suggest that 3T3-L1 adipocytes are sufficiently similar to in vivo adipocytes to provide a relevant model for investigating (a) the regulation of hormone receptors, (b) the coupling of these receptors to their physiological effector systems, and (c) the long-term effects of hormones and nutritional state on adipocyte intermediary metabolism. Although other adipocyte culture systems have been developed (56–67), few, if any, have been as thoroughly studied as 3T3 adipocytes, and few have the proven potential of 3T3 adipocytes for studies of metabolic regulation.

Glutamine occupies a key position in mammalian nitrogen metabolism. Its amide nitrogen can be utilized for the biosynthesis of several widely diverse compounds, including purine and pyrimidine nucleotides, amino sugars, and various amino acids (68). L-Glutamine is the amino acid of highest concentration in human plasma (69) and whole blood (70). Studies in man and in the rat indicate that glutamine is a vehicle of major importance for nitrogen and carbon transport from peripheral tissues to the splanchnic circulation (69–83). In adipose tissue (84) and skeletal muscle (83) the production of glutamine seems to be the major route for disposal of amino groups released during the metabolism of branched-chain amino acids. It has been estimated that in the rat, intestinal mucosal utilization of L-glutamine would result in a total turnover of the glutamine pool every 4.6 min (75). Over the past several years it has been demonstrated that L-glutamine is the preferred oxidative substrate for intestinal mucosa (75, 85), reticulocytes (86), and rapidly dividing cells in tissue culture (87–89). Glutamine synthetase, the enzyme responsible for glutamine synthesis, is therefore a pivotal enzyme in nitrogen metabolism. This enzyme links the myriad of catabolic processes by which ammonia is produced with numerous biosynthetic processes directed ultimately toward the formation of proteins, nucleic acids, and complex polysaccharides. In addition, glutamine serves as an important transport molecule for nitrogen and carbon from peripheral tissue to the splanchnic circulation. Therefore, GS represents a potentially strategic target for hormonal regulation of nitrogen metabolism.

Glutamine synthetase from *Escherichia coli* and related bacteria is regulated by a complex enzyme cascade involving adenylylation/deadenylylation and uridylylation/deuridylylation (90–92). Mammalian GS has been extensively studied in brain, liver, and kidney (93), and is also known to occur in spleen, testis, retina (94), skeletal muscle (95), and adipose tissue (96–99). Mammalian glutamine synthetases are susceptible to feedback regulation by end products of glutamine metabolism. However, little evidence would indicate that mammalian GS is regulated by any posttranslational covalent modification (91–93).

Over the past 25 years there has been considerable interest in the regulation of GS in cultured cells. Glutamine synthetase regulation has been studied in HeLa cells (100), mouse L cells (101–103), W1-38 human fibroblasts (104, 105), primary rat hepatocyte cultures (106, 107), GH1 pituitary tumor cells (108), Chinese hamster cells (109–112), hepatoma tissue culture (HTC) cells (113–121), and 3T6 cells (122). In many but not all cases, glutamine deprivation and/or glucocorticoid treatment has reportedly yielded an increase in GS activity. Growth hormone (107) but not triiodothyronine (108) may also increase GS activity.

Recently, manyfold amplification of the GS gene has been achieved in subclones of 3T6 cells (122) and Chinese hamster ovary (CHO) cells (123). Both clones were selected by growth of cells in glutamine-free medium containing gradually increasing concentrations of the GS inhibitor methionine sulfoximine. Cellular GS content was estimated to be 1% of the soluble protein in the 3T6 clone and 30% of the soluble protein in the CHO clone.

Glutamine synthetase gene amplification was estimated to be in the range of 700- to 1,400-fold in the CHO mutant clone. An amplified mutant-specific fragment of CHO nuclear deoxyribonucleic acid (DNA) was identified and employed to prepare recombinant genomic GS clones (including pGS-103) in the plasmid vector pUC-9.

During adipocyte conversion of confluent 3T3-L1 cells, we found that glutamine synthetase (GS) activity increases more than 100-fold regardless of the culture conditions employed to achieve differentiation (25). This rise in GS activity is paralleled by a rise in cellular GS content, and it is closely associated with morphological adipocyte differentiation, with an increase in the rate of triglyceride synthesis from acetate, with increased protein and DNA, and with increases in the activities of G3PD glucose-6-phosphate dehydrogenase, and hexokinase (13, 14, 25). Incubation of 3T3-L1 adipocytes with Bt_2cAMP plus theophylline for 24 hr yields a substantial decrease in cellular GS activity and content but little, if any, change in protein per culture or in the activities of hexokinase, glucose-6-phosphate dehydrogenase, or G3PD. During adipocyte conversion of confluent 3T3-L1 cells and in fully differentiated 3T3-L1 adipocytes, hydrocortisone increases GS activity and cellular content while insulin has little effect or decreases these parameters (13, 14).

In order to begin to understand the mechanisms for these observed effects, we studied the effects of Bt_2cAMP, insulin, and hydrocortisone on the biosynthesis and degradation of GS and soluble protein (13, 124). The synthesis rate was estimated from the incorporation of L-[^{35}S]methionine into GS and protein. The degradation rate was estimated from the first-order disappearance of radioactivity from GS and soluble protein in 3T3-L1 adipocytes previously incubated with L-[^{35}S]methionine. Glutamine synthetase was isolated by immunoprecipitation followed by sodium dodecylaulfate (SDS)-polyacrylamide gel electrophoresis. Protein was isolated by precipitation with trichloroacetic acid. Incorporation of L-[^{35}S] methionine into GS was compared with incorporation of L-[^{35}S] methionine into soluble protein. The resulting ratio between GS radioactivity and protein radioactivity provided both an estimate of the relative rate of GS synthesis and a control between experimental conditions, for possible differences in the time-integrated specific radioactivity of the methionine pool. Effects of experimental conditions on the cellular methionine pool were assessed by measuring the trichloroacetic acid-soluble radioactivity in cell lysates.

Recently, we selected a GS cDNA containing placmid clone (pGSRK-1) from a rat kidney cDNA library (125) with two probes: (a) a mixture of synthetic tetradecameric deoxyribonucleotides corresponding to all possible codon combinations for a pentapeptide sequence from bovine brain GS (126) and (b) a CHO genomic GS DNA clone from mutant CHO cells with multiple copies of the GS gene (123). Northern hybridization analysis of RNA from 3T3-L1 adipocytes with pGSRK-1 yields two hybridizable GS RNAs of length 3.2 and 1.6 kilobases (kb) (127, 128). Analyses of the Northern blots probed with pGSRK-1 indicate that the 3.2-kb GS-specific RNA is more abundant than the 1.6-kb GS RNA. Quantitative dot-blot hybridization analysis indicates that DEX increases GS mRNA, while both insulin and Bt_2cAMP decrease GS mRNA and/or prevent the DEX-mediated increase (128). Thus, data suggest that there are at least two GS mRNAs in 3T3-L1 adipocytes and that they are regulated in parallel by DEX, insulin, and (Btu_2cAMP).

REGULATION OF GLUTAMINE SYNTHETASE IN 3T3-L1 ADIPOCYTES

Growth and Morphology

3T3-L1 cells were obtained from Dr. Howard Green (Massachusetts Institute of Technology) and are now routinely cultured in this laboratory. Unless otherwise indicated, cells were plated near confluence (2 to 4 $\times$ 10^4 cells/cm^2) and grown to confluence in Dulbecco's modified Eagle's (DME) medium containing 10% calf serum (CS), 100 units/ml penicillin, and 100 μg/ml streptomycin. Confluent 3T3-L1 cultures maintained in the presence of insulin acquired the adipocyte phenotype. By contrast, 3T3 L1 cultures maintained for the same time interval in the absence of added insulin retained the preadipocyte morphology. This is illustrated in Figure 8-1, which shows phase-contrast micrographs of 3T3-L1 cells maintained for 14 days after confluence in the absence (left) or presence (right) of 2 μg/ml added insulin (magnification is 200$\times$). As is indicated below, adipocyte conversion of confluent 3T3-L1 cells can be more rapidly achieved by incubation with DEX and MIX or by incubation with insulin and hydrocortisone. Complete details of methods employed are given below (See Methods).

Analytical Methods

In our studies (13, 14, 25, 124), at the indicated times after confluence, cultures were washed two or three times with phosphate-buffered saline (PBS), scraped into buffer, and frozen at -80°C prior to assays. Thawed samples were clarified by centrifugation (15,600 $\times$ g, 4°C, 10 min), and aliquots were taken for assay of enzyme activities, protein, DNA, and radioactivity. Details of analytical methods are given below (See Experimental Procedures).

Glutamine Synthetase Activity and Adipocyte Differentiation

In the experiment shown in Figure 8-2 (25), 3T3-L1 cells were plated in 20-cm^2 dishes and maintained after confluence (day zero) in the absence (▲) or presence (●) of 2 μg/ml added insulin. At the indicated times duplicate cultures were harvested for analyses of enzyme activities and protein. Duplicate cultures were also incubated with 5 mM [^{14}C]acetate to determine the rate of triglyceride synthesis. Glutamine synthetase activity increased more than 100-fold during the insulin-mediated adipocyte conversion of confluent 3T3-L1 cells (Fig. 8-2*A*). The rise in GS activity closely paralleled acquisition of the adipocyte morphology. In addition, the increase in GS activity was closely associated with the increase in protein (Fig. 8-2*C*), and the increases in the activities of G3PD (Fig. 8-2*D*), glucose-6-phosphate dehydrogenase (Fig. 8-2*E*), and glutamate dehydrogenase (Fig. 8-2*F*). Neither morphological adipocyte conversion nor biochemical evidence of dif-

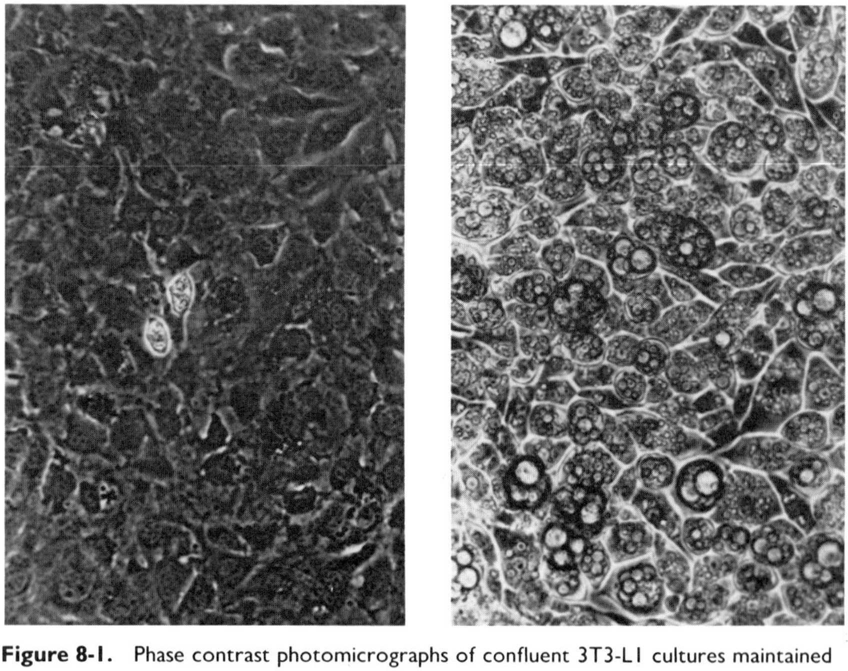

Figure 8-1. Phase contrast photomicrographs of confluent 3T3-L1 cultures maintained for 14 days in the absence (left) or present (right) of 2 μg/ml added insulin. (magnification is 200 ×). The culture medium, containing 10% CS, was replaced with fresh medium every other day.

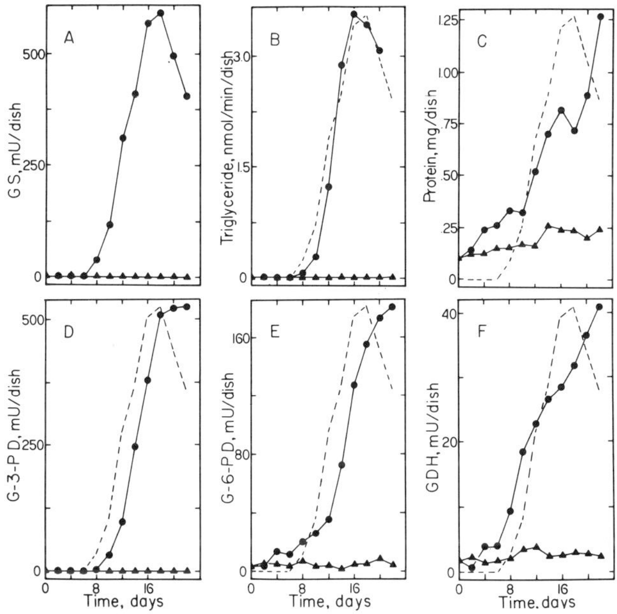

Figure 8-2. Temporal relationship in 3T3-L1 cultures between GS activity, the rate of triglyceride synthesis from acetate, soluble protein, GPDH activity, glucose-6-phosphate dehydrogenase activity, and glutamate dehydrogenase activity. Cultures were plated in 60 mm (20 cm^2) dishes at a density of 2.1 x 10^4 cells/cm^2, grown to confluence in the absence of added insulin, and maintained after confluence (Day 0) in standard medium in the absence (▲) or presence (●) of 2 μg/ml added insulin. Cultures were fed with 5 ml of fresh medium every 2 days. At the indicated times duplicate cultures were harvested for enzyme activities and protein, and duplicate cultures were incubated with 5 mM [^{14}C]acetate for determination of the rate of triglyceride synthesis. Thawed samples harvested for enzyme activity and proteins were sonicated and clarified by centrifugation prior to assays. (*A*) GS, milliunits per dish; (*B*) triglyceride synthesis from acetate, nmol/min/dish; (*C*) soluble protein, mg/dish; (*D*) (G-3-PD) activity, milliunits per dish; (*E*) glucose-6-phosphate dehydrogenase (G-6-PD) activity, milliunits per dish; (*F*) glutamate dehydrogenase (GDH) activity, milliunits per dish. Data points are the averages of values from duplicate cultures. Variation of duplicate values from the mean was 10% or less for the parameters measured. Duplicate assays performed on a single sample differed by less than 5%. Each culture was analyzed for all of the parameters indicated except triglyceride synthesis. For reference purposes the curve for GS activity in insulin-treated cultures (from panel *A*) is repeated as a broken line in panels *B, C, D, E,* and *F* with the same ordinate scale as in panel A. In addition, the maximum ordinate value for each parameter measured is plotted at the same distance above the abscissa. (*From Miller and Carrino, 1981, p. 491.*)

ferentiation occurred in confluent 3T3-L1 cells maintained for 22 days in the absence of added insulin. For reference, the curve for GS activity in insulin-treated cultures (from panel *A*) is repeated as a broken line in the other panels with the same ordinate scale as in panel *A*.

Culture Conditions Required for Adipocyte Differentiation

We find that adipocyte differentiation of 3T3-L1 cells is associated with a rise in cellular GS regardless of the culture conditions used to achieve the differentiated phenotype (Table 8-1). Incubation of confluent 3T3-L1 cells for 48 hr with 0.25 uM DEX and 0.5 mM MIX (DEX-MIX), then for 5 days without added DEX-MIX, yielded adipocyte differentiation and a rise in GS, glucose-6-phosphate dehydrogenase, G3PD protein, and DNA in cultures maintained in medium containing 10% fetal calf serum (FCS). By contrast, differentiation failed to occur and enzymes, protein, and DNA did not increase in similarly treated cultures maintained in medium containing 10% calf serum. When cultures were maintained for 7 days in the presence of either calf serum or fetal calf serum, insulin plus hydrocortisone yielded differentiation and increases in enzyme activities, protein, and DNA. Hydrocortisone markedly accelerated insulin-mediated differentiation, but hydrocortisone alone yielded no adipocyte differentiation in cultures maintained with 10% calf serum (not shown). By contrast, insulin alone was adequate to promote differentiation in calf serum-containing medium (Fig. 8-2). Thus insulin may be required for adipocyte conversion of confluent 3T3-L1 cells maintained in the presence of calf serum.

Effects of Bt_2cAMP on GS in 3T3 Adipocytes

Incubation of 3T3 adipocytes for 22 hr with 0.5 mM Bt_2cAMP plus 0.5 mM theophylline, with 0.2 mM 8-bromo-cAMP, with 10 uM epinephrine, or with 1 μg/ml 1-24 adrenocorticotropin (ACTH) decreased GS activity by more than 60%. During the same incubation period these compounds had no effect on protein or on the activities of glucose-6-phosphate dehydrogenase or hexokinase.

Immunotitration of GS activity was performed in order to determine the relationship between GS activity and GS content in 3T3-L1 cells during differentiation and in 3T3-L1 adipocytes during treatment with Bt_2cAMP plus theophylline or with 8-Br-cAMP plus theophylline (13). The time course for GS specific activity is shown in Figure 8-3*A* for cultures maintained after confluence (Day 0) with 10 μg/ml insulin (——) and for cultures

Table 8-1. Effect of Hormones and Serum on Enzymes, Protein, and DNA in Confluent 3T3-L1 Cells[a]

Culture conditions[b]									
DEX (μM)	*MIX (mM)*	*Insulin (μg/ml)*	*Cortisol (μg/ml)*	*10% Serum*	*Glutamine Synthetase (mU/dish)*[c]	*Glucose-6-phosphate Dehydrogenase (mU/dish)*[c]	*Glycerol-3-phosphate Dehydrogenase (mU/dish)*[c]	*Protein (μg/dish)*[c]	*DNA (μg/dish)*[c]
0.25	0.5	—	—	FCS	1,392 ± 170	67 ± 4	377 ± 58	610 ± 50	29 ± 1
0.25	0.5	—	—	CS	57 ± 3	12 ± 2	N.D.[d]	323 ± 10	16 ± 2
—	—	1.0	0.1	FCS	1,182 ± 121	148 ± 26	403 ± 50	979 ± 50	32 ± 4
—	—	1.0	0.1	CS	1,010 ± 52	103 ± 13	246 ± 13	1,177 ± 34	42 ± 3

[a] 3T3-L1 cells were plated in 20 cm^2 culture dishes at a density of 2×10^4 cells/cm^2 and grown to confluence in standard culture medium containing 10% calf serum (cs) and no added hormones. Confluent cultures were subsequently maintained for 7 days as indicated. Dexamethasone (DEX) and 1-methyl-3-isobutylxanthine (MIX) were present only for the first 48 hr after confluence. Subsequently, cultures treated with DEX and MIX were maintained in medium containing no added hormone or phosphodiesterase inhibitor. Insulin and hydrocortisone (cortisol) were present during the entire 7 days after confluence. (From Miller and Carrino, 1981)

[b] FCS, fetal calf serum

[c] Data are mean ± SD for three replicate cultures. Duplicate assays of the same culture differed by less than 5%. Each culture was analyzed for all of the indicated parameters.

[d] N.D., not detectable.

treated, beginning on day 12, with 0.5 mM Bt_2cAMP plus 0.5 mM theophylline (— —) or with 0.2 mM 8-Bromo-cAMP plus 0.5 mM theophylline (------). Sonicates from cultures designated by unique symbols in Figure 8-3*A* were subjected to immunotitration. The GS-specific activity in these

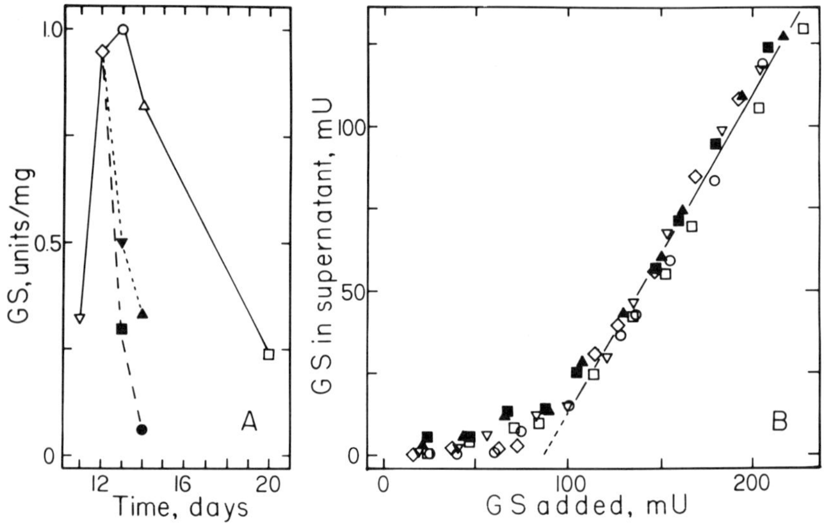

Figure 8-3. Immunotitration of GS activity in clarified sonicates prepared from 3T3-L1 adipocytes. 3T3-L1 cells were plated in 50-cm^2 dishes at a density of 3×10^4 cells/cm^2 and maintained in medium containing no added insulin. On the 5th day after plating (Day 0 of insulin treatment), cells had reached a density of 5.74×10^4 cells/cm^2, and thereafter all cultures were maintained in medium containing added insulin at 10 μg ml. On Day 12 of insulin treatment, some of the cultures were shifted to medium containing insulin at 10 μg/ml and either 0.5 mM Bt_2cAMP plus 0.5 mM theophylline or 0.2 mM 8-bromo-cyclic AMP plus 0.5 mM theophylline. The culture medium was changed every 2 days until Day 12. Thereafter it was changed daily. Cultures treated with insulin alone (——) were harvested 11 (▽)12 (◇), 13 (○), 14 (△), or 20 (□) days after initiation of insulin treatment. Insulin-treated cultures also treated with Bt_2cAMP plus theophylline (— —) were harvested on Days 13 (■) and 14 (●), 24 and 48 hr after initiation of treatment with Bt_2cAMP plus theophylline. Insulin-treated cultures also treated with 8-bromo-cyclic AMP plus theophylline (------) were harvested on Days 13 (▼) and 14 (▲), 24 and 48 hr after initiation of treatment with 8-bromo-cyclic AMP plus theophylline. Immunotitration was performed with sonicates from cultures designated symbolically by ▽, ◇, ○, □, ■, and ▲. *A*, GS specific activity. Data points are the means of values obtained from duplicate cultures; the variation between duplicate cultures was 10% or less for the parameters measured. *B*, immunotitration of GS activity. (*Miller and Carrino, 1980, p. 5495.*)

sonicates varied over a four-fold range. Immunotitration was performed by incubating increasing amounts of each cell sonicate with a constant amount (2 μl) of GS antiserum. Total GS (Fig. 8-3*B*, abscissa) was determined in incubation mixtures containing sonicate and antiserum. (It is noteworthy that under the conditions of immunotitration our GS antiserum did not significantly inhibit GS activity.) The mixture was then sedimented to separate antibody-bound GS from free GS. Supernatant activity (free GS) is indicated on the ordinate.

The amount of GS activity specifically precipitated by the added antiserum (Fig. 8-3*B*, abscissa intercept) was determined for each sonicate by extrapolation to the abscissa of the line through the linear portion of the titration curve (ordinate values greater than 25 milliunits). This abscissa intercept provides an estimate of the activity contained in the immunoprecipitated GS. Although only one line is shown, a separate line can be drawn for each sonicate evaluated. Since these lines did not differ significantly in slope or intercept, only one line is shown. Thus, the immunotitration equivalence point was virtually the same for all sonicates tested, indicating that observed changes in GS activity are due to changes in the cellular content of GS molecules.

Specificity of Immunoprecipitation of ^{35}S-labeled GS

Glutamine synthetase antibody was used for isolation of radiolabeled GS from 3T3-L1 adipocytes (13). Immunoprecipitates were prepared from clarified lysates of 3T3 adipocytes that had been incubated with L-[^{35}S]methionine. When subjected to sodium dodecyl sulfate (SDS)-polyacrylamide slab gel electrophoresis, the washed immunoprecipitates were found to contain as the major radioactive component a protein with a mobility identical to that of purified rat liver GS (see ref. 13,d Fig. 6). When immunoprecipitates were prepared from increasing amounts of the same [^{35}S]methionine-labeled lysate, radioactivity in the GS recovered from the SDS-polyacrylamide gel was directly proportional to the amount of lysate used. Immunoprecipitates were also prepared from lysates of ^{35}S-labeled 3T3-C2 cultures. (The 3T3-C2 cells contained no detectable GS activity.) These lysates were first supplemented with purified rat liver GS. When such 3T3-C2 immunoprecipitates were subjected to SDS-polyacrylamide gel electrophoresis and analyzed by autoradiography of the dried gels, no radioactive protein had the mobility of GS. Thus, our immunoprecipitation procedure, coupled with SDS-polyacrylamide gel electrophoresis, allows isolation and quantification of radiolabeled GS.

Bt_2cAMP Decreases GS Biosynthesis

We studied the effect of Bt_2cAMP plus theophylline on the synthesis of GS and protein in 3T3 adipocytes (13). Incubation of the adipocytes with Bt_2cAMP plus theophylline resulted in a rapid decline in the apparent rate of GS synthesis with little effect on the synthesis of soluble protein. In the experiment shown in Fig. 8-4, 3T3 adipocytes were maintained for 25 hr with 1 μg/ml insulin and 0.1 μg/ml hydrocortisone (●) or with 1 μg/ml insulin, 0.1 μg/ml hydrocortisone, 1 mm Bt_2cAMP, and 0.5 mM theophylline (▲). Glutamine synthetase activity in Bt_2cAMP-treated cultures decreased with a half-time of 14 to 18 hr (Fig. 8-4*A*). The GS biosynthetic rate was estimated from the incorporation of L-[^{35}S]methionine into GS isolated by immunoprecipitation followed by SDS-polyacrylamide gel electrophoresis. The Bt_2cAMP-mediated fall in GS activity was accompanied by a much more prompt (half-time about 4 hr) and precipitous decline in the rate of GS biosynthesis. (Fig. 8-4*B* and 4*C*). By 4 hr of Bt_2cAMP treatment, GS activity had declined by less than 5% compared with control cultures. After the same interval, the apparent synthesis rate for the enzyme had declined by more than 50%. By 13 hr of Bt_2cAMP treatment, GS activity had declined by about 45% and the apparent GS biosynthetic rate had declined by more than 75%. By contrast, this treatment had little if any effect on soluble protein per culture (Fig. 8-4*D*) and this treatment had a substantially smaller effect on the incorporation of [^{35}S]methionine into soluble protein (Fig. 8-4*E* and *F*). In a separate experiment Bt_2cAMP plus theophylline was found to decrease the half-life of GS from 24.5 hr in control cultures to 16 hr in Bt_2cAMP-treated cultures (13).

Hydrocortisone Increases GS Biosynthesis

In the experiment shown in Figure 8-5 (124) confluent 3T3-L1 adipocytes were incubated for the indicated times with the following: no additions (○), 1 μg/ml insulin (■), 0.1 μg/ml hydrocortisone (▲) or 1 μg/ml insulin plus 0.1 μg/ml hydrocortisone (●). These cultures had been maintained for the preceding 8 days in the absence of added hormones. Beginning at the indicated times, four replicate cultures from each treatment group were incubated for 2 hr with 2 ml of fresh medium containing the same additions and also 28 microcuries (μCi)/ml L-[^{35}S]methionine. Cultures were harvested upon completion of this 2-hr incubation. Treatment with hydrocortisone with or without insulin resulted in a gradual rise in GS activity (4.9-fold by 32 hr) (Fig. 8-5*A*) and a rapid increase in the incorporation of L-[^{35}S]methionine into GS (approximately fivefold by 4 hr) (Fig. 8-5*B*). By contrast, hydrocortisone had little, if any, effect on the incorporation of

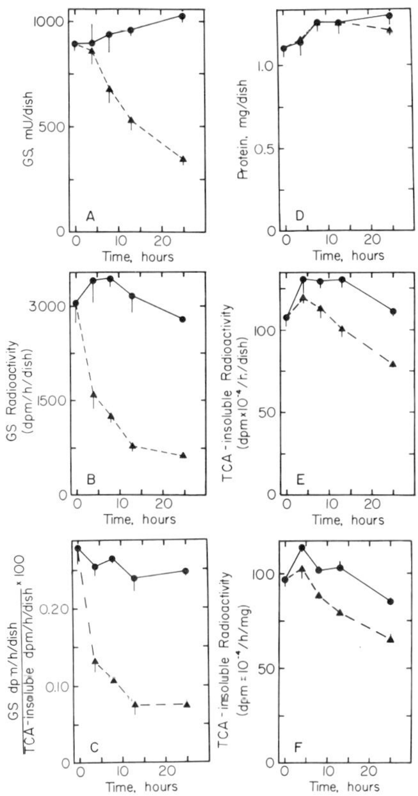

Figure 8-4. Effect of Bt_2cAMP plus theophylline on the synthesis of GS and protein in 3T3-L1 adipocytes. Confluent 3T3-L1 cultures were maintained in the presence of insulin at 1 μg/ml and hydrocortisone at 0.1 μg/ml for 12 days, during which time a substantial fraction of the cells acquired the adipocyte phenotype. During this period, fresh medium was added every 2 days. On Day 13 of treatment with insulin and hydrocortisone, fresh medium was added containing either insulin at 1 μg/ml and hydrocortisone at 0.1 ug/ml (●) as before, or at insulin 1 μg/ml, hydrocortisone at 0.1 μg/ml, 1.0 mM Bt_2cAMP, and 0.5 mM theophylline (▲). At the indicated times, the culture medium was removed by aspiration and the cultures were incubated (37°C) for 60 min with 2 ml of medium of the same composition as indicated above but also containing 26.3 μCi/ml L-[^{35}S]methionine. After incubation with L-[^{35}S]methionine, cultures were harvested and samples were stored at -80°C prior to analysis. Each sample was analyzed for all measured parameters. *A*, GS activity; *B*, GS radioactivity; *C*, radioactivity, as a fraction of trichloroacetic acid (TCA)-insoluble radioactivity; *D*, soluble protein; *E* and *F*, TCA-insoluble radioactivity. Each data point is the mean plus or minus the standard deviation for three or four replicate cultures. (*From Miller and Carrino, p. 5496, 1980.*)

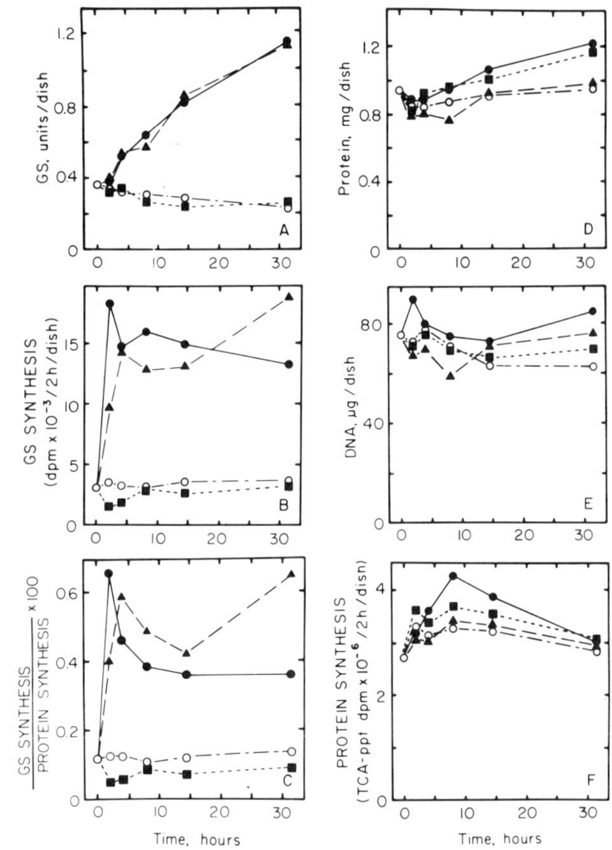

Figure 8-5. Incorporation of L-[^{35}S]methionine into GS and soluble protein in 3T3-L1 adipocytes. Confluent 3T3-L1 cells were incubated for 48 hr with medium containing 10% FCS, 0.25 µM DEX, 0.5 mM MIX, and 1 µg/ml added insulin. Thereafter, cultures were maintained for 8 days with medium containing 10% FCS and no other additions. Beginning 8 days after treatment with DEX, MIX, and insulin, cultures were incubated with the following: ○, no additions; ■, 1 µg/ml insulin; or ▲, 0.1 µg/ml hydrocortisone; or ●, 1 µg/ml insulin and 0.1 µg/ml hydrocortisone. Beginning at the indicated times four replicate cultures from each treatment were incubated for 2 hr with 2 ml of fresh medium containing 28 µCi/ml L-[^{35}S]methionine (138 µgCi/µmol). This medium contained the same additions indicated above. Cultures were harvested upon completion of the 2-hour incubation with L-[^{35}S]methionine. Each culture was analyzed for all indicated parameters. Data points are the mean for four replicate cultures. TCA-ppt, trichloroacetic acid-precipitable; GS synthesis, incorporation of L- [^{35}S]methionine into immunoprecipitable GS. Protein synthesis, incorporation of L-[^{35}S]methionine into TCA-ppt material. (*From Miller et al., p. 541, 1983.*)

[^{35}S]methionine into soluble protein (Fig. 8-5*F*). After 31.5 hr the ratio between incorporation of [^{35}S]methionine into GS and into soluble protein was 0.14 ± 0.03% (mean ± S.D.) in control cultures, 0.09 ± 0.02% in insulin-treated cultures, 0.65 ± 0.04% in hydrocortisone-treated cultures, and 0.35 ± 0.03% in cultures treated with insulin plus hydrocortisone (Fig. 8-5*C*). Radioactivity in the cellular methionine pool (estimated from trichloroacetic acid-soluble radioactivity in cell lysates) was not affected by treatment with hydrocortisone but it increased by 1.4- to 1.8-fold during insulin treatment (data not shown).

Insulin Decreases GS Biosynthesis in Hydrocortisone-Treated 3T3 Adipocytes

In the experiment shown in Figure 8-6 (128) confluent 3T3 adipocytes were treated with 0.1 μg/ml of hydrocortisone for 20 hr before beginning the indicated treatments. Beginning at time zero in the figure, cultures were incubated with fresh medium containing the following: no added hormones (○), 1 μg/ml insulin (■), 0.1 μg/ml hydrocortisone (▲), or 1 μg/ml insulin plus 0.1 μg/ml hydrocortisone (●). At the indicated times cultures were pulse labeled for 2 hr with medium containing 37 μCi/ml [^{35}S]methionine (185 mCi/mmol). Treatment with insulin had little effect on GS activity per dish but it decreased fractional GS synthesis (the ratio between [^{35}S]methionine incorporated into GS and [^{35}S]methionine incorporated into soluble protein) whether or not hydrocortisone remained present (Fig. 8-6). The continued presence of hydrocortisone yielded an increase in GS fractional synthesis from 0.8 to 1.0% during the first 4 hr. By contrast, withdrawal of hydrocortisone yielded a decline in GS synthesis that was accelerated somewhat by the addition of insulin. After 4 hr of treatment, the ratio between radioactivity incorporated into GS and radioactivity incorporated into soluble protein was 1.0% in hydrocortisone-treated cultures, 0.34% in insulin plus hydrocortisone-treated cultures, 0.18% in insulin-treated cultures, and 0.35% in cultures maintained with no added hormones. In addition to GS activity and GS radioactivity (Fig. 8-6*A* and *B*), protein and protein radioactivity (Fig. 8-6*D* and *E*), and relative GS biosynthetic rate (Fig. 8-6*C*), each culture was also analyzed for trichloroacetic acid-soluble radioactivity (*F*), DNA (*G*), G3PD activity (*H*), and glucose-6-phosphate dehydrogenase activity (*I*). It is noteworthy that no culture condition had a large effect on DNA. By contrast, insulin with or without hydrocortisone increased trichloroacetic acid-precipitable and -soluble radioactivity in parallel and it also increased protein and the activities of GPDH and glucose-6-phosphate dehydrogenase.

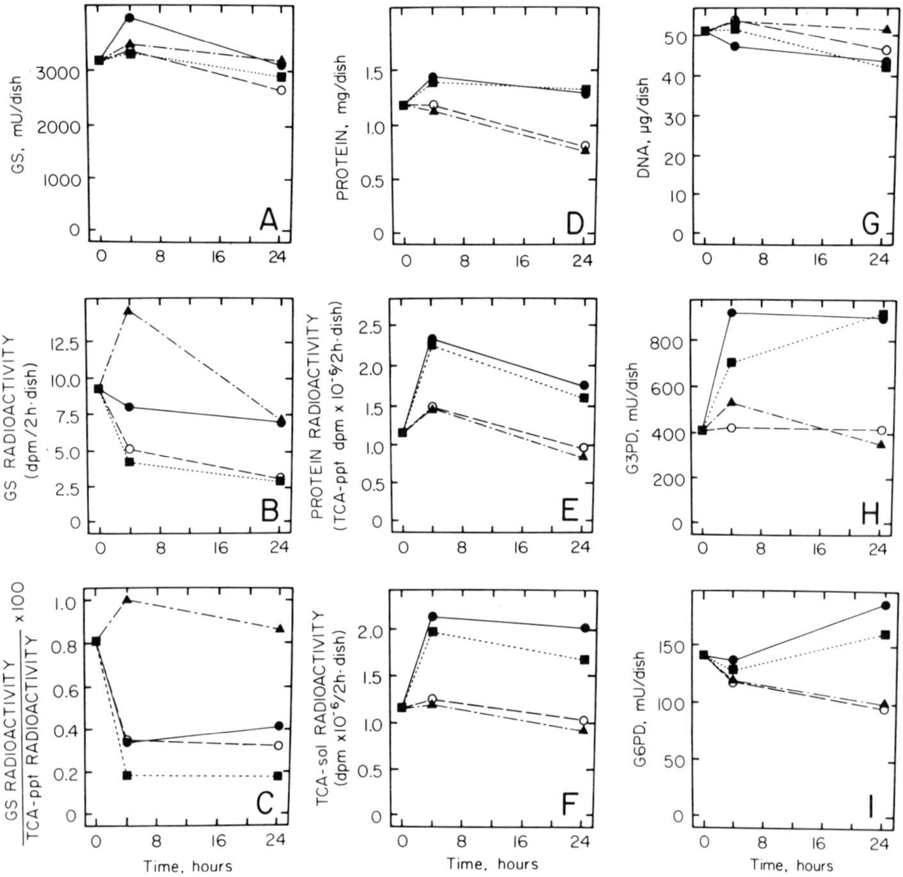

Figure 8-6. Incorporation of L-[^{35}S]methionine into GS and soluble protein in hydrocortisone-treated 3T3-L1 adipocytes. Confluent 3T3-L1 cells were treated with 0.25 µM DEX, 0.5 mM MIX, and 1 µg/ml insulin as described in the legend of Figure 8–5. Thereafter, the cultures were maintained for 7 days with medium containing no added hormone. Beginning 7 days after treatment with DEX, MIX, and insulin, all cultures were treated with medium containing 0.1 µg/ml added hydrocortisone. After 20 hr of incubation with hydrocortisone (zero time in the figure), the medium was removed by aspiration, cultures were washed once with 5 ml of phosphate-buffered saline, and fresh medium was added containing the following: ○, no additions; ■, 1 µg/ml insulin; ▲, 0.1 µg/ml hydrocortisone; or ●, 1 µg/ml insulin and 0.1 µg/ml hydrocortisone. Beginning at the indicated times, four replicate cultures from each treatment were incubated for 2 hr with 2 ml of fresh medium containing 37 µCi/ml L-[^{35}S]methionine (185 µCi/µmol). This medium contained the same additions indicated above. Cultures were harvested at the end of the 2-hour incubation with L-[^{35}S]methionine. Data points are the mean for four replicate cultures. TCA-ppt, trichloroacetic acid-precipitable; TCA-sol, trichloroacetic acid soluble; G3PD, glycerol-3-phosphate dehydrogenase; G6PD, glucose-6-phosphate dehydrogenase. (*Miller and Burns, 1985, p. 70.*)

Table 8-2. Half-Lives of Glutamine Synthetase and Soluble Protein in 3T3-L1 Adipocytes

Culture conditions		*Glutamine synthetase*			*Soluble protein*		
Insulin μg/ml	*Hydro-cortisone μg/ml*	*t½ (hr)*	*Relative t½*	r^2	*t½ (hr)*	*Relative t½*	r^2
0	0	23.8	(1.00)	0.95	34.4	(1.00)	0.95
1	0	31.9	(1.34)	0.93	61.3	(1.78)	0.91
0	0.1	20.6	(0.87)	0.91	33.3	(0.97)	0.97
1	0.1	30.7	(1.29)	0.95	57.6	(1.67)	0.85

Note: Confluent 3T3-L1 cells were incubated with dexamethasone, methylisobutylxanthine and insulin as described in the legend to Fig. 4. Thereafter, the cultures were maintained for 5 days with medium containing no added hormone. Beginning 5 days after treatment with dexamethasone, methylisobutylxanthine and insulin, all cultures were incubated for 18 hours with fresh medium containing 0.1 μg/ml hydrocortisone and 10 μCi/ml L-[^{35}S]methionine (50 μCi/umole). After 18 hours of incubation this medium was aspirated, all cultures were washed twice with 5 ml of phosphate buffered saline, and 5 ml of fresh medium were added containing 5 mM L-methionine and either no added hormone, 1.0 μg/ml insulin, 0.1 μg/ml hydrocortisone or/both 1.0 μg/ml insulin and 0.1 μ/ml hydrocortisone. Four replicate cultures from each treatment group were harvested at zero, 1.5, 3, 6, 12, 24 and 36 hours after beginning the indicated treatment. Cell lysates were analyzed for radioactivity in immunoprecipitable glutamine synthetase and in trichloroacetic acid-precipitable material. At the time of initiation of treatments (zero time), radioactivity in glutamine synthetase and soluble protein (mean for 4 replicates) was 8544 dpm per dish and 1.44×10^6 dpm per dish, respectively. Plots of the logarithm of radioactivity as a function of time were subjected to linear regression analyses. Each data point plotted was the mean value for four replicate cultures. t½, half-life. r^2, correlation coefficient from linear regression analysis of seven data points. The first order degradation rate constant K_d equals 0.693 divided by t½. (From reference # 124)

Degradation of GS and Protein. In order to estimate the first-order degradation rate constants for GS and for soluble protein, we monitored the disappearance of ^{35}S-radioactivity from immunoprecipitable GS and soluble protein in 3T3-L1 adipocytes previously incubated with L-[^{35}S]methionine (124). The half-lives of both GS and soluble protein were increased, compared with controls, in the presence of insulin alone by approximately 34 and 78%, respectively (Table 8-2; r^2 is the correlation coefficient for a linear regression analysis of seven data points from a plot of the logarithm of GS radioactivity as a function of time). They were similarly increased in the presence of both insulin and hydrocortisone. By contrast, hydrocortisone alone had little effect on the half-life of soluble protein and it decreased the half-life of GS by 13%. The half-lives for GS and soluble protein, respectively, were 24 and 34 hr in control cultures, 32 and 61 hr in insulin-treated cultures, 21 and 33 hr in hydrocortisone-treated cultures, and 31 and 58 hr in cultures treated with both insulin and hydrocortisone. When [^{3}H]leucine was substituted for [^{35}S]methionine, similar half-lives

Table 8-3. Quantitative Dot Blot Analysis of Total RNA from 3T3-LI Adipocytes

	Culture Condition	*GS mRNA Content*	
a.	Control	0.40	(1.00)
b.	Dexamethasone (DEX)	5.70	(14.30)
c.	Insulin	<0.02	(<0.05)
d.	Bt_2cAMP plus theophylline	<0.02	(<0.05)
e.	DEX plus insulin	1.85	(4.63)
f.	DEX plus Bt_2cAMP plus theophylline	1.00	(2.50)

Note: Cultured 3T3-L1 adipocytes (five 50 cm^2 dishes in each treatment group) were incubated for 6 hours in the presence of the following: (a) no additions; (b) 3μM dexamethasone; (c) 1 μg/ml insulin; (d) 0.5 mM dibutyryl cAMP plus 0.5 mM theophylline; (e) 3 μM dexamethasone, plus 1 μg/ml insulin; (f) 3 μM dexamethasone, plus 0.5 mM Bt_2cAMP, plus 0.5 mM theophylline. Total RNA was isolated from the cultured adipocytes. Dot blot analysis with total RNA (1 to 5 μg) or ploy(A−)RNA (1 to 5 μg) was performed on a single nitrocellulose filter. An autoradiograph of dot blots was obtained after hybridization of RNA samples with the ^{32}P-radiolabeled probe, pGSRK-1. Slopes were obtained from a graphical analysis of densitometer tracing peak heights plotted as a function of the mass of total RNA spotted. The slope (1.0) for poly(A-)RNA was subtracted from each value. These differences are recorded in the table as "GS mRNA content." Differences less than 0.02 are below the limit of reliable detection. The values in parenthesis are normalized to the control value.

were obtained for both GS and soluble protein. In these experiments 5 mM leucine was substituted for 5 mM methionine during the "chase" period. These degradation data, together with estimates of enzyme synthesis rate, indicate that insulin and hydrocortisone regulate GS content in 3T3-L1 adipocytes primarily at the level of enzyme synthesis.

GS mRNA in 3T3-L1 Adipocytes. The effects of DEX, insulin, and Bt_2cAMP on GS mRNA were evaluated by quantitative dot-blot analysis (Table 8-3) (128). 3T3-L1 adipocytes were incubated for 6 hours with or without added hormones and Bt_2cAMP. An autoradiograph of the dot-blots obtained after hybridization with [^{32}P]pGSRK-1 was quantified by densitometric analysis. Graphical analysis of densitometer tracing peak heights for 1 to 5 μg of total RNA from each sample and from poly(A-) RNA yielded a linear relationship between the mass of RNA spotted and the corresponding peak height for at least 4 of 5 RNA dots. The slope for the poly(A-)RNA (1.0 densitometric unit per microgram RNA) was subtracted from that obtained from each sample and these differences are recorded in Table 8-3.

The data demonstrate a DEX-mediated increase in GS mRNA that reaches a value that is 14-fold above the control value. In addition, the data indicate that insulin or Bt_2cAMP plus theophylline markedly decrease GS mRNA. Furthermore, either insulin or Bt_2cAMP plus theophylline at least partially prevent the DEX-mediated increase in GS mRNA.

CONCLUSION

Our data indicate that in 3T3-L1 adipocytes the GS biosynthetic rate is decreased by Bt_2cAMP or insulin, while it is increased by hydrocortisone. By contrast, the first-order degradation rate constant for GS is increased by Bt_2cAMP, decreased by insulin, and affected little, if at all, by hydrocortisone. In order to determine the mechanisms by which hormones and cAMP regulate the biosynthesis of GS it will be necessary (a) to analyze the transcription rate and stability for GS mRNA and (b) to isolate the GS gene and analyze its structure, organization, and regulatory regions.

EXPERIMENTAL PROCEDURES

Equipment

Major special equipment used in this laboratory for studies of GS in cultured 3T3-L1 adipocytes includes the following: Baker Biogard biohazard hood; Wedco laminar-flow CO_2 incubator; Zeiss inverted phase-contrast microscope; Coulter model ZF particle counter; Linde liquid nitrogen freezer; and Millipore positive pressure filtration apparatus. Other required equipment includes the following: pH meter; slab gel electrophoresis apparatus and power supply; liquid scintillation spectrometer; centrifuges; spectrophotometer; and automatic pipetting devices.

Materials

Starter cultures of 3T3-L1 and 3T3-C2 cells were generously provided by Dr. Howard Green, Department of Physiology and Biophysics, Harvard University. These cells are now available from the American Tissue Type Culture Collection. Porcine cystalline zinc insulin was obtained from Eli Lilly Company. L-[^{35}S]methionine was from New England Nuclear Corporation. All other chemicals were of the purest grade commercially available. Most were from Sigma Chemical Company or Grand Island Biological Company. Calf serum and fetal calf serum were from Flow Laboratories. Each batch of serum must be tested for toxicity. This is achieved by comparing 3T3-L1 cell growth in each new serum batch with cell growth in serum known to be nontoxic. 3T3-L1 cells have a doubling time of about 20 hours in nontoxic serum. Toxic serum not only inhibits growth but it may cause growing cells to detach from the culture dish surface. Corning plastic disposable culture dishes are used routinely.

Dulbecco's modified Eagle's medium is commercially available from Gibco and Flow. However, we prepare the medium without glutamine from reagent grade chemicals. Nonsterile stock solutions from which the medium is prepared include the following 20 × (20-fold concentrated) salts, 25-fold concentrated vitamins without biotin, and 50 × amino acids. Solid D-glucose (4.5 g/l), Sodium bicarbonate ($NaHCO_3$) (26.2 mM for 5% carbon dioxide (CO_2) or 44.0 mM for 10% CO_2), and D-biotin (1 μg/ml) are dissolved in the completed medium just prior to filter sterilization (0.22 μm Millipore-type GS filter). The composition of stock solutions is given below. Complete medium of the desired composition is prepared by adding to the sterile 1 × DME medium (pH 7.4), serum (10% final concentration), 200 mM glutamine in 150 mM NaCl (0 to 4 mM final concentration), 10,000 u/ml penicillin, 10 mg/ml streptomycin in 150 mM NaCl (100 u/ml penicillin and 100 μg/ml streptomycin, final concentration), and hormones or potential effectors.

The composition of concentrated stock solutions is as follows: *20 × salts:* 36 mM Calcium chloride ($CaCl_2$) 5 mM Ferric nitrate ($Fe(NO_3)_3$) 107 mM KCl, 16.2 mM magnesium sulfate ($MgSO_4$) 2.2 M NaCl, 18.2 mM monobasic sodium phosphate (NaH_2PO_4) and 300 mg/l water-soluble phenol red. *25 × vitamins,* pH 8.0: 200 mg/l D-calcium pantothenate, 100 mg/l choline chloride, 100 mg/l folic acid, 180 mg/l i-inositol, 100 mg/l nicotinamide, 100 mg/l pyridoxal HCl, 40 mg/l riboflavin, and 100 mg/l thiamine HCl. The final pH of the stock vitamin solution is adjusted to pH 8.0. *50 × amino acids,* pH 1.1 to 1.2: 20 mM L-arginine, 10 mM L-cystine, 20 mM L-glycine, 10 mM L-histidine, 40 mM L-isoleucine, 40 mM L-leucine, 40 mM L-lysine, 10 mM L-methionine, 20 mM L-phenylalanine, 20 mM L-serine, 40 mM L-threonine, 4 mM L-tryptophan, 20 mM L-tyrosine, and 40 mM L-valine. The pH of the 50 × amino acid stock solution is adjusted to 1.1 to 1.2 with hydrochloric acid (HCl). Glucose, $NaHCO_3$, D-biotin, L-glutamine, and antibiotics are added separately (see above).

Concentrated stock solutions of insulin, hydrocortisone, theophylline, and methylisobutylxanthine have the following composition: 0.5 mg/ml insulin in 1 mM HCl (The absorbance at 278 nm of a 1 mg/ml solution of porcine insulin at pH 2.5 to 3.0 is 1.05); 0.1 mg/ml hydrocortisone in 50% ethanol; 50 mM Bt_2cAMP, pH 7.0, in water; 25 mM theophylline in water; and 50 mM MIX in 50% ethanol. These solutions were filter sterilized (0.22 micron Millipore type GS filter) and stored at −20°C prior to use.

Cell Culture Methods

Except where indicated, methods were identical to those previously described (13, 14, 25, 123). 3T3-L1 cells were grown to confluence in DME medium containing 10% CS, 100 u/ml penicillin, 100 μg/ml streptomycin, 1 μg/ml D-biotin, 8 μg/ml calcium pantothenate, and either 26.2 mM or

44.0 mM $NaHCO_3$. Thereafter, the cultures were maintained with the same medium, but in some cases 10% FCS replaced the CS. Cultures were grown and maintained at 37°C in 20-cm^2 (60 mm) or 50-cm^2 (100 mm) culture dishes in a humidified atmosphere composed of either 5% CO_2 and 95% air with 26.2 mM $NaHCO_3$ or 10% CO_2 and 90% air with 44.0 mM $NaHCO_3$. Unless otherwise indicated the culture medium was removed by aspiration and fresh medium was added every 2 days. Confluence, assessed by phase-contrast microscopy, denotes the earliest time when the culture dish surface was completely covered by cells. This corresponds to a cell density of 3 to 5 x 10^4 cells/cm^2. For experiments, cultures were plated at a density of 2 to 4 x 10^4 cells/cm^2. This yielded uniform plating of multiple cultures. The undiluted CS used for medium preparation contained insulin at less than 0.3 ng/ml and hydrocortisone at less than 40 ng/ml. The undiluted FCS contained insulin at less than 0.25 ng/ml and hydrocortisone at less than 5 ng/ml.

Adipocyte conversion (differentiation) of confluent 3T3-L1 cultures was achieved either by treatment with 1 μg/ml insulin plus 0.1 μg/ml hydrocortisone or by treatment with DEX and MIX according to the method of Rubin and co-workers (9). The latter method involves incubation of confluent 3T3-L1 cultures for 48 hr with medium containing 10% fetal calf serum, 0.25 uM DEX and 0.5 mM MIX. Subsequently, the cultures were maintained for 4 to 6 days with medium containing 10% fetal calf serum and no added hormone. This treatment yielded adipocyte conversion by more than 85% of the cells. This method has the advantage that the adipocytes are not chronically exposed to supraphysiological concentrations of insulin during adipocyte conversion. The extent of adipocyte conversion was assessed by phase-contrast microscopy. Cultures were harvested as described previously (13, 25). At the time of harvesting the culture medium was removed by aspiration; the cultures were washed twice with 5 ml of phosphate-buffered saline; and the cells were scraped into 1 ml of buffer (20 mM imidazole HCl, pH 7.0, 150 mM KCl, 0.1 mM ethylenediamine tetraacetate (EDTA), frozen, and stored at −80°C for up to 2 weeks. Harvested samples were vigorously mixed and clarified by centrifugation (15,600 × *g;* 4°C; 10 min). The supernatant (clarified cell lysate) was analyzed for enzyme activities, radioactivity, and immunoprecipitable GS. The pellet was analyzed for DNA.

Assays

Glutamine synthetase activity was assayed by the gamma-glutamyl transfer assay. Glycerol-3-phosphate dehydrogenase (G3PD) activity was determined from the initial rate of nicotinamide adenine dinucleotide (NADH) oxidation and glucose-6-phosphate dehydrogenase (G6PD) activity was determined from the initial rate of NADP nicotinamide adenine dinucleotide phosphate reduction. One unit of activity is that amount catalyzing the production of 1 μmol/min product under the specified assay conditions (13, 14). Protein was quantified by the method of Lowry and colleagues (129). DNA was assayed by the method of Burton (130) as described previously (25).

The rate of synthesis of GS and protein in cultured 3T3-L1 cells was estimated from the incorporation of L-[^{35}S]methionine into the enzyme and protein during a 2-hr incubation period. The degradation rate constants for GS and protein were estimated from the first-order disappearance of radioactivity from GS and protein in 3T3-L1 adipocytes previously incubated with L-[^{35}S]methionine. Glutamine synthetase labeled with L-[^{35}S]methionine was isolated by immunoprecipitation followed by SPS-polyacrylamide gel electrophoresis (13). Protein was isolated by trichloroacetic acid precipitation. Trichloroacetic acid-precipitable and trichloroacetic acid-soluble radioactivity in clarified cell lysates were determined as described previously (13).

In pulse labeling experiments incorporation of L-[^{35}S]methionine was compared with incorporation of L-[^{35}S]methionine into trichloroacetic acid-precipitable material. The resulting ratio between GS radioactivity and protein radioactivity provides a control, between experimental conditions, for possible differences in the time-integrated specific radioactivity of the methionine pool (131). Such an analysis assumes that both GS and soluble protein are synthesized from the same methionine pool and that changes in the methionine concentration or specific radioactivity in this pool will yield parallel changes in the incorporation of L-[^{35}S]methionine into the enzyme and soluble protein (131).

Isolation of total RNA, preparation of the radiolabeled GS cDNA probe by nick translation and quantitative dot-blot hybridization procedures are described in reference 132 and detailed procedures were performed according to the methods of Maniatis and co-workers (133).

SUMMARY

Confluent 3T3-L1 cells have the capacity to acquire many of the morphological and biochemical characteristics of adipocytes. Glutamine synthetase, the enzyme responsible for glutamine synthesis, is a pivotal enzyme in nitrogen metabolism because it links the multitude of catabolic processes by which ammonia is produced with numerous biosynthetic reactions directed ultimately toward the formation of proteins, nucleic acids, and complex polysaccharides. Glutamine synthetase specific activity and cellular content (determined by immunotitration) increase by more than 100-fold during adipocyte conversion of the 3T3-L1 cells regardless of the culture conditions employed to achieve differentiation. In 3T3-L1 adipocytes Bt_2cAMP decreases GS activity and content by more than 70% and hydrocortisone increases GS activity and content. We studied the effects of Bt_2cAMP, insulin, and hydrocortisone on the synthesis and degradation of GS and soluble protein. The biosynthetic rate was estimated from the incorporation of L-[^{35}S]methionine into GS and protein during a 2-hour incubation period. The degradation rate was estimated from the first-order disappearance of radioactivity from GS and soluble protein in 3T3 adipocytes previously incubated with L-[^{35}S]methionine. Glutamine synthetase was isolated by immunoprecipitation followed by SDS-polyacrylamide gel electrophoresis. Protein was isolated by precipitation with trichloroacetic acid. Incorporation of L-[^{35}S]methionine into GS was compared with incorporation of L-[^{35}S]methionine into trichloroacetic acid-precipitable material. The resulting ratio between GS radioactivity and protein radioactivity provided both an estimate of the relative rate of GS biosynthesis and a control, between experimental conditions, for possible differences in the time-integrated specific radioactivity of the methionine pool. Our data indicate that in 3T3-L1 adipocytes the GS biosynthetic rate is decreased by Bt_2cAMP or insulin, while it is increased by hydrocortisone. By contrast, the first-order GS degradation rate constant is increased by Bt_2cAMP, decreased by insulin, and affected little, if at all, by hydrocortisone. Recently, we found that Northern hybridization analysis of RNA from 3T3 L1 adipocytes with a rat GS cDNA clone (pGSRK-1) yield two hybridizable GS RNAs of lengths 3.2 and 1.6 kb. Analyses of the Northern blots indicate that the 3.2-kb GS-specific RNA is more abundant than the 1.6-kb GS RNA. Quantitative dot-blot hybridization analysis indicates that dexamethasone increases GS mRNA while both insulin and dibutyryl cAMP decrease GS mRNA and/or prevent the dexamethasone-mediated increase. Our data suggest that there are at least two GS mRNAs in 3T3-L1 adipocytes and that they are regulated in parallel by dexamethasone, insulin, and Bt_2cAMP.

REFERENCES

1. H. Green and O. Kehinde, 1974, Sublines of mouse 3T3 cells that accumulate lipid, *Cell* **1**:113–116.
2. H. Green and M. Meuth, 1974, An established pre-adipose cell line and its differentiation in cultures, *Cell* **3**:127–133.
3. H. Green and O. Kehinde, 1975, An established preadipose cell line and its differentiation in culture. II. Factors affecting the adipose conversion, *Cell* **5**:19–27.

4. H. Green and O. Kehinde, 1976, Spontaneous heritable changes leading to increased adipose conversion in 3T3 cells, *Cell* **7**:105–113.
5. J. C. Mackall, A. K. Student, S. E. Polakis, and M. D. Lane, 1976, Induction of lipogenesis during differentiation in a "preadipocyte" cell line, *J. Biol. Chem.* **251**:6462–6464.
6. O. M. Rosen, C. J. Smith, A. Hirsch, E. Lai, and C. S. Rubin, 1979, Recent studies of the 3T3-L1 adipocyte-like cell line, *Recent Prog. Horm. Res.* **35**:477–499.
7. T. R. Russel and R. Ho, 1976, Conversion of 3T3 fibroblasts into adipose cells: Triggering of differentiation by prostaglandin $F_2\alpha$ and 1-methyl-3-isobutylxanthine, *Proc. Natl. Acad. Sci. (USA)* **73**:4516–4520.
8. I. H. Williams and S. E. Polakis, 1977, Differentiation of 3T3-L1 fibroblasts to adipocytes. The effect of indomethacin, prostaglandin El and cyclic AMP on the process of differentiation, *Biochem. Biophys. Res. Commun.* **77**:175–186.
9. C. S. Rubin, A. Hirsch, C. Fung, and O. M. Rosen, 1978, Development of hormone receptors and hormonal responsiveness in vitro. Insulin receptors and insulin sensitivity in the preadipocyte and adipocyte forms of 3T3-L1 cells, *J. Biol. Chem.* **253**:7570–7578.
10. M. Morikawa, T. Nixon, and H. Green, 1982, Growth hormone and the adipose conversion of 3T3 cells, *Cell* **29**:783–789.
11. S. Keay and S. E. Grossberg, 1980, Interferon inhibits the conversion of 3T3-L1 mouse fibroblasts into adipocytes, *Proc. Natl. Acad. Sci. (USA)* **77**:4099–4103.
12. T. Murray and T. R. Russel, 1980, Inhibition of adipose conversion in 3T3-L2 cells by retinoic acid, *J. Supramol. Struct.* **14**:255–266.
13. R. E. Miller and D. A. Carrino, 1980, Dibutyryl cyclic AMP decreases glutamine synthetase in cultured 3T3-L1 adipocytes, *J. Biol. Chem.* **255**:5490–5500.
14. R. E. Miller, R. W. Hackenberg, and H. Gershman, 1978, Regulation of glutamine synthetase in cultured 3T3-L1 cells by insulin, hydrocortisone, and dibutyryl cyclic AMP, *Proc. Natl. Acad. Sci. (USA)* **75**:1418–1422.
15. J. C. Mackall and M. D. Lane, 1977, Role of pyruvate carboxylase in fatty acid synthesis: Alterations during preadipocyte differentiation, *Biochem. Biophys. Res. Commun.* **79**:720–725.
16. W. Kuri-Harcuch and H. Green, 1977, Increasing activity of enzymes on pathway of triacylglycerol synthesis during adipose conversion of 3T3 cells, *J. Biol. Chem.* **252**:2158–2160.
17. L. S. Wise and H. Green, 1978, Studies of lipoprotein lipase during the adipose conversion of 3T3 cells, *Cell* **13**:233–242.
18. W. Kuri-Harcuch, L. S. Wise, and H. Green, 1978, Interruption of the adipose conversion of 3T3 cells by biotin deficiency: differentiation without triglyceride accumulation, *Cell* **14**:53–59.
19. C. S. Rubin, E. Lai, and O. M. Rosen, 1977, Acquisition of increased hormone sensitivity during in vitro adipocyte development, *J. Biol. Chem.* **252**:3554–3557.
20. R. A. Coleman, B. C. Reed, J. C. Mackall, A. K. Student, M. D. Lane, R. M. and Bell, 1978, Selective changes in microsomal enzymes of triacylglycerol phosphatidylcholine, and phosphatidylethanolamine biosynthesis during differentiation of 3T3-L1 preadipocytes, *J. Biol. Chem.* **253**:7256–7261.
21. P. M. Spooner, S. S. Chernick, M. M. Garrison, and R. O. Scow, 1979, Development of lipoprotein lipase activity and accumulation of triacyglycerol in differentiating 3T3-L1 adipocytes. Effects of prostaglandin $F_2\alpha$, 1-methyl-3-isobutylxanthine, prolactin and insulin, *J. Biol. Chem.* **245**:1305–1311.
22. P. M. Spooner, S. S. Chernick, M. M. Garrison, and R. O. Scow, 1979, Insulin regulation

of lipoprotein lipase activity and release in 3T3-L1 adipocytes. Separation and dependence of hormonal effects of hexose metabolism and synthesis of RNA and protein, *J. Biol. Chem.* **254:**10021–10029.

23. L. S. Wise and H. Green, 1979, Participation of one isozyme of cytosolic glycerophosphate dehydrogenase in the adipose conversion of 3T3 cells, *J. Biol. Chem.* **254:**273–275.
24. O. M. Rosen, C. J. Smith, C. Fung, and C. S. Rubin, 1978, Development of hormone receptors and hormone responsiveness in vitro. Effect of prolonged insulin treatment on hexose uptake in 3T3-L1 adipocytes, *J. Biol. Chem.* **253:**7579–7583.
25. R. E. Miller and D. A. Carrino, 1981, An association between glutamine synthetase activity and adipocyte differentiation in cultured 3T3-L1 cells, *Arch. Biochem. Biophys.* **209:**486–503.
26. S. O. Freytag and M. F. Utter, 1980, Induction of pyruvate carboxylase apoenzyme and holoenzyme in 3T3-L1 cells during differentiation, *Proc. Natl. Acad. Sci. (USA)* **77:**1321–1325.
27. G. H. Weiss, O. M. Rosen, C. S. Rubin, 1980, Regulation of fatty acid synthetase concentration and activity during adipocyte differentiation. Studies on 3T3-L1 cells, *J. Biol. Chem.* **255:**4751–4757.
28. A. Student, R. Y. Hsu, and M. D. Lane, 1980, Induction of fatty acid synthetase synthesis in differentiating 3T3-L1 preadipocytes, *J. Biol. Chem.* **255:**4745–4750.
29. R. A. Coleman and R. M. Bell, 1980, Selective changes in enzymes of the sn-glycerol 3-phosphate and dihydroxyacetone-phosphate pathways of triacylglycerol biosynthesis during differentiation of 3T3-L1 preadipocytes, *J. Biol. Chem.* **255:**7681–7687.
30. M. Kawamura, D. F. Jensen, E. V. Wancewicz, L. L. Joy, J. D. Khoo, and D. Steinberg, 1981, Hormone-sensitive lipase in differentiated 3T3-L1 cells and its activation by cyclic AMP-dependent protein kinase, *Proc. Natl. Acad. Sci. (USA)* **78:**732–736.
31. A. Y.-C. Liu, 1982, Differentiation-specific increase of cAMP-dependent protein kinase in the 3T3-L1 cells, *J. Biol. Chem.* **257:**298–306.
32. F. E. Frerman, J. L. Sabran, J. L. Taylor, and S. E. Grossberg, 1983, Leucine catabolism during the differentiation of 3T3-L1 cells. Expression of a mitochondrial enzyme system, *J. Biol. Chem.* **258:**7087–7093.
33. T. Murray and T. R. Russell, 1980, Acquisition of an insulin-sensitive activity of adenosine-3′,5′-monophosphate phosphodiesterase during adipose conversion of 3T3-L2 cells, *Eur. J. Biochem.* **107:**217–224.
34. M. L. Elks, V. C. Manganiello, and M. Vaughan, 1983, Hormone-sensitive particulate cAMP phosphodiesterase activity in 3T3-L1 adipocytes. Regulation of responsiveness by dexamethasone, *J. Biol. Chem.* **258:**8582–8587.
35. E. Lai, O. M. Rosen, and C. S. Rubin, 1981, Differentiation-dependent expression of catecholamine-stimulated adenylate cyclase. Roles of the beta-receptor and G/F protein in differentiating 3T3-L1 adipocytes, *J. Biol. Chem.* **256:**12866–12874.
36. S. S. Hoffmann and G. M. Kolodny, 1977, Insulin receptors in 3T3 fibroblasts. Relationship to growth phase, transformation and differentiation into new cell types, *Exp. Cell Res.* **107:**293–299.
37. B. C. Reed, S. H. Kaufmann, J. C. Mackall, A. K. Student, and M. D. Lane, 1977, Alterations in insulin binding accompanying differentiation of 3T3-L1 preadipocytes, *Proc. Natl. Acad. Sci. (USA)* **74:**4876–4880.
38. T. H. Chang and S. E. Polakis, 1978, Differentiation of 3T3-L1 fibroblasts to adipocytes. Effect of insulin and indomethacin on the levels of insulin receptors, *J. Biol. Chem.* **253:**4693–4696.

39. F. A. Karlsson, C. Grunfeld, C. R. Kahn, and J. Roth, 1979, Regulation of insulin receptors and insulin responsiveness in 3T3-L1 fatty fibroblasts, *Endocrinology* **104:**1383–1392.
40. O. M. Rosen, G. H. Chia, C. Fung, and C. S. Rubin, 1979, Tunicamycin-mediated depletion of insulin receptors in 3T3-L1 adipocytes, *J. Cell Physiol.* **99:**37–42.
41. B. C. Reed and M. D. Lane, 1980, Insulin receptor synthesis and turnover in differentiating 3T3-L1 preadipocytes, *Proc. Natl. Acad. Sci. (USA)* **77:**285–289.
42. B. C. Reed, J. Moss, P. H. Fishman, and M. D. Lane, 1980, Loss of choleragen receptor and ganglioside upon differentiation of 3T3-L1 preadipocytes, *J. Biol. Chem.* **255:**1711–1715.
43. B. M. Spiegelman and H. Green, 1980, Control of specific protein biosynthesis during the adipose conversion of 3T3 cells, *J. Biol. Chem.* **255:**8811–8818.
44. R. S. Sidhu, 1979, Two-dimensional electrophoretic analyses of proteins synthesized during differentiation of 3T3-L1 preadipocytes, *J. Biol. Chem.* **254:**11111–11118.
45. G. V. Ronnett, V. P. Knutson, and M. D. Lane, 1982, Insulin-induced down-regulation of insulin receptors in 3T3-L1 adipocytes. Altered rate of receptor inactivation, *J. Biol. Chem.* **257:**4285–4291.
46. G. V. Ronnett, G. Tennekoon, V. P. Knutson, and M. D. Lane, 1983, Kinetics of insulin receptor transit to and removal from the plasma membrane, *J. Biol. Chem.* **258:**283–290.
47. G. J. Smith, P. J. Wehksnora, J. F. Warner, C. S. Rubin, and O. M. Rosen, 1979, Insulin-stimulated protein phosphorylation in 3T3-L1 preadipocytes, *Proc. Natl. Acad. Sci. (USA)* **76:**2725–2729.
48. C. J. Smith, C. S. Rubin, and O. M. Rosen, 1980, Insulin-treated 3T3-L1 adipocytes and cell-free extracts derived from them incorporate ^{32}P into ribosomal protein S6, *Proc. Natl. Acad. Sci. (USA)* **77:**2641–2645.
49. G. D. Swergold, O. M. Rosen, and C. S. Rubin, 1982, Hormonal regulation of the phosphorylation of ATP citrate lyase in 3T3-L1 adipocytes. Effects of insulin and isoproterenol, *J. Biol. Chem.* **257:**4207–4215.
50. L. M. Petruzzelli, S. Ganguly, C. J. Smith, M. H. Cobb, C. S. Rubin, and O. M. Rosen, 1982, Insulin activates a tyrosine-specific protein kinase in extracts of 3T3-L1 adipocytes and human placenta, *Proc. Natl. Acad. Sci. (USA)* **79:**6792–6796.
51. H. Green and O. Kehinde, 1979, Formation of normally differentiated subcutaneous fat pads by an established preadipose cell line, *J. Cell Physiol.* **101:**169–172.
52. B. M. Spiegelman and H. Green, 1981, Cyclic AMP-mediated control of lipogenic enzyme synthesis during adipose differentiation of 3T3 cells, *Cell* **24:**503–510.
53. B. M. Spiegelman and S. R. Farmer, 1982, Decrease in tubulin and actin gene expression prior to morphological differentiation of 3T3 adipocytes, *Cell* **29:**53–60.
54. B. M. Spiegelman, M. Frank, and H. Green, 1983, Molecular cloning of mRNA from 3T3 adipocytes. Regulation of mRNA content for glycerophosphate dehydrogenase and other differentiation-dependent proteins during adipocyte development, *J. Biol. Chem.* **258:**10083–10089.
55. D. Steinberg, S. E. Mauyer, J. C. Khoo, E. A. Miller, R. E. Miller, B. Fredholm, and R. Eichner, 1975, Hormonal regulation of lipase, phosphorylase, and glycogen synthetase in adipose tissue, *Adv. Cyc. Nucl. Res.* **5:**549–568.
56. R. O. Adebonojo, 1975, Studies on human adipose cells in culture: Relation of cell size and multiplication to donor age, *Yale J. Biol. Med.* **48:**9–16.
57. F. O. Adebonojo, 1975, Monolayer cultures of disaggregated human adipocytes, *In Vitro* **11:**50–54.

58. D. Dixon-Shanies, J. Rudeck, and J. L. Knittle, 1975, Observations on the growth and metabolic functions of cultured cells derived from human adipose tissue, *Proc. Soc. Exp. Biol. Med.* **149**:541–545.
59. D. Dixon-Shanies and J. L. Knittle, 1976, Effect of hormones on cyclic AMP levels in cultured human cells, *Biochem. Biophys. Res. Commun.* **68**:982–987.
60. R. L. R. Van, C. E. Bayliss, and D. A. K. Roncari, 1976, Cytological and enzymological characterization of adult human adipocyte precursors in culture, *J. Clin. Invest.* **58**:699–704.
61. D. A. Roncari and R. L. R. Van, 1978, Promotion of human adipocyte precursor replication by 17 beta-estradiol in culture, *J. Clin. Invest.* **62**:503–508.
62. R. Negrel, P. Grimaldi, and G. Ailhaud, 1978, Establishment of preadipocyte clonal line from epididymal fat pad or ob/ob mouse that responses to insulin and to lipolytic hormones, *Proc. Natl. Acad. Sci. (USA)* **75**:6054–6058.
63. P. Grimaldi, R. Negrel, J.-P. Vincent, and G. Ailhaud, 1979, Differentiation of ob 17 preadipocytes to adipocytes. Effect of insulin on the levels of insulin receptors and on the transport of alpha-aminoisobutyrate, *J. Biol. Chem.* **254**:6849–6852.
64. D. A. Roncari, H. Wang, and D. S. Desai, 1980, Cyclic AMP metabolism and lipolysis in cultured human adipocyte precursors, *Can. J. Biochem.* **53**:201–205.
65. M. de la Llera, J. M. Glick, and G. Rothblat, 1981, Mechanism of triglyceride accumulation in rat preadipocyte cultures exposed to very low density lipoprotein, *J. Lipid Res.* **22**:245–253.
66. C. Vannier, H. Jansen, R. Negrel, and G. Ailhaud, 1982, Study of lipoprotein lipase content in Ob17 preadipocytes during adipose conversion. Immunofluorescent localization of the enzyme, *J. Biol. Chem.* **257**:12387–12393.
67. R. Sager and P. Kovac, 1982, Pre-adipocyte determination either by insulin or by 5-azacytidine, *Proc. Natl. Acad. Sci. (USA)* **79**:480–484.
68. E. R. Stadtman, 1973, A note on the significance of glutamine in intermediary metab olism, in *The Enzymes of Glutamine Metabolism,* S. Prusiner, and E. R. Stadtman, eds., Academic Press, New York, pp. 1–6.
69. P. Felig, J. Wahren, I. Karl, E. Cerasi, R. Luft, and D. M. Kipnis, 1973, Glutamine and glutamate metabolism in normal and diabetic subjects, *Diabetes* **22**:573–576.
70. O. P. Ganda and N. B. Ruderman, 1976, Muscle nitrogen metabolism in chronic hepatic insufficiency, *Metabolism* **25**:427–435.
71. E. B. Marliss, T. Aoki, T. Pozefsky, A. S. Most, and F. G. Cahill Jr., 1971, Muscle and splanchnic glutamine and glutamate metabolism in postabsorptive and starved man, *J. Clin. Invest.* **50**:814–817.
72. N. B. Ruderman and P. Lund, 1972, Amino acid metabolism in skeletal muscle. Regulation of glutamine and alanine release in the perfused rat hindquarter, *Isr. J. Med. Sci.* **8**:295–302.
73. P. Felig, J. Wahren, and L. Raf, 1973, Evidence of inter-organ amino-acid transport by blood cells in humans, *Proc. Natl. Acad. Sci. (USA)* **70**:1775–1779.
74. N. B. Ruderman and M. Berger, 1974, The formation of glutamine and alanine in skeletal muscle, *J. Biol. Chem.* **249**:5500–5506.
75. H. G. Windmueller and A. E. Spaeth, 1974, Uptake and metabolism of plasma glutamine by the small intestine, *J. Biol. Chem.* **249**:5070–5079.
76. P. Felig, 1975, Amino acid metabolism in man, *Annu. Rev. Biochem.* **44**:933–955.
77. A. J. Garber, I. E. Karl, and D. M. Kipnis, 1976, Alanine and glutamine synthesis and release from skeletal muscle. I. Glycolysis and amino acid release, *J. Biol. Chem.* **251**:826–835.

78. A. J. Garber, I. E. Karl, and D. M. Kipnis, 1976, Alanine and glutamine synthesis and release from skeletal muscle. II. The precursor role of amino acids in alanine and glutamine synthesis, *J. Biol. Chem.* **251**:836–843.
79. A. J. Garber, I. E. Karl, and D. M. Kipnis, 1976, Alanine and glutamine synthesis and release from skeletal muscle. IV. Beta-Adrenergic inhibition of amino acid release, *J. Biol. Chem.* **251**:851–857.
80. I. E. Karl, A. J. Garber, and D. M. Kipnis, 1976, Alanine and glutamine synthesis and release from skeletal muscle. III. Dietary and hormonal regulation, *J. Biol. Chem.* **251**:844–850.
81. T. W. Chang and A. L. Goldberg, 1978, The origin of alanine produced in skeletal muscle, *J. Biol. Chem.* **253**:3677–3684.
82. T. W. Chang and A. L. Goldberg, 1978, The metabolic fates of amino acids and the formation of glutamine in skeletal muscle, *J. Biol. Chem.* **253**:3685–3695.
83. A. L. Goldberg and T. W. Chang, 1978, Regulation and significance of amino acid metabolism in skeletal muscles during fasting, *Fed. Proc.* **37**:2301–2307.
84. M. E. Tischler and A. L. Goldberg, 1980, Leucine degradation and release of glutamine and alanine by adipose tissue, *J. Biol. Chem.* **255**:8074–8081.
85. H. G. Windmueler and A. E. Spaeth, 1980, Respiratory fuels and nitrogen metabolism in vivo in small intestine of fed rats. Quantitative importance of glutamine, glutamate, and aspartate, *J. Biol. Chem.* **255**:107–112.
86. S. Rapoport, J. Rost, and M. Schultze, 1971, Glutamine and glutamate as respiratory substrates of rabbit reticulocytes, *Eur. J. Biochem.* **23**:166–170.
87. M. Donnelly and I. E. Scheffler, 1976, Energy metabolism in respiration-deficient and wild type Chinese hamster fibroblasts in culture, *J. Cell Physiol.* **89**:39–52.
88. H. R. Zielke, P. T. Ozand, J. T. Tildon, D. A. Sevdalian, and M. Cornblath, 1976, Growth of human diploid fibroblasts in the absence of glucose utilization, *Proc. Natl. Acad. Sci. (USA)* **73**:4110–4114.
89. L. J. Reitzer, B. M. Wice, and D. Kennell, 1979, Evidence that glutamine, not sugar, is the major energy source for cultured HeLa cells, *J. Biol. Chem.* **254**:2669–2676.
90. E. R. Stadtman and A. Ginsberg, 1974, The glutamine synthetase of *Escherichia coli:* Structure and control, in *The Enzymes,* Vol. 10, P. D. Boyer, ed., Academic Press, New York, pp. 755–897.
91. E. R. Stadtman and P. B. Chock, 1978, Interconvertible enzyme cascades in metabolic regulation, *Curr. Top. Cell. Regul.* **13**:53–95.
92. P. B. Chock, S. G. Rhee, E. R. Stadtman, 1980, Interconvertible enzyme cascades in cellular regulation, *Annu. Rev. Biochem.* **49**:813–843.
93. A. Meister, 1974, Glutamine synthetase of mammals, in *The Enzymes,* Vol. 10, P. D. Boyer, ed., Academic Press, New York, pp. 699–754.
94. P. Lund, 1970, A radiochemical assay for glutamine synthetase, and activity of the enzyme in rat tissues, *Biochem. J.* **118**:35–39.
95. K. Iqbal and J. G. Ottaway, 1970, Glutamine synthetase in muscle and kidney, *Biochem. J.* **119**:145–156.
96. R. E. Miller, 1975, Adipose tissue and skeletal muscle glutamine synthetase in normal and diabetic rats, *Diabetes* **24** (Suppl. 2):416 (abstract).
97. R. E. Miller, R. W. Hackenberg, and H. Gershman, 1978, Regulation of glutamine synthetase in cultured 3T3-L1 cells by insulin, hydrocortisone, and dibutyryl cyclic AMP, *Proc. Natl. Acad. Sci. (USA)* **75**:1418–1422.
98. L. Arola, A. Palou, X. Remesar, and M. Alemany, 1981, Glutamine synthetase activity in the organs of fed and 24-hours fasted rats, *Horm. Metab. Res.* **13**:199–202.

99. X. R[illegible]nesar, L. Arola, A. Palou, and M. Alemany, 1982, Glutamine synthetase activity in rat tissues during pregnancy and lactation, *Horm. Metab. Res.* **14**:419–421.
100. R. DeMars, 1958, The inhibition by glutamine of glutamyl transferase formation in cultures of human cells, *Biochim. Biophys. Acta* **27**:435–436.
101. J. Paul and P. Fottrell, 1963, Mechanism of D-glutamyltransferase repression in mammalian cells, *Biochim. Biophys. Acta* **67**:334–336.
102. P. R. Barnes, D. Youngberg, and P. A. Kitos, 1971, Factors affecting the production of glutamine in cultured mouse cells, *J. Cell. Physiol.* **77**:135–144.
103. P. R. Barnes, R. T. Hersh, and P. A. Kitos, 1974, Regulation of glutamine synthetase in L cell by cortisol and L-glutamine, *In Vitro* **9**:230–238.
104. D. Viceps and V. J. Cristofalo, 1975, Glutamine synthetase activity in WI-38 cells, *J. Cell Physiol.* **86**:9–12.
105. D. Viceps-Madore and V. J. Cristofalo, 1978, Age-associated changes in glutamine synthetase activity in WI-38 cells, *Mech. Aging Dev.* **8**:43–50.
106. R. Gebhardt, P. Bellemann, and D. Mecke, 1978, Metabolic and enzymatic characteristics of adult rat liver parenchymal cells in non-proliferating primary monolayer cultures, *Exp. Cell Res.* **112**:431–441.
107. R. Gebhardt and D. Mecke, 1979, The role of growth hormone, dexamethasone and triiodothyronine in the regulation of glutamine synthetase in primary cultures of rat hepatocytes, *Eur. J. Biochem.* **100**:519–525.
108. H. H. Samuels, D. Klein, F. Stanley, and J. Casanova, 1978, Evidence for thyroid hormone-dependent and independent glucocorticoid action in cultured cells. Studies on the induction of growth hormone and glutamine synthetase in GH1 cells and tyrosine aminotransferase in Reuber II-35 cells, *J. Biol. Chem.* **253**:5895–5898.
109. D. C. Tiemeier and G. Milman, 1972, Chinese hamster liver glutamine synthetase. Purification, physical and biochemical properties, *J. Biol. Chem.* **247**:2272–2277.
110. D. C. Tiemeier and G. Milman, 1972, Regulation of glutamine synthetase in cultured Chinese hamster cells. Induction and repression by glutamine, *J. Biol. Chem.* **247**:5722–5727.
111. D. C. Tiemeier, D. Smotkin, and G. Milman, 1973, Regulation of glutamine synthetase in Chinese hamster cells, in *The Enzymes of Glutamine Metabolism,* S. Prusiner, and E. R. Stadtman, eds., Academic Press, New York, pp. 145–166.
112. G. Milman, L. S. Portnoff, and D. C. Tiemeier, 1975, Immunochemical evidence for glutamine-mediated degradation of glutamine synthetase in cultured Chinese hamster cells, *J. Biol. Chem.* **250**:1393–1399.
113. R. G. Kulka, G. M. Thompkins, and R. B. Crook, 1972, Clonal differences in glutamine synthetase activity of hepatoma cells. Effects of glutamine and dexamethasone, *J. Cell. Biol.* **54**:175–179.
114. R. G. Kulka and H. Cohen, 1973, Regulation of glutamine synthetase activity of hepatoma tissue culture cells by glutamine and dexamethasone, *J. Biol. Chem.* **248**:6738–6743.
115. A. Freikoph and R. G. Kulka, 1975, Specificity of the glutamine-binding site involved in the regulation of glutamine-synthetase activity in hepatoma tissue-culture cells, *Eur. J. Biochem.* **56**:483–492.
116. G. Arad, A. Freikopf, and R. G. Kulka, 1976, Glutamine-stimulated modification and degradation of glutamine synthetase in hepatoma tissue culture cells, *Cell* **8**:95–101.
117. A. Freikopf and R. G. Kulka, 1976, Involvement of protein synthesis in the post-translational control of glutamine synthetase activity of cultured hepatoma cells, *Biochem. Biophys. Res. Commun.* **72**:1195–1200.

118. G. Arad and R. G. Kulka, 1978, Effects of glutamine, methionine sulfone and dexamethasone on rates of synthesis of glutamine synthetase in cultured hepatoma cells, *Biochem. Biophys. Acta* **544:**153–162.
119. R. B. Crook and G. M. Tompkins, 1978, Effect of glutamine on the degradation of glutamine synthetase in hepatoma tissue-culture cells, *Biochem. J.* **176:**47–52.
120. R. B. Crook, M. Louie, T. F. Deuel, and G. M. Tompkins, 1978, Regulation of glutamine synthetase by dexamethasone in hepatoma tissue culture cells, *J. Biol. Chem.* **253:**6125–6131.
121. E. B. Thompson, D. K. Granner, T. Gelehrter, J. Erickson, and F. L. Hager, 1979, Unlinked control of multiple glucocorticoid-induced processes in HTC cells. *Mol. Cell. Endocrinol.* **15:**135–150.
122. A. P. Young and G. M. Ringold, 1983, Mouse 3T6 cells that overproduce glutamine synthetase, *J. Biol. Chem.* **258:**11260–11266.
123. P. Sanders and R. Wilson, 1984, Amplification and cloning of the Chinese hamster glutamine synthetase gene, *EMBO. J.* **3:**65–71.
124. R. E. Miller, S. R. Pope, J. W. DeWille, and D. M. Burns, 1983, Insulin decreases and hydrocortisone increases the synthesis of glutamine synthetase in cultured 3T3-L1 adipocytes, *J. Biol. Chem.* **258:**5405–5413.
125. D. M. Burns, B. Bhandari, and R. E. Miller, 1985, Selection of a rat glutamine synthetase cDNA clone, *Biochem. Biophys. Res. Commun.* **134:**146–151.
126. R. E. Miller, and D. M. Burns, 1985, Regulation of glutamine synthetase in cultured 3T3-L1 adipocytes by insulin, hydrocortisone, and cyclic AMP, *Curr. Top. Cell. Regul.* **26:**65–78.
127. B. K. Bhandari, R. C. Hoffman, D. M. Burns, and R. E. Miller, 1985, Glutamine synthetase in cultured 3T3-L1 adipocyte is encoded by more than one mRNA. 13th International Congress of Biochemistry, p. 672. (Abstract)
128. B. Bhandari, D. M. Burns, R. C. Hoffman, and R. E. Miller, 1986, Glutamine synthetase mRNA in cultured 3T3-L1 adipocytes: Complexity, content and hormonal regulation, *Molecular and Cellular Endocrinology,* in press.
129. O. H. Lowry, N. J. Rosebrough, A. L. Farr, and R. J. Randall, 1951, Protein measurement with the Folin phenol reagent, *J. Biol. Chem.* **193:**265–275.
130. K. Burton, 1968, Determination of DNA concentration with diphenylamine, *Methods Enzymol.* **12B:**163–169.
131. R. T. Schimke, 1975, Methods for analysis of enzyme synthesis and degradation in animal tissues, *Methods Enzymol.* **40E:**241–266.
132. B. Bhandari and R. E. Miller, 1985, Glycerol-3-phosphate dehydrogenase mRNA content in cultured 3T3-L1 adipocytes: Regulation by dibutyryl cyclic AMP, *Biochem. Biophys. Res. Commun.* **131:**1193–1197.
133. T. Maniatis, E. F. Fritsch, and J. Sambrook, 1982, *Molecular Cloning, A Laboratory Manual,* Cold Spring Harbor Laboratory, Cold Spring Harbor, New York, p. 545.

9

Determination of Adipose Tissue Cellularity

Aubrey Lee Cartwright
U.S.D.A. Poultry Research Laboratory

Adipose tissue, once regarded as a static tissue, is now recognized as a dynamic organ and an important participant in the control of body metabolism. Investigations involving adipose tissue present unique problems since depot fat grows more substantially by cell hypertrophy than other tissues. The adipocyte is the specific cell type of adipose tissue and is the storage site of triglyceride. The greatest portion of the adipocyte volume is thus occupied by a metabolically neutral lipid droplet. Metabolically active cytoplasm constitutes a regressively smaller portion of the adipocyte's mass as the cell's volume increases. Therefore, it is prudent that metabolic data on adipose tissue be expressed as a function of adipocyte number.

Comparisons of depots indicate that adipose tissue is not a homogeneous tissue but one that has characteristics specific for functions other than storage of energy (1). Some depots resist fatty acid metabolism after severe starvation, while others have physical and mechanical restraints to cell hypertrophy (2). Even adipose tissue sites most frequently studied are different in cell composition, mean cell size, and age of maturity. Adipocyte size itself dramatically influences the metabolism of the cell (Fig. 9-1). This mul-

---

The views expressed here are those of the author and do not reflect the policies of the United States Department of Agriculture. Mention of a trade name, proprietary product, or specific equipment does not constitute a guarantee or warranty by the United States Department of Agriculture and does not imply its approval to the exclusion of other products that may be suitable.

tiformity of function and morphology requires differences in responsiveness to modulators of the biochemistry of the cell. The discovery of how adipose tissues respond to different physiological states or chemical influences requires determination not only of cell number but also of cell size.

Cell number and size data have become prerequisites that replace mere knowledge of the lipid content of a sample of adipocytes. The amount of body fat may be determined or estimated by a variety of means (4), but cellularity is more difficult to quantify.

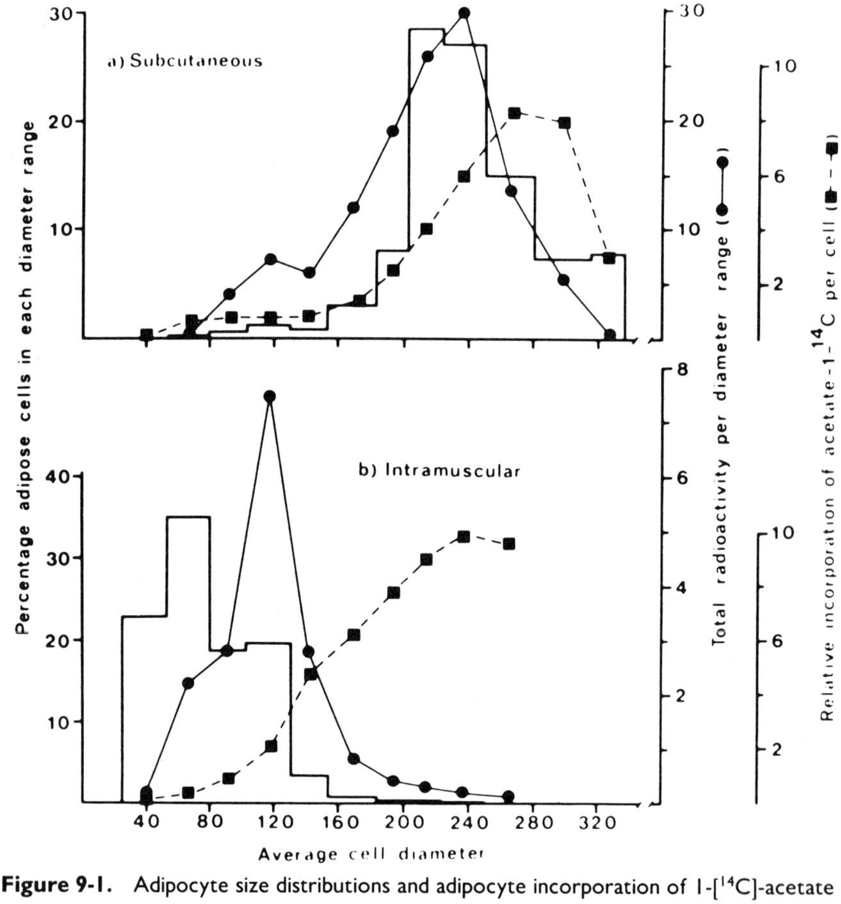

Figure 9-1. Adipocyte size distributions and adipocyte incorporation of 1-[^{14}C]-acetate into lipid are shown. The adipocytes from subcutaneous and intramuscular ovine tissue were incubated as tissue slices for 2 hr. Note the difference in adipocyte size between the depots, and the increased incorporation of radioactivity in larger adipocytes. (*From Hood and Thornton, 1980; reproduced with permission of Journal of Lipid Research.*)

APPROACHES TO DETERMINING CELLULARITY

Adipocyte number can be estimated through either of two modes of rationalization. In the first case, one places confidence in the accuracy of mean adipocyte volume determinations. Additionally, one assumes that essentially all of the tissue lipid, volume, or mass is due to adipocytes. Therefore the total adipocyte number is calculated from the total lipid, volume, or mass of the tissue and the mean adipocyte. Alternatively, one assumes total recovery and accurate quantification of the number of cells in an analyzed piece of adipose tissue. Therefore the quantity of cells in the piece of tissue is directly proportional to the depot mass. Mean adipocyte size can be estimated indirectly from cell number and tissue mass as in the previous strategy, or it can be determined by direct measurement. Incorrect assumptions or false values resulting from incorrect assumptions can seriously affect the accuracy of cellularity determinations.

Adipose tissue is composed of adipocytes of different sizes. It has been assumed that the adipocytes are normally distributed about a mean diameter. Thus we have mathematical formulas for transforming mean cell diameter to mean cell volume (5, 6) and correcting apparent diameter in histological sections (7) by considering the variance of raw size data. Adipocyte size distribution can in fact be skewed or biphasic (1, 2, 8–17). Indeed, a normally distributed adipocyte population may be unusual. If this is the case, the basic assumption for determination of mean adipocyte size, upon which the accuracy of cell number is founded, is in error. Direct determination of cell volume from size distributions makes transformations to determine volume unnecessary. An optimal system would not only arrive at a mean adipocyte volume in this manner but would also yield a quantitative value for adipocyte number.

Four methods are used to obtain data from which adipocyte size and number are estimated. Several modifications of the methods are found in the literature, but each requires as a basis deoxyribonucleic acid (DNA) determination, adipocyte isolation with collagenase, microscopic examination of adipose tissue or fixed adipocytes, or electronic quantification. The technique chosen to fulfill experimental requirements is dictated by the relative importance of the advantages and disadvantages of a particular method, and the availability of resources.

The methods have been evaluated and critically compared in a comprehensive review (4). Other papers present validations of specific methods (6, 17, 18, 19) and evidence of improvement (20–23). Comparisons of these techniques are also published (8, 24, 25) as well as a review of investigations that utilized techniques to observe preadipocyte cellularity (26).

DNA Method

The simplest extrapolation of adipose tissue cellularity requires quantification of DNA. This method assumes that the DNA content of adipose tissue is directly proportional to the number of adipocytes in the depot and that a fixed amount of DNA is associated with a cell. The adipocyte, the distinguishing cell type of adipose tissue, is assumed to comprise a proportionally large amount of the cells. An extension of these assumptions is that the total DNA of adipose tissue is the product of the number of adipocytes and the amount of DNA in a cell. Furthermore, the mean mass of an adipocyte is estimated from the total number of adipocytes and the mass of the tissue. Analytical accuracy is dependent upon the DNA determination and the veracity of the assumptions.

In the past the completeness of extraction of DNA from adipose tissue and isolated adipocytes has been suspect (5, 27), but recently laboratory procedures for DNA determination have received appropriate attention and should be less of a serious concern. Quantitative determinations of DNA specifically developed and evaluated for white and brown adipose tissue are available (27, 28). However, the accuracy of DNA determinations cannot correct inappropriate assumptions inherent in the method.

The number of nuclei per adipocyte might not be absolutely consistent with these assumptions.* However, the constancy of nuclear material within the cells of a species is not a serious concern. The flaw is the assumption that adipocytes are a major portion, or even a constant proportion, of the cells.

Quantitative evidence by Rakow (29) indicates that the number of non-adipocyte supportive cells (stromal-vascular cells) is inconstant and comprises the major portion of cells in adipose tissue. Adipocyte and stromal-vascular cellularity were compared in lean mice, and in genetically and chemically obese mice. Adipocyte size was measured by microscopic examination, and total adipocyte number was calculated as a proportion of fat pad mass. Additional cytophotometric and pulse labeling examinations showed that as mean adipocyte size and adipose tissue mass increased, proliferation of stromal cells increased. Adipocyte number did not change among lean or obese mice, but a threefold increase in stromal cell DNA occurred. Stromal-vascular DNA is not only a variable component of adipose tissue but is also the major source of DNA. In the guinea pig a forty-fold ratio of stromal cells to adipocytes was observed (30). It is generally

*J. Novakofski, 1984, Adipocyte development and differentiation, presented as part of the Adipose Tissue Differentiation and Growth Minisymposium of the Seventy-Sixth Annual Meeting of the American Society of Animal Science. Photomicrographs indicated that a small percentage of adipocytes becomes polynuclear.

agreed that the DNA method will cause an underestimation of adipocyte volume since 60 to 98% of DNA is from stromal-vascular cells (1, 29–32). Measurement of tissue DNA is not an accurate basis for estimation of adipocyte cellularity.

Even though tissue DNA is not an adequate index of adipocyte number, DNA is a useful measure of cell number in solutions of isolated adipocytes (33). In this case adipocytes are freed from the interfering stromal cells. This is accomplished by employing a technique characteristic of the second method used to estimate cellularity, the collagenase method.

Collagenase Method

Several procedures for estimating cellularity evolved from the classical technique of Rodbell (31) for preparing isolated adipocytes. Adipocytes are freed from surrounding stroma by enzymatic digestion of minced adipose tissue in a physiological solution of collagenase. The free spherical adipocytes with a low specific gravity are easily separated from tissue residue. Adipocytes rapidly float to the surface of the solution, while other cell types and tissue debris settle to the bottom of the tube.

The isolation technique was modified specifically for determination of adipose cellularity by DiGirolamo and co-workers (19) and was improved to minimize adipocyte lysis (8, 34). An aliquot of the suspension of isolated adipocytes is placed on a slide, or a lake of solution is prepared by using silicone grease as a dam. Adipocytes float to the surface of the solution, where they are measured microscopically with the aid of a micrometer eyepiece, or any one of several improvisations developed to allay the tedious sizing process (35, 36, 37).

The collagenase technique requires equipment that is readily available and has advantages over the DNA estimation of cellularity. A major advantage is that the adipocytes are directly measured, yielding accurate diameter, volume, and surface area estimations of the adipocytes in solution, and a reasonable means by which to estimate total adipocyte number in a depot. Direct measurement of intact adipocytes also alleviates the dimensional problems associated with histological thin sections. If availability of tissue is not limiting, it is advantageous to isolate cells from a large sample to avoid isolation of adipocytes not representative of the depot. The adipocytes can be used in conjunction with metabolic and morphological studies.

The collagenase method has serious disadvantages that affect both quantitative recovery and size distribution. Incomplete digestion, aspiration loss, and cell lysis are sources of the errors. March and Hansen (38) monitor recovery of adipocytes from the tissue by lipid extraction of isolated adipocytes. But this method does not adjust for changes in size distribution.

Incomplete digestion is a function of agitation, amount of connective tissue (4), and collagenase quality (8). Adipocytes retained in connective tissue and cell clumps as a result of incomplete digestion are separated from isolated adipocytes by aspiration. During aspiration of debris, small adipocytes are disproportionately lost because their density is greater than the density of large adipocytes. Their slow flotation rate hinders their separation from tissue debris.

The third source of error is lysis. Lysis accounts for a considerable loss of cells even when great care is exercised (24, 39) and appears to predominately affect larger adipocytes (6, 22, 40). The effect of shaking and species is dramatically depicted in Figure 9-2. Agitation of samples during incubation can potentiate lysis (Fig. 9-2a). Emphasis should be placed on determining the correct amount of agitation to produce maximal yield of adipocytes and minimal lysis during isolation and incubation.

Rat adipocytes are not as labile as cells from pigs, mice, and chickens (unpublished observations). Figure 9-2b compares the disappearance of rat and pig adipocytes over 2 hours after cell isolation. Isolation and subsequent incubation of mouse adipocytes resulted in rapid lysis similar to that observed in pig adipocytes. Lysis was coincident with a decrease in mean adipocyte diameter (41). The decreased mean diameter resulted from smaller cells contributing a disproportionate amount to the number of surviving adipocytes as larger cells lysed. The effect of lysis of the more stable rat adipocyte on size distribution was investigated by DeMartinis (17). Fixation procedures were used to arrest adipocyte lysis immediately after collagenase digestion. Comparisons between population distributions of cells isolated by collagenase and cells fixed in situ with osmium tetroxide showed that collagenase digestion significantly affected adipocyte size distribution (Fig. 9-3). Mean adipocyte diameter of cells larger than 35 μm was significantly larger for osmium-fixed cells (92.8 $\pm$ 17.1 μm) than for collagenase-treated cells (72.6 $\pm$ 9.7 μm). The proportion of very small fat cells was also significantly smaller for collagenase-treated cells (5, 9) than for osmium-fixed cells (9, 4). Even though progress has been made to improve the technique (8, 34), lysis is difficult to consistently control. Lysis affects mean cell size and renders cellularity determination of a depot erroneous. Furthermore, lysis differentially affects adipocyte size, thereby rendering comparisons between experimental treatments, which might affect adipocyte size, suspect.

Intact Tissue

Direct microscopic examination of adipose tissue offers the best method to minimize the effects of cell loss during cellularity determinations. Cells are analyzed from fixed histological sections (7 to 200 μm thick) or from intact

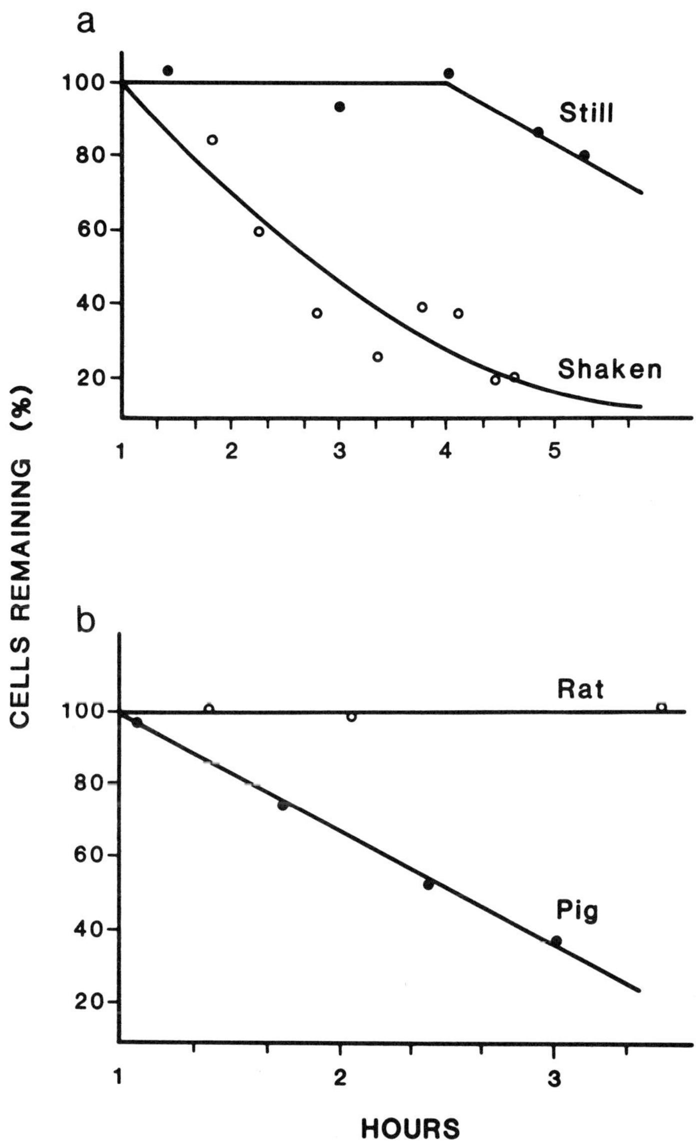

Figure 9-2. The effect of agitation and species on adipocyte lysis over time is apparent in this figure by Gurr and Kirtland. The abscissa represents the collagenase-isolated adipocytes remaining as a percentage of cells obtained after a 1-hour digestion period. The effect of shaking (*a*) and the effect of species (*b*) on adipocyte disappearance are depicted. (*From Gurr and Kirtland, 1978, p. 354; figure reproduced with permission of International Journal of Obesity.*)

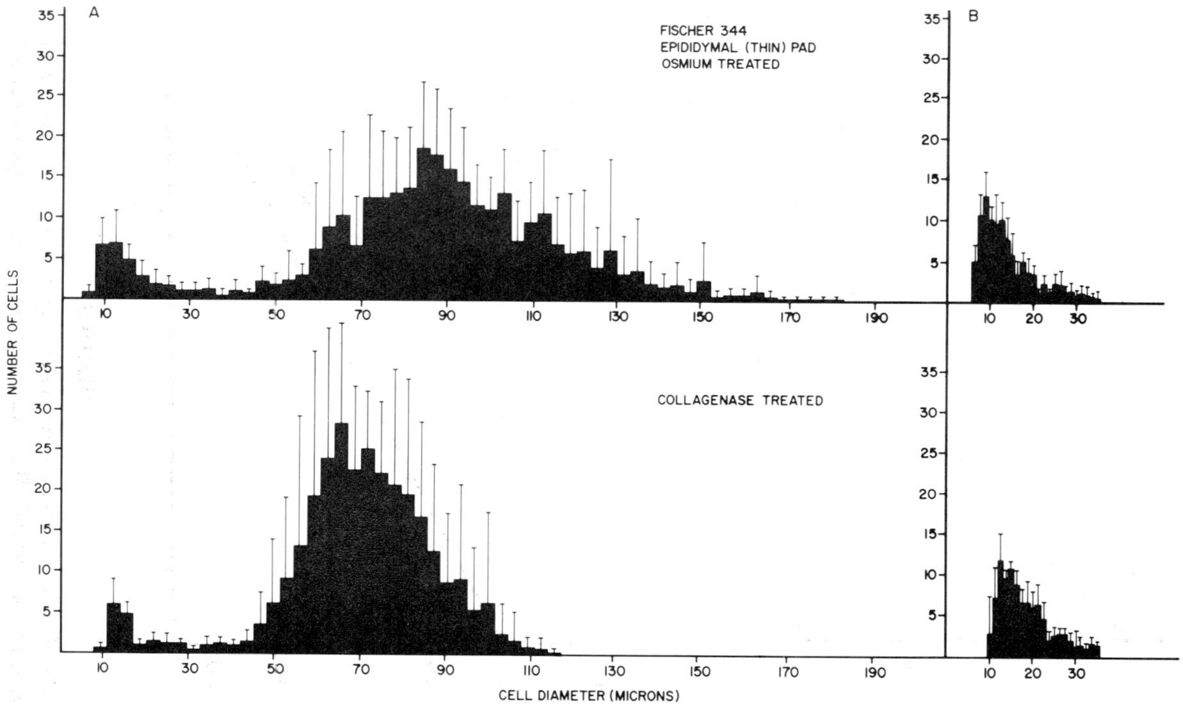

Figure 9-3. *A*. Histograms of the average of adipocyte diameters of cells isolated from rat epididymal depot either by a modified osmium fixation method or by the collagenase method. *B*. Histograms of the average distribution of adipocyte diameters 35 μm or less in diameter isolated by each method. (*From DeMartinis, 1985, p. 354; figure reproduced with permission of American Journal of Physiology.*)

tissue pieces. The means of estimating cell size depend upon whether thin or thick sections, or intact tissue pieces are used.

Both thin and thick sections are prepared from fixed tissues in paraffin blocks. The sections are then stained appropriately for adipocyte measurement (7, 18, 24). Adipocyte ghosts are sized by microscopic observation of either the cell cross sections (thin sections) or through measurement of the maximal diameter of ghosts (thick sections).

Thin sections are not thick enough to contain whole adipocytes. The histological preparation is therefore an array of random cross sections of adipocytes. Only a fraction of the cross sections present the maximum diameter. The apparent mean diameter of the adipocytes is recorded and corrected for statistical deviations from the actual diameter. This phenomenon and methods necessary to arrive at the mean diameter are discussed by Bjrulf (18) and Lemmonier (7, 42).

Thick sections (18) require analysis similar to that required of thin sections. However, thick sections contain a distribution of intact adipocytes. The ability of the microscope to focus within the tissue is utilized to locate the plane that displays an adipocyte's maximum diameter. The observer can be satisfied that false apparent diameters of adipocytes are not a complication and apppropriate volume calculations are made directly.

Bjrulf's (18) comparison of data collected from thick and thin sections (1959) showed that the two methods are usually in agreement. Lemmonier determined apparent cell size of thin sections by counting the number of adipocytes within a known area of the slide and assuming that: (a) the adipocyte shape approximated a perfect sphere; and (b) adipocytes covered the entire measurement area. Mean adipocyte size was derived by using a simplified mathematical formula that considered the variance of cell size and planar cuts of spheres. Ashwell and co-workers (24) tested Lemmonier's method against the isolated adipocyte method to suggest a constant correction factor that could be used to adjust the apparent diameter of fixed adipocytes. The correction factor, 1.1, was in agreement with Lemmonier's (42) theoretical correction factor of 1.15.

Like data collection from thick paraffin sections, the frozen-cut method proposed by Sjostrom and co-workers (20) allows direct measurement of adipocytes. The frozen-cut method is similar in principle to direct observation of adipose tissue pieces in physiological solution (5) or in glycerol (25). In Sjostrom's procedure small pieces of tissue are collected and briefly fixed in 35% formaldehyde. The fixed tissue is frozen and cut into 200-μm sections and allowed to float freely in the lake of Ringers solution on a glass slide. The largest diameter of each intact cell in the section is recorded. Minor modifications of the method have been made to render analysis of large numbers of samples more accurate and less tedious (4).

The microscopic methods are low-cost techniques of high accuracy that use equipment usually available in laboratories. Fixed slides or photomicrographs of samples allow cellularity analysis with a minimal amount of tissue and provide permanent records of an experiment. Greater detail and more information can be accumulated from sections since the adipocytes are observed within the tissue matrix. The number of cells measured (usually 100), can be practically increased through use of the method of Lemmonier (7). It is more rapid since counting cells (at least 1000) is less time-consuming than measuring diameters. Freeze-cut sections offer the specific advantage of immediate results since tissue preparation time is minimal.

The advantages, as well as the disadvantages, of direct tissue observation are associated with discrete detailed examinations of samples. Tissue fixing and sectioning cause some shrinking and distortion (4, 18, 20), but the effects are considered insignificant (4, 7). However, when thin sections are used, mathematical corrections are needed to arrive at actual values (7, 24). To make these corrections, subjective assumptions about the relative contribution of supporting tissue, the representative nature of the sample, and the symmetry of the adipocyte must be made. Many of these disadvantages are not encountered by the electronic quantification technique.

Electronic Quantification

The final method uses electronic means to quantify adipocyte cell number and volume. Briefly, osmium tetroxide is used to rapidly fix adipocytes within the adipose tissue matrix. The connective tissue and tissue debris are then removed. The remaining fixed lipid droplets are suspended in saline, and counted and sized electronically. The original method and validation (6) was later improved (20, 21, 22). A modification of this method has successfully been used to quantify the cellularity of brown adipose tissue (43). Detailed procedures for white adipose tissue will be described in the methodology section of this chapter.

Cell number, mean cell diameter, or mean cell volume of a tissue can be determined either directly or indirectly, depending on the electronic equipment available. Direct determinations are completely objective, with cell volume being a measure of volume displacement. The volume is therefore not affected by the polyhedral cell shape or the subjective judgments of the observer. Mathematical manipulation and assumptions are unnecessary. Quantitative cell counts of large numbers of cells are obtained that minimize sampling errors and the possibility of analyzing areas of tissue unrepresentative of the depot.

The technique does have limits beyond which precautions should be taken. Data on small adipocytes are occasionally affected by spurious counts caused by debris. Measures to circumvent this problem are discussed in more detail later. Other disadvantages are a consequence of the expense of the equipment and the osmium tetroxide, and the length of time required for sample preparation. Swelling of osmium-fixed adipocytes was originally feared (6), but no evidence of this has been found (20).

Both the quantitative abilities of the electronic technique and the resilience of fixed adipocytes have been utilized to investigate the effects of cell size on metabolism. Isolated adipocytes fixed in osmium tetroxide can be counted and sized for estimations of cell number and morphology (6, 44, 45, 46). Furthermore, the properties of fixed adipocytes allow cell lipid integrity to be retained through rigorous separation procedures from adipose tissue slices (3, 47). Effects of adipocyte size on the chemistry of the adipocyte can be investigated while deleterious effects of cell isolation are avoided.

Summary

Direct microscopic observation of intact tissue or fixed tissue, and electronic quantification of osmium-fixed adipocytes are clearly superior methods for determining adipocyte cellularity of adipose tissues. Determination of the DNA content of adipose tissue does not yield data directly related to adipocyte number. DNA is an accurate index of adipocyte number only after adipocytes have been dissociated from stromal-vascular cells. The collagenase method yields isolated adipocytes well suited for metabolic studies, but problems associated with recovery and the fragility of cells render the method of questionable value for cellularity studies. Cell loss, directly affecting estimations of mean adipocyte size, invalidates extrapolation of cell number from size data. The osmium tetroxide and electronic quantification technique was developed to circumvent the cell lysis problems of the collagenase method (6). Animals with small adipocytes were utilized for comparison of the collagenase and the osmium techniques to minimize these problems. The observation that osmium-fixed cells were larger than collagenase-isolated cells sized by other methods resulted in the conclusion that osmium fixation caused the cells to swell. This observation has not been reported by the investigators (20) and was probably due to false data resulting from collagenase treatment.

Examination and electronic quantification of intact tissue offer specific advantages, the relative importance of which depends on experimental objectives. Fixed adipose tissue sections enable a researcher to examine adipocyte cellularity (7, 24) and enzymology (23) as well as to evaluate the fine structure of adipose tissue. The apparent sizes of the adipocytes are not unacceptably affected by fixing and sectioning processes (18). Freeze-cut sections allow rapid measurement of adipocyte size distributions which are similar to those obtained from electronic analysis (20), but little additional information can be obtained.

The method of DeMartinis (17), in which osmium-fixed adipocytes are sized microscopically, offers a more accurate measure of the very small adipocyte population. Quantification of this population is more difficult by other methods. Ideally, cellularity determinations would be a product of the electronic quantification method and the in situ examination of the adipocyte. Electronic quantification, however, makes possible collection of direct, objective, quantitative determinations of both cell number and cell size.

METHODOLOGY AND APPROACH

This section considers specific methodology of fixing adipose tissue, separating adipocytes from tissue debris, and subsequently counting and sizing adipocytes electronically. In addition, advantages and quality assurance precautions of the technique are discussed. Because an understanding of the principles of electronic counting are needed before other concerns can be appreciated, the electronic principles and equipment will be described first.

Electronic Counting and Sizing

Principles of electronic counting and sizing are simple (48, 49, 50). Fixed adipocytes are evenly dispersed in an electrolyte solution and drawn singly into an aperture in a nonconductive glass tube (Fig. 9-4). Since an electric current has been created across the aperture, passage of nonconductive particles displaces a volume of electrolyte and interferes with current flow. This change in conductance creates a pulse that is detected by the counter and a count is registered. The total number of pulses collected during a counting session can be used to determine the total number of adipocytes in a defined piece of tissue if certain quantitative values are known. Since the counts are generated by displacement of electrolyte by the adipocyte, the size of the pulse is proportional to the volume of the adipocyte. So both the number and the size distribution of an adipocyte population can be obtained electronically.

The equipment necessary for efficient utilization of time and labor in counting and sizing are the counter and the multichannel analyzer (Fig. 9-5). The counter alone can be used to obtain cell number and size data, but the multichannel analyzer dramatically increases productivity and reduces error.

The equipment with which the author is most familiar is produced by Coulter Electronics, Inc., and Particle Data, Ltd. Equipment from both companies is comparable. Detailed descriptions in this section are of Coulter Electronics' equipment (Coulter counter, model ZBI; multichannel analyzer, Channelyzer C-1000*), but minor modifications of procedure easily adapt other equipment to the purpose.

The counter consists of two main sections: the sample stand (which contains the aperture tube, electrodes, manometer, and vacuum pump) and the electronic counter. The sample stand must be equipped with a mixer with which homogeneous suspensions of fixed adipocytes are maintained while counting. A mixer attachment is available for newer models, but a fluid-driven magnetic stirrer can be mounted to the beaker platform.

Quantitative information is obtained with the counter because a known volume of suspension passes through the aperture during the counting process. The manometer and vacuum pump are responsible for drawing a precisely measured volume of cell suspension through the sensing zone off the aperture tube. A stopcock control assembly regulates exposure of the mercury-filled manometer to a vacuum. When the vacuum is opened to

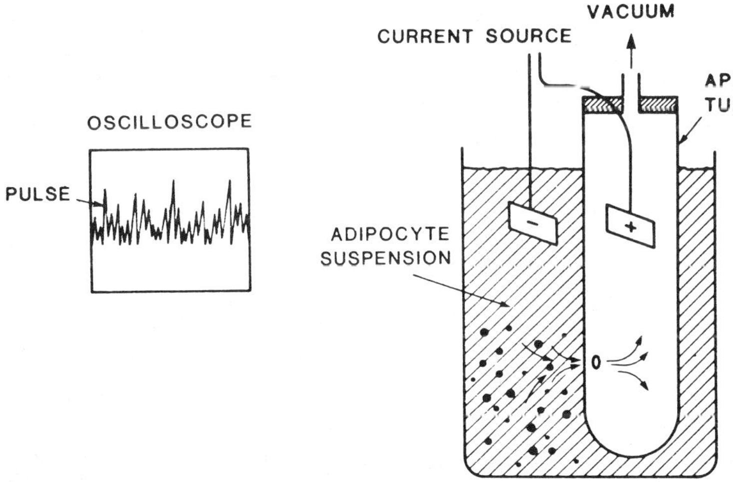

Figure 9-4. Diagram of the adipocyte suspension flow through the counter aperture unit (right) and the pulses generated by their passage (left).

*Coulter Electronics has recently developed the Channelized 256. Many calculations required for calibration of the Channelizer C-1000 are performed by microprocessers in the 256 model.

the manometer, mercury is pulled above a pair of electrodes. When the vacuum reservoir is closed, the mercury returns to equilibrium and contacts the electrodes, which start and stop the electronic counter from accumulating counts. Various manometer sizes are available, but the 2-ml manometer is best suited to aperture sizes appropriate for counting adipocytes.

The electronic counter contains equipment for displaying counts, and dials for setting amplification, aperture current, and size exclusion. Counts can be read quantitatively from a numeric readout, or generated pulses may be visualized during the counting process on an oscilloscope. The size of the pulses is the result of the current supplied by the counter and can be adjusted by altering the amplification and the aperture current. A gain adjustment can also be used on some models to fine-tune pulse size. Therefore, the effective range and sensitivity of the counter are adjustable for each aperture tube size.

Different-sized pulses make it possible to measure cell volume if pulses that are too large or too small are excluded. This exclusion is achieved by using the upper and lower thresholds, and the aperture current switches. Upper and lower threshold settings are controlled by individual graduated dials to delimit particular size ranges.

Figure 9-5. Coulter counter model ZBI, which houses the counter and sample stand (left), and Coulter Channelyzer C-1000 multichannel analyzer (right).

Precise size increments are associated with each graduation of the threshold settings by counting a standard and using the formula:

$$K = \frac{V}{AIT},$$

where V = mean volume of standard particles, A = 1/amplification setting, I = 1/aperture current setting, T = the lower threshold setting at which the standard particles are registered.

Once K is known, the volume represented at each increment of the threshold settings for each A and 1 is used to count adipocytes of a certain size. Counts of known size increments can be obtained with the counter alone by repeatedly counting a cell suspension, and changing threshold and current settings.

The multichannel analyzer is designed to receive pulses from the counter and separate them by size into several size ranges simultaneously. With the counter alone many counting volumes are required to obtain size distribution data with relatively small numbers of counts in each size range. However, all cells in a solution are counted and sized in a single step with a multichannel analyzer.

The Coulter Channelyzer C-1000 is capable of analyzing pulses from the full range of the counter, as delimited by threshold settings, into 100 size ranges. These data can be viewed as a frequency distribution on an oscilliscope, as counts from individual channels, or as counts from integrated groups of channels.

The volumes of particles that are registered in each channel are calculated by the formula:

$$\text{Volume} = (\text{CN} \times \frac{\text{WW}}{100} + \text{BC}) \times \text{TF},$$

where CN = channel number, WW = window width setting, BC = base channel threshold setting, TF = threshold factor determined when counter is calibrated.

Groups of channels can be integrated to contain a predetermined volume range for analysis.

Calibration is a critical procedure. Once the counter threshold and the multichannel analyzer channels are aligned (as described in the manufacturer's manual), the two machines can be calibrated. The calibration procedure must be performed for each aperture tube, and at each setting of the matching switch or gain trim that is used.

Calibration requires the availability of a precise standard. The standard particles should be similar in size to that of samples for analysis. Fortunately, there are several commercial sources of accurate particle standards. Since resistance determines the detected volume of the particles, differences in particle composition could affect the apparent size of fixed adipocytes. Fixed cells can be used as standards for calibration. Known size ranges of adipocytes obtained by fractionation and commercially prepared polystyrene standards have given the same response in this laboratory (unpublished observations).

The calibration procedure involves counting standard particles and defining the peak accumulation channel on the multichannel analyzer. With this information and the mean volume of the standard particle, the threshold factor may be calculated from the equation:

$$TF = (CN \times \frac{WW}{100}) + BC$$

Once the threshold factor is known for one amplification and aperture current setting, threshold factors for other settings are calculated. Thus, the sizing range of one aperture tube is expanded. Sizing flexibility is expanded further when different sized aperture tubes, for which threshold factors have been determined, are available.

Other equipment is available for expanding the range and the data handling capacity of the system. The most useful pieces of equipment are the plotter, the logarithmic range expander (LRE), and computer-assisted data handlers. The plotter gives a rapid hard copy of size distribution data but is not easily adaptable to quantitative analysis. Theoretically, the LRE will increase the dynamic range of the multichannel analyzer from 100:1 to 10,000:1. Automatic data handling systems collect, calculate, and display data in preplanned formats automatically. While this equipment may not appreciably improve data quality, it will allow an experimenter to record and analyze data more rapidly.

Adipocyte Preparation

Most apparatuses for fixation and subsequent isolation of adipocytes from tissue debris are readily available. A component of the fixing solution, osmium tetroxide, and the precision nylon screen material for filtering might be special needs.

Osmium tetroxide is necessary for rapid fixation of adipose tissue. Other fixatives may be used, but adipocyte preparation is usually cumbersome and results are less accurate than the osmium fixation methods developed by

Hirsch and Gallian (6) and Etherton and co-workers (22). One must bear in mind that osmium is highly volatile and toxic, so work with osmium or solutions of osmium must always be performed under a fume hood.

Filtering apparatus consists of an aspirating flask, filtering screens, and screen-holding materials. Several 35- to 40-mm lengths of 20- to 25-mm diameter plastic conduit and connectors are connected in series to hold 40-mm squares of 240-, 64-, and 25-μm nylon mesh screening material. A neoprene stopper, cut to accommodate the conduit, is fitted into an aspirating flask. Cells are filtered in the filtering apparatus with the aid of an aspirated vacuum and washed repeatedly to assure proper fractionation of the cells for counting.

Copious amounts of normal saline are used during sample preparation and counting. An additional detergent saline solution is made by adding 0.1% Triton X-100 to normal saline. Care should be exercised to ensure solubilization of the detergent before use. If a large number of samples are to be analyzed, it is best to prepare carboys of both saline and detergent saline. Solutions should be rendered dust-free by connecting a cartridge filter to the saline-filled carboys, or by prefiltering the solutions with a 0.45μm filter. Prefiltering is advisable if the solutions are stored for an extended period.

Procedure

Adipose tissue is a fragile tissue. The range of volumes of adipocytes is extensive. Forces that are imparted to large cells may cause them to rupture preferentially. One must remember that since volume increases as a cubed function of diameter, the volume that large adipocytes contribute to the tissue is disproportionate to their number. Proper respect must be paid to the fragility of adipocytes if accurate cellularity data are sought.

Excised fat pads are weighed, and 50- to 100-mg samples are removed with sharp scissors for osmium fixation. Oil droplets that might inhibit fixation of cells are removed from the tissue surface by two brief rinsings in saline at 37°C. A scintillation vial receives the sample, and 3 ml collidine-HCl buffer* is washed onto the tissue to remove remaining oil immediately before 5 ml osmium tetroxide solution† is added. The vial is capped§ and stored at room temperature for 2-4 days.

---

*Prepare 50 mM collidine-HCl buffer (pH 7.4) by diluting 25 ml 0.2 M collidine and 22.5 ml 0.1 N HCl (hydrochloric acid) to 100 ml with 0.154 M NaCl (sodium chloride).

†Prepare by dissolving 5 g osmium tetroxide in 100 ml of 50 mM collidine-HCl buffer.

§Foil-lined caps should not be used, because the corrosive nature of osmium will result in sample contamination.

Fixed adipocytes are separated from tissue debris, and connective tissue is solubilized. Osmium tetroxide solution is separated from fixed tissue by filtration through 25-μm nylon screen. Fixed tissue is washed onto the 25-μm screen with copious amounts of saline. Isolated tissue and 10 ml saline are added to a scintillation vial and sealed. After 24 hr, the tissue is re-collected on a 25μm screen and added to a scintillation vial with 10 ml 8 M urea in 0.154 M NaCl. Samples are stored in capped vials at room temperature for several days and agitated occasionally.

Fixed adipocytes are separated into two size ranges by sequential filtering through 240-, 64-, and 25μm nylon screens.* Fixed oil droplets and tissue debris that collect upon the 240μm screen are discarded.† Large and small adipocytes are isolated upon 64- and 25-μm screens, respectively. Each fraction is rinsed into tared 400-ml beakers with detergent saline solution. A total volume of approximately 300 ml, which may be measured gravimetrically, is added to each beaker.

Adipocyte number and size distribution are now measured electronically. Several 2-ml aliquots of continuously stirred cell suspensions are sized and counted. Small adipocyte fractions are counted and sized by using a 280-μm aperture. A 400-μm aperture is used to count adipocytes retained by the 64-μm screens.

Adipocyte size distribution can be analyzed in 10μm increments. Diameter distributions are the sum of distributions of the large and small cells. The total cell number of the fat pad is estimated directly by using tissue weight and adjustments for suspension volume. Average cell diameter, volume, and surface area are calculated geometrically from the accumulated distributions, assuming that the normal polyhedral shape of the adipocyte approximates a sphere.

Small samples of isolated adipocytes can also be quantified by a modification of this method. A volume of isolated adipocytes added to a 1% solution of osmium tetroxide in a ratio of 1:1 (v:v) is fixed within minutes and can be analyzed electronically within the hour, or at a later time. A coefficient of variation of cell number determinations should be substantially less than 5% by this technique (unpublished observations).

Data Interpretation

The total number of adipocytes in a tissue sample or in an entire adipose depot is of obvious use when results of biochemical studies should be ex-

*Filtering with 240- and 25-μm screens and counting with a 400-μm aperture is all that is required if an LRE is available.

†The amount of fixed oil and debris should be minimal and utilized as an indicator of technique quality.

pressed as a function of adipocyte number. Likewise, the need for precise knowledge of the number of isolated adipocytes for adjustment of suspension concentration is vital in the study of receptors and in studying the biochemical response of isolated adipocytes to a concentration of an effector. But knowledge of size distribution allows the investigator to estimate physical characteristics of the cells, to be more confident of the data, and to draw inferences during analysis.

An easy comparison of adipocyte size in different tissues can be made on the basis of mean cell size, which may be expressed as diameter, volume, or surface area. *The values are not interconvertible.* Since volume increases as a cube and surface area increases as a square of cell diameter, the mean of neither of these values can be used to estimate a mean volume of the other. These values must be calculated individually from raw data. The value to use in an estimation of size may not be critical, but nevertheless, an investigator should carefully consider the physics that would be involved in the response measured by the study.

Size distributions can alert an investigator to possible errors in analysis based upon adipocyte number and mean size comparisons. Data collections that have become routine can develop a lack of consideration of complete size distributions. Size distributions can indicate the presence of a biphasic population or the incomplete collection of quantitative data. In these cases mean cell size measurements lose validity. Even if a case can be made for using a mean value, it is prudent to present a distribution graph for consideration by others.

Size distributions can be expressed quantitatively or as a percentage of the tissue quantity. Most frequently, distributions in graphic form use adipocyte diameter as the abscissa. This approach is desirable since plots based on diameter approach a Gaussian distribution. The ordinate, depending on the emphasis of the author, may be number, volume, or a frequency of either when expressed as a proportion of the total. The latter will present size distributions in which differences in size are more easily seen. However, a distribution of total cells also supplies the number of adipocytes in a tissue of interest. Examples of several methods of graphic display are presented in recent publications (16, 51).

Statistical analysis of the curves described above presents a problem since different curves usually cross. Cumulative frequency distributions, when compared at particular percentiles, can be used as a measure of statistical differences, but more difficult problems are faced when a curve becomes biphasic. DeMartinis has used the goodness of fit chi square test to compare either two distributions or for goodness of fit to a uniform distribution (13, 17). Perhaps an appropriate solution is to supply standard errors of the mean at all points in a graphic comparison in addition to other statistical comparisons. Differences in size distributions are usually apparent.

Advantages and Precautions

This method exceeds the specific performance capabilities of other counting and sizing methods* (light blocking, scalling, diffraction, elutriation, sedimentation, sieving, and microscopy) for determination of adipose tissue cellularity. There are advantages to the osmium fixation and electronic quantitation of adipocytes that are related to the nature of the fixation process, the size of the sample, the capacity for data accumulation, and the objective nature of analysis. The method has the capacity for the resolution, range, sensitivity, and mildness necessary for accurate determination of adipocyte size and number. But limitations of the method must be known and precautionary measures must be taken to ensure collection of valid data and to minimize errors.

Advantages. One characteristic ascribed to adipocytes is their tendency to lyse under mild perturbation. Fragility seems to be a function of cell size, but species differences also influence adipocyte stability. A vicious cycle probably also exists whereby lysis during adipocyte isolation potentiates further lysis. An advantage to determining cellularity through osmium fixation is that cells are fixed in the tissue matrix. Penetration of osmium tetroxide into the tissue and the resultant fixation of the cells is rapid. Cell lysis and disruption of the normal cell shape is minimized (Fig. 9-6). Even extremely fragile adipose tissue such as that of the chicken can be successfully fixed if proper respect for tissue fragility is observed.

An additional advantage of the method is that large samples may be analyzed in which thousands of cells can be counted and sized rapidly. Various methods that utilize microscopy suffer from the disadvantages of measuring a one-dimensional characteristic that must then be statistically adjusted to a theoretically correct value, measuring a relatively small number of cells by a tedious process, or both. While electronic counting works well with large samples, small samples may be analyzed as well.

The statistical advantages of equipment that can detect and size 5,000 counts/sec is apparent. In practice counting, rates are much lower than the maxima since the concentrations of cells in solution are kept at several hundred cells per milliliter of electrolyte solution. Nevertheless, the large numbers of cells that form the data base for estimating cell concentrations and size distributions dramatically improve repeatability and accuracy.

---

*Considerable progress has been made in particle sizing using laser technology and image analysis; however, the ability of these new instruments to determine quantitatively cell number per unit volume of a suspension is unacceptable for our purposes.

Precautionary Concerns. Several disadvantages are largely due to electronic or physical contamination, but detection of "true adipocytes" in the lower end of the range is also a concern. The size range limit that makes fractionation of cells necessary introduces a chance of error as well.

The utilizable range of this method is constrained by the equipment available. Without the proper range of aperture sizes the effective sizing range is a limiting factor.* It is usually necessary to fractionate cells during the filtering process so that cells can be analyzed with different-sized apertures to obtain the full size distribution. This practice is especially critical when adipose tissues of different sizes are compared.

Fractionation of cells into two or more size ranges will require recombination of data to reform the size distribution of cells in the tissue. Data are lost if fractionation is not complete or if sample loss or contamination results. Inclusion of cells that are too large or too small for a particular counting range results in lost data and therefore alters the accuracy of the

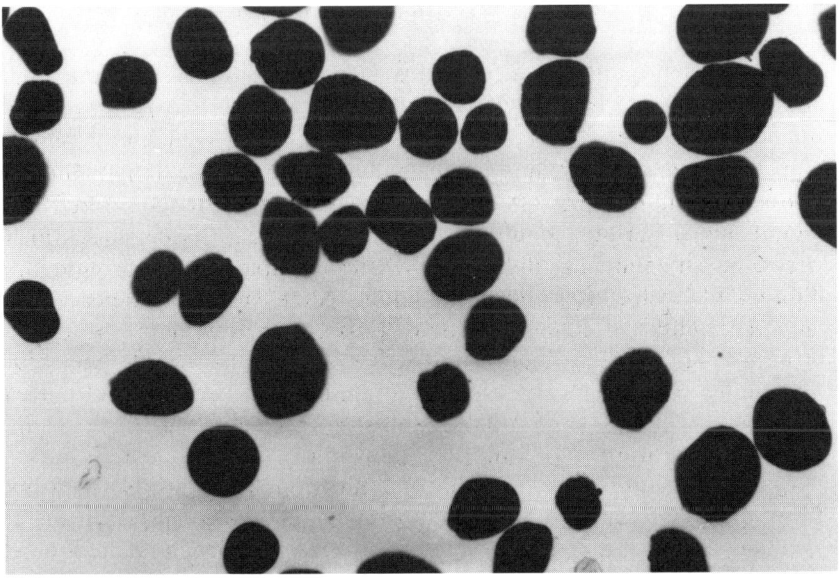

Figure 9-6. Osmium tetroxide-fixed rat adipocytes, × 120. Note the irregular shapes similar to those of adipocytes in situ.

*An LRE will allow counting and sizing of adipocytes without fractionation of adipocytes and requires only a 400 μm aperture.

process. When data are recombined the final product contains the sum of the errors of each counting session. If the counting and sizing range of one aperture tube can be expanded with a logorithmic range expander, unnecessary fractionation can be avoided. Nevertheless, extreme care must be exercised during the final stages of tissue processing.

Other possible sources of error are affected by the characteristics of both the adipocyte preparation and the adipocytes. Very young or lean animals have a large proportion of small adipocytes, and these adipocytes may be counted with proper preparation and equipment. However, one must note that preparations of small adipocytes contain more debris. This debris registers as adipocytes. Currently, no means is available to rid a sample of unwanted interference of this type. It is believed that the fixation process causes dissolution of the cell membrane and cytoplasm. Thus the electronic quantification method actually measures the fixed lipid droplet and not the adipocytes per se. Fixation of a multilocular adipocyte may therefore liberate several small droplets that either will not be counted or will register as individual adipocytes. The frequency of multilocular cells is low for white adipose tissue analysis but should be a vital concern during analysis of brown adipose tissue. Cells larger than 20 μm usually are not affected by these errors.

Sample quality may be affected by storage conditions. The best results are obtained when adipocytes are counted expeditiously after processing. But sample storage is necessary if equipment is not available or samples must be acquired over time. Mersmann and MacNeil (52) compared freshly processed osmium-fixed adipose tissue with adipose tissues stored for 1 month under various conditions. Fixation of frozen tissue caused an increase in cell volume, as did storage of fresh samples in post-osmium urea. Storage of fresh samples in post-osmium saline caused an increase in the apparent number of adipocytes. Samples stored in osmium or post-osmium detergent saline did not significantly affect cell number or cell volume, but the number of particles less than 35 μm in diameter was elevated in detergent saline cells. In view of these results samples can be stored in osmium for a short period if they cannot be analyzed in a timely manner.

Spurious counts due to sample contamination are controlled by proper filtration of solutions and by minimizing contact of samples with the environment. Dust is a contamination that can be virtually eliminated. Fibers are troublesome unless care is exercised to prevent the screens from shedding fibers during the filtering process. Bubbles should not be created by stirring the dense fixed adipocytes into a homogeneous suspension since they are counted as adipocytes. Careful regulation of stirring speed eliminates the problem of air bubbles. If necessary, glycerol is used to increase the viscosity of the solution to aid adipocyte suspension (20). Electronic error, when encountered, can be a more difficult obstacle to overcome.

Error associated with background electronic noise is negligible when properly controlled. Background occurs in the lower channels coincident with the counting of osmium-fixed cells. This noise can be controlled with the lower threshold setting. If a multichannel analyzer is available, the lower channels may be omitted from the analysis. One problem with omitting lower channels is that range can be severely curtailed. Since the distribution recorded is proportional to the volume of the cells, the lower channels contain data from cells with the smallest diameters. When lower channel capacity is lost, the effective range of the instrument is dramatically decreased. The logorithmic range expander can be used to extend the available range of sizes, but if noise is a considerable problem the range expander will also expand noise.

The shadow effect is a problem associated with solution flow allowing reentry of previously counted cells to return to the sensing zone of the aperture. Spurious counts are subsequently registered. Increasing the viscosity of the electrolyte solution by adding glycerol has been suggested as a method to prevent recirculation of particles. However, for the aperture sizes used in the analysis of adipocytes, the necessary viscosity of the electrolyte solution would be impractical. In fact, a solution of 50% glycerol has an insignificant effect on the shadow effect of adipocyte counting (unpublished observation). The shadow effect is insignificant if the aperture tube is frequently flushed free of residual adipocytes.

Interference from laboratory machinery is occasionally a problem. Cen trifuges, noisy motors, nonmotorized high-power equipment, or lighting may be sources of spurious counts. Relocation of counting equipment to a more favorable environment may be necessary.

A serious and less visible source of error is the coincident count. These false pulses result when more than one cell at a time enters the sensing zone. This necessarily gives a false reading and may alter the size and numbers counted. These spurious counts are dependent on the concentration of the cells in suspension and the size of the aperture being used. Formulas* for determining coincidence:

$$P = 2.5 \times (D/100)^3 \times 500/V,$$

*A high-resolution aperture or "long-bore aperture" may not be appropriate for use in sizing adipocytes. Nevertheless, the formula for determining coincidence for an aperture tube of this type is: P = 3.8 × (D/100)3 × 500/V.

where P = correction factor, D = aperture diameter (μm), V = manometer volume (μl),

$$\text{and } N'' = P \times (N'/1000)^2,$$

where N″ = correction count added, N′ = count, P = correction factor give us a means to correct the numeric error that is created. The corrected count is N′ + N″. Caution should be exercised to guarantee that N″ should not exceed 15% of N′. Even though coincidence does not substantially affect the frequency distribution, coincidence may affect the volume distribution.

Coincidence has two effects: loss of particle counts, and abnormally large pulses due to coincident passage. Lost data will retain the normal frequency distribution, since passage is statistical. But spurious counts will collect in larger size ranges and skew size distributions. Efforts should be made to reduce cell suspension concentrations so that coincident counts decrease to acceptable levels.

In summary, the advantages of electronic counting of osmium-fixed cells are many since the technique enables the investigator to rapidly obtain accurate, reliable data on adipocyte number, volume, and surface area of populations of adipocytes. The disadvantages are primarily cautionary in nature and may be overcome with proper respect for tissue integrity and the limitations of the counting and sizing apparatus.

REFERENCES

1. C. E. Allen, D. C. Beitz, D. A. Cramer, and R. G. Kauffman, 1976, *Biology of Fat in Meat Animals,* Research Division, College of Agriculture and Life Sciences, University of Wisconsin-Madison.
2. C. E. Allen, 1976, Cellularity of adipose tissue in meat animals, *Fed. Proc.* **35**:2302–2307.
3. R. L. Hood and R. F. Thornton, 1980, A technique to study the relationship between adipose cell size and lipogenesis in a heterogeneous population of adipose cells, *J. Lipid Res.* **21**:1132–1136.
4. M. I. Gurr and J. Kirtland, 1978, Adipose tissue cellularity: A review. 1. Techniques for studying cellularity, *Int. J. Obes.* **2**:401–427.
5. R. B. Goldrick, 1967, Morphological changes in the adipocyte during fat deposition and mobilization, *Am. J. Physiol.* **212**:777–782.
6. J. Hirsch and E. Gallian, 1968, Methods for the determination of adipose cell size in man and animals, *J. Lipid Res.* **9**:110–119.
7. D. Lemmonier, 1972, Effect of age, sex, and site on the cellularity of the adipose tissue in mice and rats rendered obese by a high-fat diet, *J. Clin. Invest.* **51**:2907–2915.
8. U. Smith, L. Sjostrom, and P. Björntorp, 1972, Comparison of two methods for determining human adipose cell size, *J. Lipid Res.* **13**:822–824.
9. R. L. Hood and C. E. Allen, 1973, Cellularity of bovine adipose tissue, *J. Lipid Res.* **14**:605–610.

10. H. J. Mersmann and L. J. Brown, 1973, Adaptation of swine (*Sus domesticus*) adipose tissue to increased lipogenesis: Expression of data, *Int. J. Biochem.* **4**:503–510.
11. H. J. Mersmann, J. R. Goodman, and L. J. Brown, 1975, Development of swine adipose tissue: Morphology and chemical composition, *J. Lipid Res.* **16**:269–279.
12. J. Robelin, 1981, Cellularity of bovine adipose tissues: Developmental changes from 15 to 65 percent mature weight, *J. Lipid Res.* **22**:452–457.
13. F. D. DeMartinis and A. Francendese, 1982, Very small fat cell populations: Mammalian occurrence and effect of age, *J. Lipid Res.* **23**:1107–1120.
14. K. J. Smith, J. M. Leatherwood, and E. J. Eisen, 1983, Effects of preweaning and postweaning feed restriction on the development of polygenic obese mice, *Growth* **47**:5–52.
15. T. M. Plucinski, R. K. Bruner, J. M. Leatherwood, and E. J. Eisen, 1984, Effects of feeding pattern and dietary regimen on growth and adipose tissue cellularity in polygenic obese mice, *J. Anim. Sci.* **59**:350–360.
16. K. L. Rogers, T. D. Etherton, and P. M. Kris-Etherton, 1984, Biphasic diameter distribution of adipocytes from lean and obese rats, *Growth* **48**:331–338.
17. F. DeMartinis, 1985, Very small fat cell populations determined by a modified osmium tetroxide-urea method, *Am. J. Physiol.* **249**:C89–C96.
18. P. Bjrulf, 1959, Atherosclerosis and body build with special reference to size and number of subcutaneous fat cells, *Acta Medica Scand. Suppl.* **349**:1–99.
19. M. DiGirolamo, S. Mendlinger, and J. W. Fertig, 1971, A simple method to determine fat cell size and number in four mammalian species, *Am. J. Physiol.* **221**:850–858.
20. L. Sjostrom, P. Björntorp, and J. Vrana, 1971, Microscopic fat cell size measurements on frozen-cut adipose tissue in comparison with automatic determinations of osmium-fixed fat cells, *J. Lipid Res.* **12**:521–530.
21. M. P. Stern and F. Conrad, 1975, An automated direct method for measuring adipocyte cell size, *Clin. Chim. Acta* **65**:29–37.
22. T. D. Etherton, E. H. Thompson, and C. E. Allen, 1977, Improved techniques for studies of adipocyte cellularity and metabolism, *J. Lipid Res.* **18**:552–557.
23. G. J. Hausman, 1981, Techniques for studying adipocytes, *Stain Technol.* **56**:149–154.
24. M. Ashwell, P. Priest, M. Bondoux, C. Sowter, and C. K. McPherson, 1976, Human fat cell sizing: A quick, simple method, *J. Lipid Res.* **17**:190–192.
25. G. Enzi, E. M. Inelmen, F. F. Rubaltelli, V. Zanardo, and L. Favaretto, 1982, Postnatal development of adipose tissue in normal children on strictly controlled calorie intake, *Metabolism* **31**:1029–1034.
26. G. J. Hausman, D. R. Campion, and R. J. Martin, 1980, Search for the adipocyte precursor cell and factors that promote its differentiation, *J. Lipid Res.* **21**:657–670.
27. J. Rafael and P. Vsiansky, 1981, Quantitative determination of nucleic acids in brown and white adipose tissue, *Anal. Biochem.* **115**:158–162.
28. P. B. Curtis-Prior, T. Hanley, and N. J. Temple, 1975, A colorimetric method for the determination of deoxyribonucleic acid in adipose tissue, *Analyst* **100**:105–110.
29. L. Rakow, 1974, Cellularity of the different cell compartments in white adipose tissue of mice in chronic starvation, refeeding and two types of obesity, in *The Regulation of the Adipose Tissue Mass,* J. Vague and J. Boyer, eds., *Excerpta Medica,* Amsterdam 140–144.
30. J. Kirtland and M. I. Gurr, 1979, Adipose tissue cellularity: A review. 2. The relationship between cellularity and obesity, *Int. J. Obes.* **3**:15–55.
31. M. Rodbell, 1964, Metabolism of isolated fat cells: I. Effects of hormones on glucose metabolism and lipolysis, *J. Biol. Chem.* **239**:375–380.
32. J. Hirsch and P. W. Han, 1969, Cellularity of rat adipose tissue: Effects of growth, starvation and obesity, *J. Lipid Res.* **10**:77–82.

33. E. W. Askew, R. L. Huston, C. G. Plopper, and A. L. Hecker, 1975, Adipose tissue cellularity: Response to exercise and cortisol treatment, *J. Clin. Invest.* **56**:521–529.
34. T. D. Etherton and C. S. Chung, 1981, Preparation, characterization, and insulin sensitivity of isolated swine adipocytes: Comparison with adipose tissue slices, *J. Lipid Res.* **22**:1053–1059.
35. E. W. Askew and A. L. Hecker, 1976, Adipose tissue cell size and lipolysis in the rat: Response to exercise intensity and food restriction, *J. Nutr.* **106**:1351–1360.
36. M. Lavau, C. Susini, J. Knittle, S. Blanchet-Hirst, and M. R. C. Greenwood, 1977, A reliable photomicrograph method for determining fat cell size and number: Application to dietary obesity, *Proc. Soc. Exp. Biol. Med.* **156**:251–256.
37. P. M. Clarkson, W. Kroll, J. Wai, and G. Kamen, 1981, A comparison of four methods to measure fat cell size, *Am. J. Clin. Nutr.* **34**:2287–2290.
38. B. E. March and G. Hansen, 1977, Lipid accumulation and cell multiplication in adipose bodies in white leghorn and broiler-type chicks, *Poult. Sci.* **56**:886–894.
39. S. C. Jamdar and L. J. Osborne, 1982, Glycerolipid biosynthesis in rat adipose tissue. 10. Changes during a starvation and re-feeding cycle, *Biochim. Biophys. Acta* **713**:647–656.
40. L. B. Salans, J. L. Knittle, and J. Hirsch, 1968, The role of adipose cell size and adipose tissue insulin sensitivity in the carbohydrate intolerance of human obesity, *J. Clin. Invest.* **47**:153–165.
41. A. L. Cartwright, 1982, Insulin resistance of muscle tissue and adipose cellularity in polygenic obese and monobese mice and diet effects, Ph.D. Thesis, North Carolina State University, Raleigh.
42. D. Lemmonier, 1981, Adipose tissue development in normal and obese animals, in *Adipose Tissue in Childhood,* F. P. Bonnet, ed., CRC Press, Cleveland, Ohio, pp. 57–80.
43. O. L. Tulp, R. Frink, and E. Danforth, Jr., 1982, Effect of cafeteria feeding on brown and white adipose tissue cellularity, thermogenesis, and body composition in rats, *J. Nutr.* **112**:2250–2260.
44. S. W. Cushman and L. B. Salans, 1978, Determination of adipose cell size and number in suspension of isolated rat and human adipose cells, *J. Lipid Res.* **19**:269–273.
45. J. E. Foley, A. L. Laursen, O. Sonne, and J. Gliemann, 1980, Insulin binding and hexose transport in rat adipocytes. Relation to cell size, *Diabetologia* **19**:234–241.
46. R. Vasilatos, T. D. Etherton, and P. J. Wangsness, 1983, Preparation of isolated bovine adipocytes: Validation of use from studies characterizing insulin sensitivity and binding, *Endocrinology* **112**:1667–1673.
47. T. D. Etherton, E. D. Aberle, E. H. Thompson, and C. E. Allen, 1981, Effects of cell size and animal age on glucose metabolism in pig adipose tissue, *J. Lipid Res.* **22**:72–80.
48. W. H. Coulter, 1956, High speed automatic blood cell counter and cell size analyzer, *Proc. Natl. Electron. Conf.* **12**:1034.
49. C. F. T. Mattern, F. S. Brackett, and B. J. Olson, 1957, Determination of number and size of particles by electrical gating: Blood cells, *J. Appl. Physiol.* **10**:56–70.
50. R. F. Karuhn and R. H. Berg, 1982, *Practical Aspects of Electrozone Size Analysis,* Particle Data Laboratories, Ltd., Elmhurst, Illinois.
51. R. L. Prior, 1983, Lipogenesis and adipose tissue cellularity in steers switched from alfalfa hay to high concentrate diets, *J. Anim. Sci.* **56**(2):483–492.
52. H. J. Mersmann and M. D. MacNeil, 1986, Variables in estimation of adipocyte size and number with particle counter, *J. Anim. Sci.* **62**:980–991.

10

Transplantation of White and Brown Adipose Tissue in Experimental Animals

Margaret Ashwell
Good Houskeeping Institute

OBJECTIVES OF ADIPOSE TISSUE TRANSPLANTATION

In general, there are two main reasons why any tissue is transplanted between a donor animal and a host animal. The first is to discover what effect, if any, the donor tissue has on the metabolism of the host. Should the presence of the transplanted tissue cause significant changes in host metabolism, it is usually because some factors are being released from the tissue. Good examples of this phenomenon are the amelioration of streptozotocin induction of diabetes in rats (1) and the prevention of hypothalamic obesity in rats (2) by the transplantation of syngeneic fetal pancrei. In both cases, the transplanted islets can release the necessary insulin to lower host blood sugar levels.

The second main reason for performing transplantation experiments is to see what effect the host metabolism has on the donor tissue. In this case one is looking at the effect that factors in the host's circulation and sometimes those released by the host's autonomic nervous system have on the metabolism of the donor tissue. Adipose tissue transplantation experiments usually fall into this second category. It is generally assumed that it is much more important to look at the effect of host factors on donor tissue metabolism, since the effects of factors released by transplanted adipose tissue are likely to be minimal. Perhaps the only exception to this statement are the experiments performed by Zirm (3) (quoted by Johansson (4)), who implanted pieces of brown adipose tissue from hibernating hedgehogs into mice. In these experiments, not only did the body temperature of the mice decrease but body weight increased.

Sometimes, normal adipose tissue is transplanted into a different environment as a method of studying its pattern of development or degradation. More specifically, adipose tissue that shows cellular and metabolic abnormalities, such as the adipose tissue from genetically obese mice, can be transplanted into normal animals to find out whether the "abnormalities" are retained by the donor tissue and are thus intrinsic factors or whether they can be corrected by the host environment and are thus caused by extrinsic factors (see Applications of Current Methodology).

METHODOLOGY

Transplanting White and Brown Adipose Tissue

Table 10-1 summarizes the results of some of the publications that have reported the transplantation of white adipose tissue in animals. Transplantation methodology has been used to study either the mechanism of fat cell formation or the question of whether there are intrinsic differences in the adipose tissue of genetically obese animals. This latter aspect will be covered in more detail later in this chapter. Regarding the former aspect, the methodology adopted by Clark and Clark (7) is particularly elegant since the histology of fat cells and the simultaneous behavior of blood vessels could be observed in ear chambers in the free living animal. All the authors state that the fat grafts become well vascularized. Peer (6) states that early circulation is established through end-to-end anastomosis between graft and host blood vessels and that this is followed by a penetration of host capillaries into the fat graft. None of these authors comment on the question of re-innervation of white adipose tissue grafts at their chosen transplantation sites.

Table 10-2 summarizes some of the experiments that have involved the transplantation of brown adipose tissue. There is a great variation in the viability of brown adipose tissue transplants, but this can probably be explained by the variability in innervation that takes place at the different sites of transplantation. Liebelt (9) did not investigate innervation, but it is possible that the greater viability of grafts placed in the axillary region reflects the greater likelihood of sympathetic innervation here than at the base of the ear. Both Ferren (14) and Smalley (15) have stated that innervation does not occur if brown adipose tissue is transplanted under the kid-

Table 10-1. Transplantation of White Adipose Tissue in Experimental Animals

Host	*Transplantation Site*	*Donor Tissue*	*Result/Conclusions*	*Reference*
Rabbit	Spleen and connective tissue from peritoneum of host rabbit	Fatty tissue fat cell formation tissues	Successful	5
?	Implanted subcutaneously	Free fat grafts	Grafts vascularize well 50% fat cell survival; cells increase in size; host gains weight	6
Rabbit	Transparent chambers in ear	Subcutaneous adipose tissue	Studied fat cell histology in relation to blood vessel activity	7
Obese and lean mice (littermates)	Subcutaneous fat transplantation by means of parabiosis followed by separation	Subcutaneous adipose tissue	*Obese* adipose tissue takes on characteristics in lean host; no intrinsic defect in *obese* adipose tissue	8
F_1 Hybrid mice (NH × DBA/2)	Base of ear	Inguinal WAT from either NS (obese) DBA/2 (lean)	Graft from obese donor deposited more fat than graft mice from lean donor; intrinsic defect in *obese* adipose tissue	9
Pig	Between rectus muscle and rectus sheath	Dermis plus underlying subcutaneous fat	Vascularisation good, but slow replacement of fat by fibrous tissue	10
Mice (lean and genetically obese from syngeneic species)	Under kidney capsule	Gonadal and subcutaneous adipose tissue	*Obese* adipose tissue takes on lean characteristics; no intrinsic defect in *obese* adipose tissue	11 12
Isogeneic or allogeneic rats	Spleen	Neonatal rat pancreas plus adjacent white adipose tissue	Proliferation of fat cells in isogeneic grafts but not allogeneic grafts probably due to localized trophic effect from pancreatic B cells	13

Table 10-2. Transplantation of Brown Adipose Tissue in Experimental Animals

Host	*Transplantation Site*	*Sympathetic Innervation Intact*	*Donor Tissue*[a]	*Result*	*Reference*
Balb/c mice aged 30 d	Base of ear	Not considered	ISBAT from Balb/c mice, aged 30 d	Remains viable at 30–90 d but atrophied at 130 d	9
Balb/c mice aged 30 d	Subcutaneous, at axillary region	Not considered	ISBAT from Balb/c mice aged 30 d	Remains viable 130 d	9
Mice	Kidney capsule	Assumed absent	ISBAT	Increased cell size; increased droplet size after 8–15 d transplantation	14
Adult hamsters	Kidney capsule	Assumed absent	ISBAT from hamsters (1 d to adult)	Cells remain or become multilocular	15
Adult hamsters	Intraocular	Yes, but can be denervated surgically	ISBAT from hamsters, aged 4–5 d (unilocular cells)	Cells become multilocular (process delayed by denervation)	16 17
Mice lean and genetically obese (ob/ob)	Kidney capsule	Partial at 23°C Increased at 4°C Decreased at 33°C (see text)	ISBAT from lean and obese mice	Cells take on host characteristics, only if innervation of host is matched	Ashwell, Wells and Dunnett (in preparation)

[a]ISBAT = interscapular brown adipose tissue

ney capsule although neither has experimental evidence for this statement. In contrast, Nechad (16, 17) has used the intraocular transplantation site mainly because Olson and Malmfors (18) have demonstrated that tissues grafted here can be re-innervated from the sympathetic nerves of the iris. It is interesting to note, however, that a recent investigation of the innervation of hamster pup brown adipose tissue grafts (17) using an autoradiographical technique has shown that the nerve fiber density in the brown adipose tissue grafts reaches a value of only 18% of the adult hamster value in situ. This low nerve fiber density is thought to be due to the undifferentiated state of the transplanted tissue. This particular transplantation site also has the advantage that the effects of denervation can be studied by previous ablation of the superior cervical ganglion, which innervates the ipsilateral iris.

Transplantation and Immunological Rejection

It is particularly important that a transplant must not suffer immunological rejection. The easiest way to avoid rejection is to perform autogenic transplants in which the donor tissue is taken from the host animal (5). Another way of avoiding rejection is to use a host possessing all of the transplantation antigens expressed by the grafted tissue. An F_1 hybrid is thus a suitable host for tissues from either of its inbred parents, because it has both sets of parental transplantation antigens (9). Recently, transplantation has been much simplified by the availability of animals bred by repeated backcrossing into a standard strain, and differing only at one genetic locus; for example, that determining obesity (11).

If none of the above methods is suitable, it should theoretically be possible to transplant tissue from any species into nude mice and rats (nu/nu). These mutants lack a functional thymus and are therefore unable to mount a normal immune response to a foreign graft. The nude mouse has been used as a host with human adipose tissue as the donor tissue (19, 20).

Transplantation of Adipose Tissue Under the Kidney Capsule

The basic technique for transplanting adipose tissue under the kidney capsule is illustrated in Figure 10-1. Fat is removed from a selected site of the donor animal, placed into a balanced salt solution, and cut into small pieces (5-10 mg). The anesthetized host animal is then prepared for transplantation by making a dorsolateral incision in the shaved skin to reveal the kidney (Figure 10-1*A*). A small incision is then made in the body wall so that the kidney can be carefully pulled through it and is retained by it during the transplantation. A small hole is made in the kidney capsule at the posterior pole, and the fat transplant is inserted and pushed under the capsule towards the anterior pole of the kidney (Figure 10-1*B*) with curved forceps. In double kidney experiments, fat transplants (usually from different donors) are inserted under both kidneys. The kidney is replaced (Figure 10-1*C*) and the skin wound is closed (Figure 10-1*D*). The grafts are left in place for at least 1 month, the animals are then killed and the kidney(s) plus the graft are exposed.

The kidney plus the graft and also a sample of the host adipose tissue are then put into histological fixative (Figure 10-2). The kidney is sectioned through the plane of the fat graft and stained with hematoxylin and eosin or aniline blue. Fat cell sizing is performed directly on fixed sections. The grafts are examined under the microscope, and areas are chosen that are suitable for cell sizing. Areas containing pockets of small cells are avoided, as are areas that show obvious signs of fat necrosis.

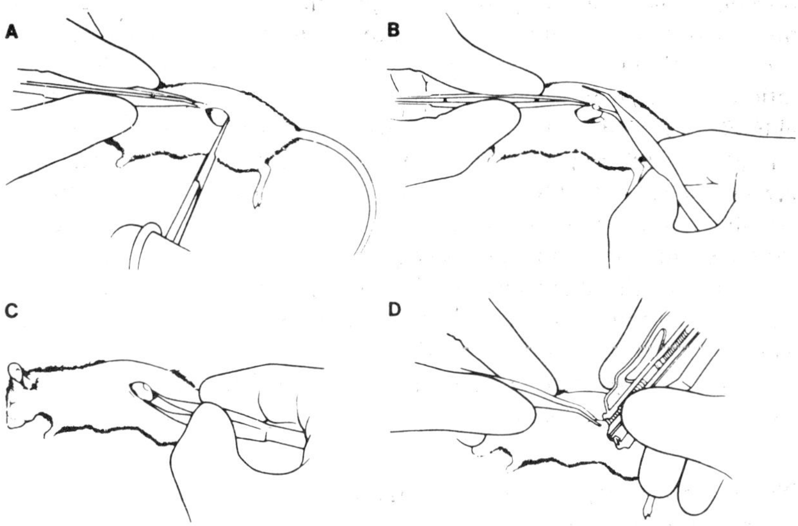

Figure 10-1. The adipose tissue transplantation technique. *A*, The anesthetized host mouse is prepared for transplantation; *B*, The adipose tissue from the donor mouse is pushed under the kidney capsule; *C*, The graft is now in place under the kidney capsule of the host mouse; *D*, The wound is closed and the mouse is left to recover from the anesthetic. (*From Ashwell and Meade, 1978, p. 466; reproduced with permission of Diabetologia.*)

The advantages of using the kidney capsule as a transplantation site are as follows:

(a) The grafted tissue is well vascularized by an extension of the kidney's normal blood supply. Histological sections of adipose tissue grafts show good evidence of vascularization, and sometimes it is possible to see blood vessels running through the graft.

(b) Even when the kidney is surrounded by much of its own perirenal adipose tissue, it is easy to identify the grafted adipose tissue because of its unique position under the capsule.

(c) It is possible to use two completely different tissues for transplantation onto the two kidneys of a single animal (providing they will not be immunologically rejected by the host) to study the effect of the common host environment on them both at the same time.

The disadvantages of using the kidney capsule as a transplantation site are as follows:

(a) The amount of grafted tissue is limited by the size of the host kidney and by the fragility of the kidney capsule. This laboratory has found that 10 mg tissue is about the maximum size that could be transplanted under a mouse kidney capsule. It apparently is possible to transplant up to 100 mg of adipose tissue under a rat kidney capsule (Ikeda, personal communication) by using a slightly different technique; that is, expelling the tissue from the end of a glass pipette. This presumably is possible with rats but

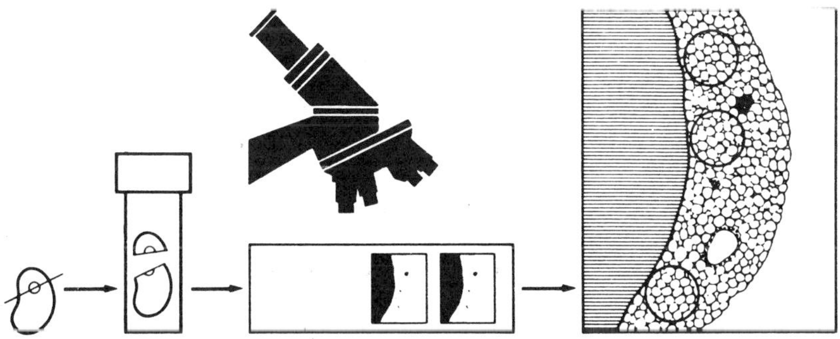

Figure 10-2. Measurement of adipocyte size in adipose tissue grafts. *A*, The kidney plus adipose tissue graft is sectioned through the plane of the graft and put into Bouin's fixative. Thin sections are prepared from the fixed tissue and stained with either hematoxylin and eosin or aniline blue. *B*, The sections are viewed under a microscope, and the cell diameter in the grafts is estimated by counting the number of cells in a known area, taking care to choose areas free of vacuoles or pockets of small cells (circled in diagram). (*From Ashwell and Meade, 1978, p. 466; reproduced with permission of Diabetologia.*

not with mice, because of the large size of the rat kidney and also the decreased fragility of its capsule.

(b) Although there is no doubt that vascularization of grafts takes place, there is considerable doubt as to whether the grafts are innervated by the autonomic nervous system. Some investigators have used the kidney capsule transplantation site with a view to studying the specific effects of denervation (21).

APPLICATIONS OF CURRENT METHODOLOGY

Is There an Intrinsic Defect in the White Adipocytes of the Genetically Obese Mouse (ob/ob)?

Background. Several of the reported differences between the adipose tissue of genetically obese rodents and their lean counterparts led one to think in the mid 1970s that factors intrinsic to the adipose tissue of obese animals predispose them to obesity; for example, the very early appearance of the increased fat cell size, the increased fat storage potential, and decreased fat breakdown potential observed in in vitro experiments (22). Other experimental approaches also suggested this: Restriction of food and the subsequent decrease in fat cell size of Zucker rats do not restore lipogenesis or lipolysis to normal (23) and streptozotocin-treated ob/ob mice have higher body fat contents than lean mice despite similar insulin levels (24).

One of the best direct tests to see whether an abnormality is caused by an intrinsic defect in a certain tissue is to transplant that tissue into a normal environment and see whether the abnormality persists. The earliest attempts at fat transplantation were made before inbred strains were available. Hausberger (8) reported that although he had little success in transplanting between very young littermates by conventional grafting methods, he could transplant subcutaneous fat from obese (ob/ob) mice to lean mice and vice versa by a process of parabiosis and separation. The survival of these mice was poor, but by examining the surviving fat grafts he showed that fat transplants assumed the typical features of the host tissue. He thus concluded that it was unlikely that the abnormal growth rate of fatty tissue in hereditary obesity was caused by properties inherent in the tissue itself.

Using a different animal model of obesity, Liebelt (9) reached the opposite conclusion. He transplanted inguinal fat from lean DBA/2 mice and obese NH mice into the ears of host NH $\times$ DBA/2 F_1 hybrid mice. Examination of the grafts by lipid extraction after 90 days revealed that the ear grafts derived from the NH (obese) parent deposited more lipid than did the grafts of the adipose tissue from the DBA/2 (lean) parent. He therefore concluded that the source of adipose tissue had more influence than its environment in deciding the character of the adipose tissue in the graft.

There Is No Intrinsic Defect in Ability to Mobilize Triacylglycerol. The first experiments using the kidney capsule method of fat transplantation were performed on C57BL/6J ob/ob mice and their lean counterparts (C57BL/6J +/+ and +/?) (11). The ring skin-grafting procedure of Billingham and Medawar (25) confirmed that all mice, lean and obese, had similar transplantation antigens and that grafts would not be immunologically rejected. Figure 10-3 summarizes the results of these experiments. The fat cells of obese donor fat decrease in size in a lean environment to a size typical of lean fat, while the cells of lean donor fat transplanted into an obese environment increase to the size typical of obese fat. When both lean and obese fat tissues are transplanted onto opposite kidneys of the same mouse, the tissues acquire cell sizes characteristic of their common host and there is no significant difference in cell size in the two grafts. These experiments showed that no intrinsic defects in the adipose tissue of ob/ob mice prevent them from mobilizing fat to decrease their cell size when they are placed in a lean environment.

The fat transplantation technique has been used in other genetically obese rodents; results are summarized in Table 10-3. In the diabetic yellow and adipose mouse (an allele of the diabetic mouse), lean fat transplanted into the corresponding obese mice or obese fat transplanted into corresponding lean mice, underwent a significant change in fat cell mass. Comparison of grafts of the same fat tissue transplanted into either obese or lean recipient mice showed significant differences in all cases. Graft cell size altered to correspond to that characteristic of the recipient. Figure 10-4 shows some photomicrographs of the grafts in diabetic mice with corresponding host and donor fat. Recently, Ikeda has used a similar fat transplantation technique to study the adipose tissue of another genetically obese rodent: the Wistar fatty rat (26). Transplantation under the rat kidney capsule has made it possible to transplant larger amounts of adipose tissue between the fatty rat and its lean counterpart. Ikeda (personal communication) has concluded not only that the large fat cells of the fatty rat decrease in a lean environment, but also that there is no decrease in total fat cell number. Conversely, the small fat cells of the lean rat increase in size when transplanted into the fatty rat, but there is no total increase in fat cell number.

There Is No Intrinsic Defect in Ability to Store Triacylglycerol. It was still possible, however, that the adipose tissue of obese mice might have an

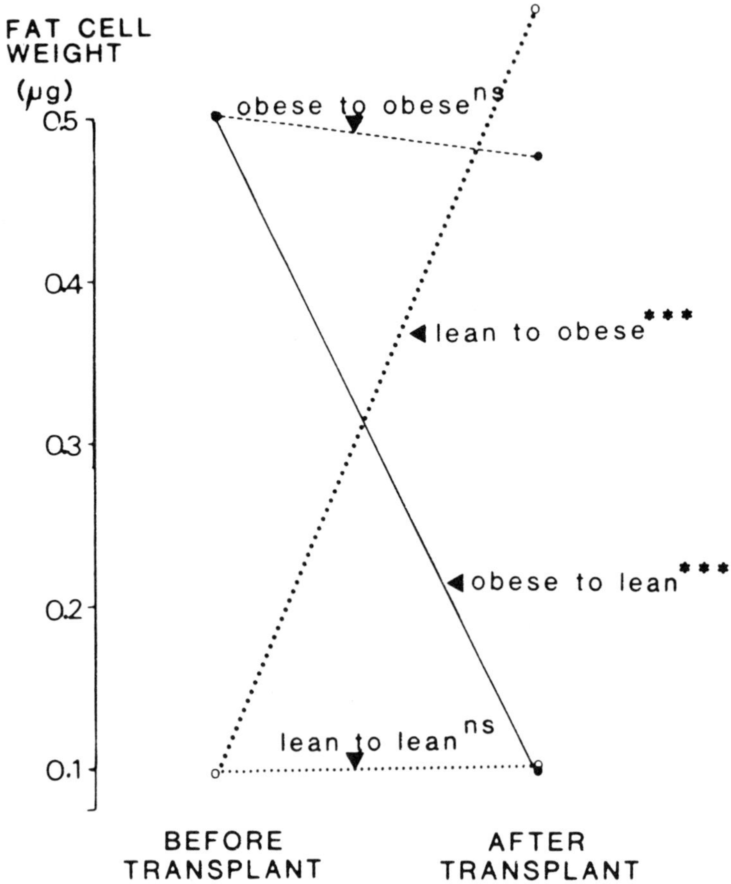

Figure 10-3. Fat cell size in grafts between genetically obese mice (ob/ob) and their lean counterparts. Epididymal fat was transplanted between genetically obese mice (C57Bl/6J ob/ob) and their lean littermates (C57Bl/6J +/+ or +/ob) by using the kidney capsule transplantation technique described in (11). After 1 month, the kidneys plus grafts were removed and the size of the fat cells in the grafts was calculated and compared with the size of the fat cells in the donor tissue. *Lean* fat cells in lean hosts and *obese* fat cells transplanted into obese hosts showed no significant change in cell weight. On the other hand, *lean* fat cells transplanted into lean hosts showed significant changes in fat cell weight. The *lean* fat cells increased to a size typical of obese fat and the *obese* fat cells decreased to a size typical of lean fat.

•—• obese obese ns = no significant difference in fat cell size
•---• obese lean *** = difference in fat cell size $p < 0.001$
○....○ lean lean ns
○—○ lean obese ***

Table 10-3. The Effect of Fat Transplantation Between Obese Mice and Their Lean Counterparts

Fat Donors		*Fat From Obese Donors (fat cell mass in grams)*			*Fat From Lean Donors (fat cell mass in grams)*			*Host Fat*	
Obese	*Lean*	*Before Transplant*	*When Removed After Period in Obese Host*	*When Removed After Period in Lean Host*	*Before Transplant*	*When Removed After Period in Obese Host*	*When Removed After Period in Lean Host*	*Obese*	*Lean*
Gold thioglucose-treated CBA mice	Untreated CBA mice	0.58	0.75	0.14	0.21	0.65	0.16	0.51	0.14
Diabetic mice C57BL/KsJ db/db		0.69	0.68	0.08	0.13	0.77	0.10	0.79	0.11
Adipose mice C57BL/6J db^{ad}/db^{ad}	Lean littermates +/?	0.69	0.52	0.07	0.09	0.60	0.07	0.66	0.09
Yellow obese mice C57BL/6J Ay/+	Lean littermates +/+	0.58	0.37	0.11	0.09	0.45	0.11	0.45	0.16

Adapted from Meade et al., 1979,

innate capacity for increased fat storage rather than a decreased capacity for fat mobilization. A further series of experiments was therefore designed using the ob/ob mouse (27). In these experiments, grafts from lean and obese donors were transplanted under the kidney capsules of lean host mice and left for 1 month. At this point, some of the host mice were treated with gold thioglucose (GTG), which is known to induce obesity mainly via hyperphagia (28), and other host mice were injected with saline as controls. The mice were then killed at monthly intervals for up to 3 months and the adipose tissue from the host mice and the grafts was examined in the normal way. In the control groups the obese donor fat cells decreased in size to that typical of their lean host. The effect of 1, 2, or 3 months' GTG treatment was to substantially increase graft and host fat cell size. In no group was there a significant difference between the cell size of "lean" and "obese" grafts. Thus, both grafts could respond similarly to the new host environment by increasing the size of their cells.

Therefore, the results reported earlier argue against the possibility that the adipose tissue of obese mice has an intrinsic defect in either fat mobilization or storage.

Is There an Intrinsic Defect in the Fatty Acid Composition of Triacylglycerol? Adipose tissue from obese mice fed on standard lab chow is characterized by a fatty acid composition with a higher proportion of palmitoleic acid and a lower proportion of linoleic acid than that of adipose tissue from lean mice (29, 30). These differences probably arise from (a) increased rates of fatty acid synthesis in the liver and adipose tissue of obese mice (31, 32), which causes a greater dilution of dietary linoleic acid and (b) the raised activity of acyl-CoA 9 desaturase (33). Experiments were con-

---

Figure 10-4. Fat cell size in grafts between diabetic mice (dg/db) and their lean counterparts.
Top left: Graft from a diabetic mouse in a lean host; mean fat cell mass = 0.111 μg. Top right: Graft from a lean mouse in a lean host; mean fat cells mass = 0.067 μg. Bottom left: Graft from diabetic mouse in overweight db/db host; mean fat cell mass = 0.829 μg. Bottom right: Graft from a lean mouse in overweight db/db host, mean fat cell mass = 0.623 μg.
Insets. Top: Adipose tissue of lean host; mean fat cell mass = 0.0918 μg. Bottom: Adipose tissue of overweight db/db host; mean fat cell mass = 0.918 μg. Left: Adipose tissue of overweight db/db donor; mean fat cell mass = 0.833 μg. Right: Adipose tissue of lean donor; mean fat cell mass = 0.115 μg. Scale marker = 100 μm.
(From Mead et al., 1979, plate 2; reproduced with permission of Proc. Soc. R. Lond.)

ducted to determine whether the increases in activity of these lipogenic enzymes were dependent on factors intrinsic or extrinsic to the adipocyte (34).

Adipose tissue transplants were performed between lean and genetically obese mice (ob/ob), and after either 1 or 2 months, the grafts and samples of perirenal and subcutaneous adipose tissue from donors and hosts were extracted with solvent. The fatty acid compositions of the lipids were then determined by gas liquid chromatography.

Grafts from lean donors in obese host mice had fatty acid compositions that resembled the perirenal adipose tissue of the hosts after 1 month, with a lower proportion of linoleic acid (L) and a higher proportion of palmitoleic acid (P) than in lean mice (i.e., a lower L:P ratio). Grafts from obese mice into lean mice had fatty acid compositions that were unchanged, partially changed, or completely resembled that in the host. The use of grafts, prelabeled by feeding the donor mouse margaric acid, indicated that a total lack of fatty acid turnover, rather than selective metabolic processes, was responsible for the failure of a few grafts from obese mice into lean mice to acquire the fatty acid composition of the perirenal adipose tissue of the host (i.e., to increase the L:P ratio). Figure 10-5 summarizes the results of these experiments, excluding those grafts in which the margaric acid content indicated that there was an abnormal turnover of fatty acids. In all cases the L:P ratio of all adipose tissue grafts was not significantly different from that of the host adipose tissue. The main conclusion from this study was that the environment in which the adipocyte finds itself is the major determinant of its content of linoleic acid and palmitoleic acid; that is, no intrinsic differences in fatty acid metabolism of adipocytes from obese mice were apparent when the cells were transplanted into lean mice.

There Is No Intrinsic Defect in Brown Adipose Tissue (BAT) Cells of the Genetically Obese Mouse

Background. Recently, attention has turned away from white adipose tissue and has focused on the brown adipose tissue of genetically obese mice. It has been shown that defective functioning of brown adipose tissue in respect to its capacity to generate heat can account for the reduced thermoregulatory energy requirements of obese mice (35, 36). The principal mechanism for thermogenesis in brown adipose tissue is through the operation of the mitochondrial proton conductance pathway; this pathway is associated with a specific membrane protein of molecular weight 32,000 commonly referred to as uncoupling protein. The amount of uncoupling protein (UCP) in brown adipose tissue mitochondria is closely correlated with the thermogenic capacity of the tissue. A sensitive radioimmunoassay (37) has been used to demonstrate that adult obese mice (ob/ob) have less uncoupling protein than their lean counterparts (38, 39). The altered mitochondrial function of the obese mouse is also reflected in the ultrastructure of the mitochondria. The cristae are less pronounced than those in lean mice and the matrix tends to be denser (40; Fig. 10-6). The differences in ultrastructure are not dissimilar to those noted for liver mitochondria prepared in different media. The *condensed* conformation of liver mitochondria (which is similar to the obese mitochondria) is transformed into the

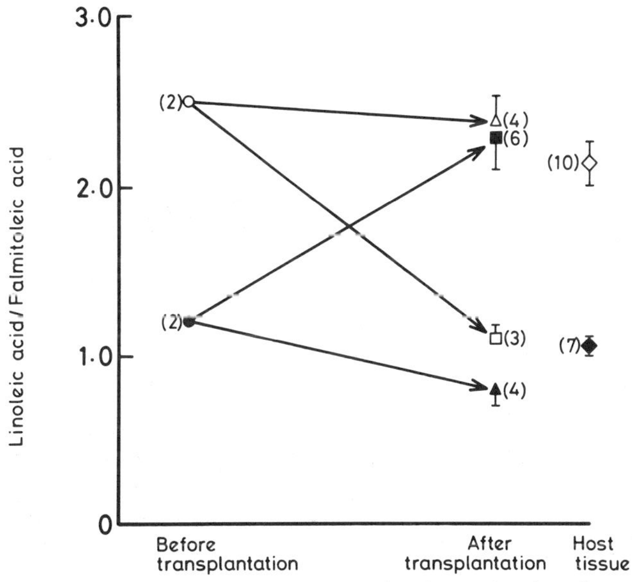

Figure 10-5. Changes in fatty acid composition (linoleic acid: palmitoleic acid ratio) in grafts between genetically obese mice (ob/ob) and their lean counterparts (summary of results).

The results are means for the number of animals stated. The bar indicates the S.E.M.

● obese donor fat;	○ lean donor fat;
▲ obese graft in obese host;	△ lean graft in lean host;
■ obese graft in lean host;	□ lean graft in obese host;
◆ obese host perirenal fat;	◇ lean host perirenal fat.

orthodox conformation (which is similar to lean mitochondria) when an uncoupling agent is present (41).

The decreased brown adipose tissue thermogenesis in obese mice has been demonstrated as early as 12 days of age (42) and occurs before any other defect reported so far for this species. The question therefore arose as to whether the abnormality in brown adipose tissue mitochondria is intrinsic to the obese mouse or whether the brown adipose tissue is receiving abnormal signals external to the tissue and the changes in the mitochondria, both in UCP content and ultrastructure, are secondary to these environmental changes.

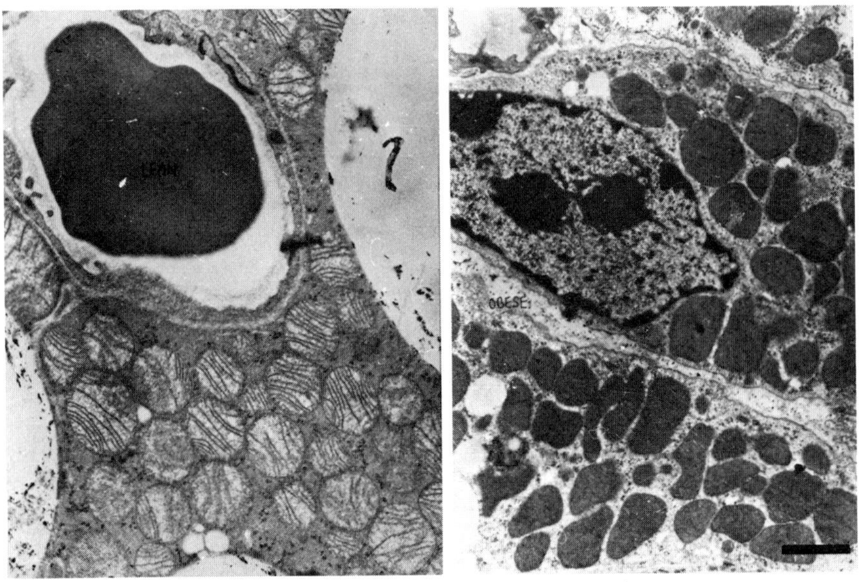

Figure 10-6. Brown adipose tissue was dissected from the interscapular depot, cut into small pieces of about 1 mm^3 and fixed in 3% glutaraldehyde solution in 0.1 M cacodylate buffer for 2 hr. After thorough washing with buffer, the tissue was postfixed with osmium and embedded in Spurrs resin. Sections were cut at 800 A and viewed in an electron microscope. Scale marker = 1 μm; m = mitochondria; l = lipid droplet.

Figures 10-6A and B show interscapular brown adipose tissue (ISBAT) from lean and obese mice, respectively, and demonstrate their typical mitochondrial ultrastructure. (*Electron microscophs by kind permission of Mr. Clive Wells, MRC Clinical Research Centre, Harrow, Middlesex U.K.*)

Transplantation of Brown Adipose Tissue Between Lean and Obese Mice. To help answer this fundamental question, interscapular brown adipose tissue (ISBAT) was transplanted between genetically obese (C57BL 6J ob/ob) mice and their lean littermates. The kidney capsule method already described for white adipose tissue (11) was used. After 6 weeks, the ISBAT grafts were removed from the kidney and fixed and stained for viewing under the light and electron microscope (Ashwell, Wells, and Dunnett, in preparation). Samples of donor and corresponding host ISBAT were prepared similarly. Initially, the mice were kept at 23°C following transplantation.

Figure 10-6 shows the essential differences between ISBAT mitochondria from lean and obese mice; however, in grafted ISBAT several intermediate forms existed. It was therefore necessary to devise a system for scoring the mitochondrial ultrastructure in grafts on a scale between 1 and 7. Thus the mitochondrial ultrastructure score (MUS) of ISBAT from lean donors was predominantly of types 1, 2, and 3 (as in Figure 10-6*a* and that from obese donor mice was predominantly of types 5, 6, and 7 (as in Figure 10-6*b*).

Figures 10-7*a* and *b* illustrate the typical outcome of the ISBAT grafts performed between lean and obese mice in terms of the resultant MUS of cells in the grafts. When lean ISBAT was transplanted into obese mice there was a partial shift towards obese type mitochondrial ultrastructure, with most of the cells having mitochondria that were neither typically obese nor lean (i.e., MUS 2–5) (Figure 10-7*a*). When obese ISBAT was transplanted into lean mice there was also only a partial shift towards lean-type mitochondrial ultrastructure (i.e., MUS 3–6) (Figure 10-7*b*). Even when grafts were left in host mice for 12 weeks rather than 6 weeks, the mitochondrial ultrastructure in the grafts was still typically midway between the characteristic lean and obese brown adipose tissue.

Although these first experiments might have been interpreted as providing some evidence for an intrinsic difference between lean and obese ISBAT mitochondria, this laboratory was reluctant to draw this conclusion without further experimentation. Whether the results of the first experiments could be explained by considering the degree of sympathetic innervation of the ISBAT transplants was questioned. The cells nearest the kidney in any graft were most inclined to show mitochondrial ultrastructure typical of the host, whereas the cells nearest the kidney capsule were inclined the least. Whether these results reflected differences in sympathetic stimulation of the cells and whether the results on mitochondrial ultrastructure were due to incomplete or inconsistent sympathetic stimulation to all the cells in the graft were questioned. Since cold acclimation is known to increase sympathetic outflow to brown adipose tissue, while warm acclimation is known to decrease it (43), it was decided to subject host mice to either warm or cold accli-

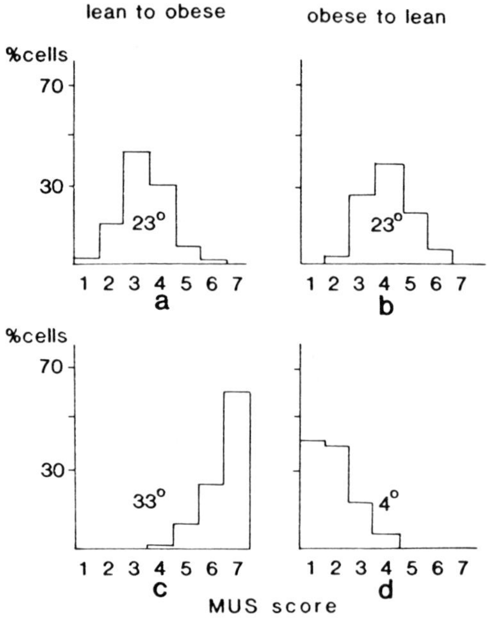

Figure 10-7. Small pieces of interscapular brown adipose tissue (ISBAT) were grafted between lean and obese mice as described in the legend to Figure 10-1. After 6 weeks at 23°C, or 6 weeks at 23°C followed by 3 weeks at 33°C (obese mice) or 4°C (lean mice) the mice were killed and the brown adipose tissue (BAT) grafts were prepared for electron microscopy as explained in the legend to Figure 10-6.

Several sections of each graft were studied, and at least 200 cells in each graft were scored for their mitochondrial ultrastructure (MUS). This scoring was done using a 7-point scale, giving the most typical lean mitochondrial ultrastructure the score of 1 and the most typical obese mitochondrial ultrastructure were scored accordingly; i.e., a cell that contained mitochondria with characteristics midway between typically lean and typically obese was scored as 4. Figure 7*A* shows the MUS profile of a typical lean ISBAT graft in an obese host kept at 23°C for 6 weeks. Figure 7*B* shows the MUS profile of a typical obese ISBAT graft in a lean host kept at 23°C for 6 weeks. Figure 7*C* shows the MUS profile of a typical lean ISBAT graft in an obese host kept at 23°C for 6 weeks and then at 33°C for 3 weeks. Figure 7*D* shows the MUS profile of a typical obese ISBAT graft in a lean host kept at 23°C for 6 weeks and then at 4°C for 3 weeks.
(*Quantitation of electron micrographs by kind permission of Mr. Clive Wells, MRC Clinical Research Centre, Harrow, Middlesex, U.K.*)

mation after the transplantation. This was done to see whether increasing or decreasing sympathetic outflow to the grafts could provide a more consistent change in mitochondrial ultrastructure in the grafts.

Transplantation with Subsequent Temperature Acclimation. Figures 10-7*c* and 10-7*d* show the typical outcome of ISBAT grafts between lean and obese mice when transplantation was followed by cold acclimation of lean mice and warm acclimation of obese mice. When lean ISBAT was transplanted into obese mice that were then warm acclimated, the typical mitochondrial ultrastructure of most cells in the graft was similar to the obese type (i.e., MUS score = 5, 6, and 7) (Fig. 10-7*c*). On the other hand, when obese ISBAT was transplanted into lean mice that were then cold acclimated, there was a nearly complete transformation to the typical mitochondrial ultrastructure of lean mitochondria (MUS score = 1, 2, and 3) (see Figure 7*d*). This complete transformation persisted even when the hosts were subsequently again housed at ambient temperatures after the period of warm or cold acclimation.

The results for this second group of ISBAT transplantation experiments can be reconciled with the view that the difference in ultrastructure, and presumably function, of the obese ISBAT mitochondria is not of primary origin. It is, however, necessary to warm or cold acclimate the host mice to demonstrate complete transformation of mitochondrial ultrastructure from lean to obese mice and vice versa. Grafts in animals kept at ambient temperatures are associated with partial transformation, probably because the resultant sympathetic stimulation is midway between that associated with lean and obese mice.

Experimental support for these conclusions has come recently from a fluorescent histochemical study of catecholamines in brown adipose tissue from lean and obese mice (44) and in brown adipose tissue grafts between lean and obese mice (Ashwell, Wells, and Dunnett, in preparation). If obese ISBAT is grafted into a lean mouse that is then cold (4°C) acclimated, there is clear evidence of the typical dual innervation of brown adipose tissue (45); that is, innervation of the blood vessels plus innervation of individual cells. On the other hand, if lean brown adipose tissue is grafted into obese mice and these obese host mice are subsequently kept at 33°C for 3 weeks, there is no evidence of any functional cellular innervation, although the blood vessels are innervated at normal. Grafts between lean and obese mice that are kept at ambient temperatures show normal innervation of the blood vessels in the graft but only partial functional cellular innervation (Fig. 10-8).

The results of several other experiments were reported while these trans-

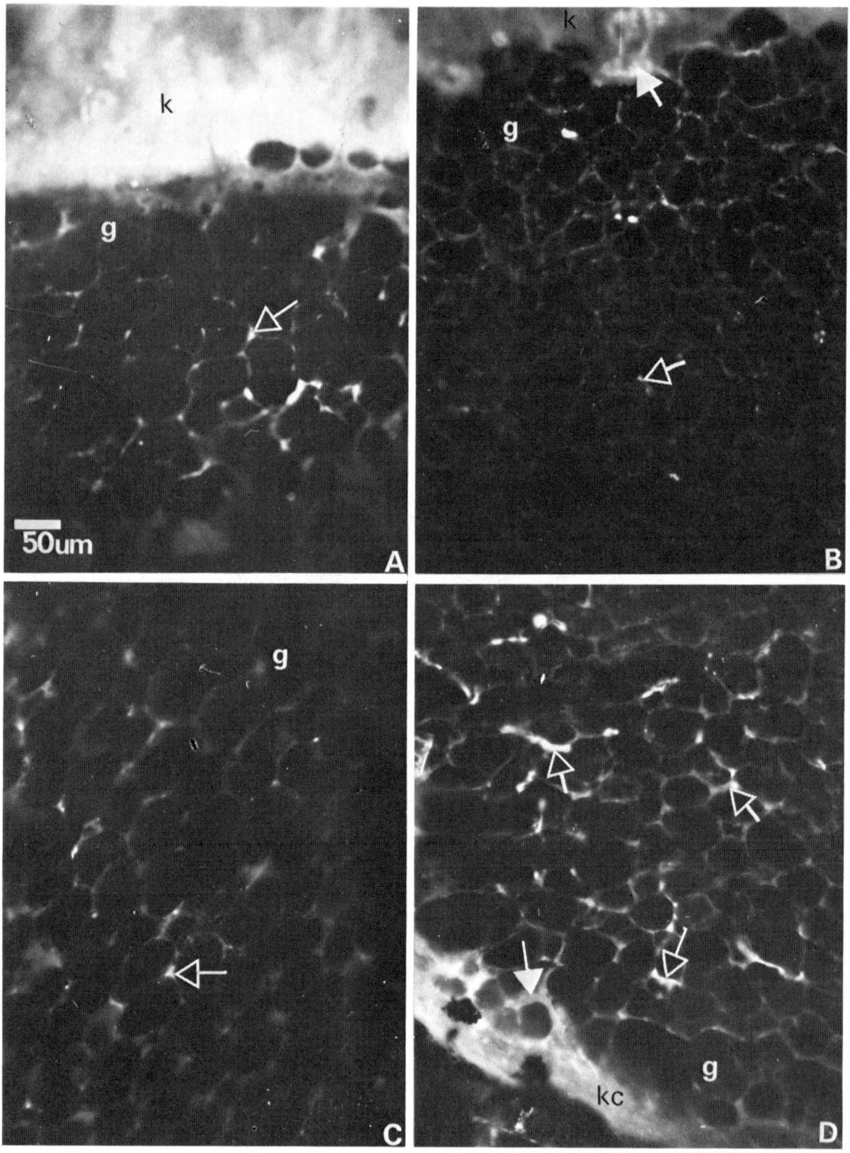
k
g
50um
A
k
g
B
g
C
g
kc
D

plantation experiments were in progress. The results are consistent with our conclusions that there is no intrinsic defect in mitochondria in the brown adipose tissue of obese mice. First, Hogan and Himms-Hagen (40) showed that obese mice acclimated at 13°C for 2 weeks increased the guanosine diphosphate (GSP) binding activity of isolated mitochondria and that the 32,000-molecular weight uncoupling protein, as estimated from polyacrylamide gels, was also increased. This increase has also been confirmed by direct radioimmunoassay of the uncoupling protein (39). Hogan and Himms-Hagen (40) also showed that the ultrastructure of ISBAT mitochondria from obese mice acclimated at 13°C tends towards the normal lean type of ultrastructure. Furthermore, measurements of noradrenaline turnover in brown adipose tissue from lean and obese mice have shown that the reduced turnover exhibited by obese mice at 23°C can be ameliorated by cold acclimation (46, 47).

The current hypothesis concerning the defect in brown adipocytes in the genetically obese mouse is that the defect lies in the reduced activity of the sympathetic nervous system. Thus, thermogenesis in the obese mouse at ambient temperature is not as fully switched on as it is in the lean mouse. This is reflected in lower GDP binding to isolated mitochondria (48) and lower mitochondrial UCP content at room temperatures (28, 39). Maximum capacity can, however, be exhibited during cold acclimation indicating that the defect can be remedied by increasing sympathetic activity and is not due to a primary defect at the mitochondrial level.

Figure 10-8. Small pieces of interscapular brown adipose tissue (ISBAT) were grafted between lean and obese mice exactly as described in the legend to Figure 10-7. The grafts were examined by fluorescence histochemistry using a combination of the techniques described in refs. 57 and 58.

A, a typical lean ISBAT graft in an obese host kept at 23°C for 6 weeks; *B*, a typical obese ISBAT graft in a lean host kept at 23°C for 6 weeks; *C*, a typical lean ISBAT graft in an obese host kept at 23°C for 6 weeks and then at 33°C for 3 weeks; *D*, a typical obese ISBAT graft in a lean host kept at 23°C for 6 weeks and then at 4°C for 3 weeks.

In all photomicrographs, the scale marker represents 100 μm. Bright green catecholamine-dependent fluorescence is indicated around blood vessels (→) and around individual cells (→) k = kidney; kc = kidney capsule; g = ISBAT graft. (*Fluorescent histochemistry technique by kind permission of Dr. Stephen Dunnett, Department of Experimental Psychology, University of Cambridge.*)

Other Applications of the Fat Transplantation Methodology

White Adipose Tissue. Some researchers have assumed the complete denervation of adipose tissue grafted under the kidney capsule in order to study the activity of adipose tissue enzymes without the intervention of the sympathetic nervous system. For example, Ikeda and co-workers (21) measured lipoprotein lipase (LPL) in adipose tissue grafts in rats and showed that LPL activity decreased even though the grafts were well vascularized and accumulated lipid. They thus concluded that LPL activity is influenced by changes in autonomic nervous tone. This particular result and conclusion contrast with that obtained by Hansson and co-workers (49), who microsurgically denerved rat epididymal adipose tissue and found no change in LPL activity or lipolysis even after 3 weeks.

It is difficult to reconcile these two results unless one assumes that the effects of surgically denervating adipose tissue and transplanting it under the kidney capsule are different. It has in fact been suggested that surgical denervation of brown adipose tissue denervates only the blood vessels and not the cells (43). Further comparisons of the two techniques and a closer examination of the innervation of transplants are obviously necessary to resolve this problem.

Apart from experiments using adipose tissue from lean and obese rodents in this laboratory, the kidney capsule method has been used to investigate such problems as concurrent fat loss and fat gain (50) and species differences in response to fat mobilization and fat storage hormones (20).

There has been a lot of interest recently in the question of preadipocyte development. Perhaps this is a new area in which transplantation techniques might be employed to elucidate the factors that stimulate recruitment and development. For example, Van and Roncari (51) implanted rat adipocyte precurser cells subcutaneously and reported their complete differentiation into a normal adipose tissue depot. In addition, Tavassoli (52) reported the intrasplenic transplantation of lipid-depleted cultured rat adipocytes. These postadipocytes accumulate lipid in vivo, providing vascularization is adequate. Loffler and co-workers (53) showed that the serum from genetically obese mice added to the culture medium of 3T3-L1 preadipocytes stimulates lipogenic enzymes in the precursor cells more than does the serum obtained from lean littermates. Preliminary experiments suggest that the adipogenic factor present in obese serum is a peptide with hormonal activity. The corollary of this experiment would be to transplant preadipocytes from lean and obese mice into identifiable sites in syngeneic mice and to follow in situ fat accumulation in the transplanted cells. In this case one would have not only the opportunity to study the effect of the adipogenic factor in obese serum on lean preadipocytes, but also to subdivide the preadipocyte population by methods such as those used by Gaben-Cogneville and co-workers (54) and to investigate which particular cells are responsive to the factor.

Brown Adipose Tissue. Transplantation methodology seems to offer great potential as an alternative for exploring the role that the sympathetic nervous system plays in the development and function of brown adipose tissue. The effects of sympathetic denervation have traditionally been studied by surgically or immunologically ablating the nerve supply to the ISBAT depot; these methods almost certainly denervate the blood vessels but they probably do not denervate the cells (43). Transplantation of brown adipose tissue under the kidney capsule (especially with subsequent warm acclimation of the host), however, seems to be a method for functionally denervating the cells of the tissue and not the blood vessels. Alternatively, intraocular transplantation of brown adipose tissue can be followed by sympathetic denervation of the host iris (17) so that the graft can be studied before and after denervation.

There is currently a lot of interest in devising drugs that can increase the thermogenic activity of brown adipose tissue (55); most of these drugs are sympathomimetics. So far, very little attention has been focused on the role of the cellular innervation of brown adipose tissue compared with the direct innervation to blood vessels. Therefore, the transplantation methodology that offers an alternative to other methods of denervation might be particularly useful when the detailed mechanisms of drug actions are investigated.

SUMMARY

This chapter is an attempt to review some of the work on adipose tissue transplantation in experimental animals. The main focus of the review is on the use of the transplantation technique to determine whether intrinsic abnormalities exist in the white and/or brown adipose tissue of genetically obese rodents; First, can an intrinsic abnormality in the cells of white adipose tissue explain why obese animals have larger fat cells than their lean counterparts, second, can an intrinsic abnormality in the cells of brown adipose tissue of obese animals explain its decreased thermogenic capacity compared with that from lean animals?

The results of transplantation experiments with lean and obese animals have argued against the existence of intrinsic abnormalities in both white and brown adipose tissues of genetically obese animals, and they are thus consistent with conclusions reached by alternative approaches to the problem.

ACKNOWLEDGMENTS

I would like to thank Dr. C. J. Meade and Dr. P. Trayhurn for making helpful comments and suggestions on the manuscript. I am grateful to Dr. S. B. Dunnett and Mr. C. Wells for allowing me to mention some of our unpublished collaborative work. Finally, I would like to thank Mrs. Audrey Broodes and her word processor for producing the manuscript so efficiently.

REFERENCES

1. J. Brown, W. R. Clark, I. G. Molnar, and Y. S. Mullen, 1976, Fetal pancreas transplantation for reversal of streptozotocin induced diabetes in rats, *Diabetes* **25**:56–64.
2. S. Inoue, G. A. Bray, and Y. S. Mullen, 1978, Transplantation of pancreatic β-cells prevents development of hypothalamic obesity in rats, *Am. J. Physiol.* **235**:E266–E271.
3. K. L. Zirm, 1956, Contribution to the knowledge of natural hibernation, *Z. Naturforsch.* **11b**:530–534.
4. B. Johansson, 1959, Brown fat: A review, *Metab. Clin. Exp.* **8**:221–240.
5. Y. Matsuoka, 1917, Transplantation of fatty tissue into the spleen, *J. Pathog. Bact.* **21**:493–501.
6. L. A. Peer, 1959, *Transplantation of Tissues.* Williams & Wilkins Co., Baltimore, Maryland.
7. E. R. Clark and E. L. Clark, 1940, Microscopic studies of the new formation of fat in living adult rabbits, *Am. J. Anat.* **67**:255–285.
8. F. X. Hausberger, 1959, Behavior of transplanted adipose tissue of hereditary obese mice, *Anat. Rec.* **135**:109–112.
9. R. A. Liebelt, 1963, Response of adipose tissue in experimental obesity as influenced by genetic, hormonal, and neurogenic factors, *Ann. N.Y. Acad. Sci.*d **110**:723–748.
10. C. P. Sawhney, T. N. Banerjee, and R. N. Chakravath, 1966, Behavior of dermal fat transplants, *Br. J. Plast. Surg.* 169–176.
11. M. A. Ashwell, C. J. Meade, P. Medawar, and C. Sowter, 1977, Adipose tissue: Contributions of nature and nurture to the obesity of an obese mutant mouse (ob/ob), *Proc. Roy. Soc. Lond. B.* **195**:343-353.
12. C. J. Meade, M. Ashwell, and C. Sowter, 1979, Is genetically transmitted obesity due to an adipose tissue defect, *Proc. Roy. Soc. Lond. B.* **205**:395–410.
13. I. G. Banks, J. M. Sloan, and J. D. Buchanan, 1982, A histological study of intrasplenic transplanted neonatal rat pancreas and of adjacent adipose tissue proliferation, *Diabetologia* **22**:128–133.
14. L. E. Ferren, 1967, Morphological differentiation of implanted brown and white fats, *Trans. Kansas Acad. Sci.* **131**:476–484.
15. R. L. Smalley, 1970, Changes in composition and metabolism during adipose tissue development, in *Brown Adipose Tissue,* O. Lindberg, ed., Elsevier, New York, pp. 73–95.
16. M. Nechad, 1980, Differentiation of the brown adipose tissue in the hamster: Evidence of a trophic role of the sympathetic innervation, Satellite of 28. *Int. Congr. Physiol. Sci. Pecs.,* Z. Szelenyi and M. Szekely, eds., Akademiai Kiado, Budapest, pp. 503–505.
17. M. Nechad and L. Olson, 1983, Development of interscapular brown adipose tissue in the hamster. II. Differentiation of transplants in the anterior chamber of the eye: Role of sympathetic innervation, *Biol. Cell.* **48**:167–174.

18. L. Olson and T. Malmfors, 1970, Growth characteristics of adrenergic nerves in the adult rat. Fluorescence histochemical and ^{3}H-noradrenaline uptake studies using tissue transplantations to the anterior chamber of the eye, *Acta Physiol. Scand. Suppl.* **384.**
19. N. Bach-Mortensen, P. Romert, and S. Ballegaard, 1976, Transplantation of human adipose tissue to nude mice, *Acta Path. Microbiol. Scand.* Sect. C **84:**283–289.
20. C. J. Meade and M. Ashwell, 1981, A quantitative study of human fat transplants in obese nude mice and the lean littermates, *Metabolism* **30:**376–379.
21. H. Ikeda, N. D. Courtney, J. D. Brunzwell, and S. D. Woods, 1984, Lipoprotein lipase activity decreases in transplanted adipose tissue, Abstract presented at Fourth International Congress of Obesity, New York, 1983 (in press).
22. M. A. Ashwell and C. J. Meade, 1979, Adipose tissue in genetically obese rodents, in *Animal Models of Obesity,* M. F. W. Festing, ed., Macmillan, London, pp. 107–130.
23. D. A. York and G. A. Bray, 1973, Genetic obesity in rats. II. The effect of food restriction on the metabolism of adipose tissue, *Metabolism* **22:**443–454.
24. C. N. Boozer and J. Mayer, 1976, Effects of long-term restricted insulin-production in obese-hyperglycemic (genotype ob/ob) mice, *Diabetologia* **12:**181–187.
25. R. E. Billingham and P. B. Medawar, 1951, The technique of free skin grafting in mammals, *J. Exp. Biol.* **28:**385–402.
26. H. Ikeda, A. Shino, T. Matsuo, H. Iwatsuka, and Z. Suzuoki, 1981, A new genetically obese-hyperglycemic rat (Wistar fatty), *Diabetes* **30:**1045–1050.
27. M. A. Ashwell and C. J. Meade, 1978, Obesity: Do fat cells from genetically obese mice (C57B1/6J ob/ob) have an innate capacity for increased fat storage?, *Diabetologia* **2:**465–470.
28. A. F. Debons, E. Krimsky, M. L. Maayan, K. Fani, and F. A. Jimenez, 1977, Gold thioglucose obesity syndrome, *Fed. Proc.* **36:**143–147.
29. H. A. Haessler and J. D. Crawford, 1965, Alterations in fatty acid composition of depot fat associated with obesity, *Ann. N.Y. Acad. Sci.* **131:**476–484.
30. M. Enser, 1979, The role of insulin in the regulation of stearic acid desaturase activity in liver and adipose tissue from obese-hyperglycamic (ob/ob) and lean mice, *Biochem. J.* **180:**551–558.
31. M. W. Bates, C. Zomzely, and J. Mayer, 1955, Acetate metabolism of obese mice, *Am. J. Physiol.* **181:**187–190.
32. E. A. Rath and S. Thenen, 1980, Influence of age and genetic background on in vivo fatty acid synthesis in obese (ob/ob) mice, *Biochim. Biophys. Acta* **618:**18–27.
33. M. Enser, 1975, Desaturation of stearic acid by liver and adipose tissue from obese-hyperglycamic mice (ob/ob), *Biochem. J.* **148:**551–555.
34. M. Enser and M. Ashwell, 1983, Fatty acid composition of triglycerides from adipose tissue transplanted between obese and lean mice, *Lipids* **18:**776–780.
35. J. Himms-Hagen, 1979, Obesity may be due to a malfunctioning of brown fat, *Can. Med. Assoc. J.* **121:**1361–1364.
36. P. Trayhurn, 1984, The development of obesity in animals: The role of genetic susceptibility, *Clin. Endocrinol. Metab.* **13:**451–474.
37. M. E. J. Lean, W. J. Branch, W. P. T. James, G. Jennings, and M. Ashwell, 1983, Measurement of rat brown adipose tissue mitochondrial uncoupling protein by radioimmunoassay: Increased concentration after cold acclimation, *Biosci. Rep.* **3:**61–71.
38. M. Ashwell, G. Jennings, and P. Trayhurn, 1983, Evidence from radioimmunoassay for a decreased concentration of mitochondrial *uncoupling* protein from brown adipose tissue of genetically obese (ob/ob) mice, *Biochem. Sci. Trans.* **11:**727–728.
39. M. Ashwell, S. Holt, G. Jennings, D. Stirling, P. Trayhurn, and D. A. York, 1985, Measurement by radioimmunoassay of the mitochondrial uncoupling protein from brown adi-

pose tissue of obese (ob/ob) mice and Zucker (fa/fa) rats at different ages, *FEBS Lett.* **179:**233–237.

40. S. Hogen and J. Himms-Hagen, 1980, Abnormal brown adipose tissue in obese (ob/ob) mice: Response to acclimation to cold, *Am. J. Physiol.* **239:**E301–309.
41. C. R. Hackenbrock, 1968, Ultrastructural basis for metabolically linked mechanical activity in mitochondria. II. Electron transport-linked ultrastructural transformations in mitochondria, *J. Cell Biol.* **57:**345–369.
42. A. E. Goodbody and P. Trayhurn, 1982, Studies on the activity of brown adipose tissue in suckling, pre-obese, ob/ob mice, *Biochim. Biophys. Acta* **680:**119–126.
43. T. Barnard, G. Mory, and M. Nechad, 1980, Biogenic amines and the trophic response of brown adipose tissue, in *Biogenic Amines in Perinatal Development,* S. Parvez and H. Parvez, eds., Elsevier, Amsterdam, pp. 391–439.
44. M. Ashwell and S. B. Dunnett, 1985, Fluorescence histochemical demonstration of catecholamines in brown adipose tissue from obese (ob/ob) and lean mice acclimated at different temperatures, *J. Autonon. Nerv.* **14:**377–386.
45. D. M. Derry, E. Schonbaum, and G. Steiner, 1969, Two sympathetic nerve supplies to brown fat of the rat, *Can. J. Physiol. Pharmacol.* **47:**57–63.
46. A. W. Knehans and D. R. Romsos, 1973, Norepinephrine turnover in obese (ob/ob) mice: Effects of age, fasting and acute cold, *Am. J. Physiol.* **244:**E567–E574.
47. G. Zaror-Behrens and J. Himms-Hagen, 1983, Cold stimulated sympathetic activity in brown adipose tissue of obese (ob/ob) mice, *Am. J. Physiol.* **244:**E361–E366.
48. J. Himms-Hagen and M. Desautels, 1978, A mitochondrial defect in brown adipose tissue of the obese (ob/ob) mouse. Reduced binding of purine nucleotides and a failure to respond to cold by an increase in binding, *Biochem. Biophys. Res. Commun.* **83:**628–634.
49. P. Hansson, T. Holmin, and P. Nilsson-Ehle, 1981, Microsurgical denervation of rat adipose tissue: Lack of effect on lipoprotein lipase, *Biochem. Biophys. Res. Commun.* **103:**1254–1257.
50. M. Ashwell and C. J. Meade, 1981, Obesity: Can some fat cells enlarge while others are shrinking?, *Lipids* **16:**475–478.
51. R. L. Van and D. A. Roncari, 1982, Complete differentiation in vivo of implanted cultured adipocyte precursors from adult rats, *Cell Tissue Res.* **225:**557–566.
52. M. Tavassoli, 1982, In vivo development of adipose tissue following implantation of lipid-depleted cultured adipocyte, *Exp. Cell. Res.* **137:**55–62.
53. G. Loffler, P. Feich, H. Hauner, and L. Herberg, 1983, An adipogenic serum factor in genetically obese rodents, *FEBS Lett.* **153:**179–182.
54. A. M. Gabewn-Cogneville, Y. Aron, G. Idriss, T. Jahchnan, S.-V. Pello, and E. Swierczenski, 1983, Differentiation under the control of insulin of rat preadipocytes in primary culture. Isolation of homogeneous cellular fractions by gradient centrifugation, *Biochim. Biophys. Acta* **762:**437–444.
55. J. R. S. Arch, A. T. Ainsworth, M. A. Cawthorne, et al., 1984, A typical β-adrenoreceptor on brown adipocytes as target for anti-obesity drugs, *Nature* **309:**163–165.
56. M. Enser and M. Ashwell, 1983, Fatty acid composition of triglycerides from adipose tissue transplanted between obese and lean mice, *Lipids* **18:**776–780.
57. I. Lorein, A. Bjorlund, and O. Lindvall, 1977, Magnesium ions in catecholamine fluorescence histochemistry. Application to the cryostat and vibratome techniques, *Histochemistry* **52:**223–239.
58. J. C. de la Torre and J. W. Surgeon, 1976, A methodological approach to rapid and sensitive monoamine histofluorescence using a modified glyoxylic acid technique: The SPG method, Histochemistry **49:**81–93.

Index